Let's Ponder!

What is the Mystery of the Creation?

Kindly break-open the horizon, O lord!
Let me find out — what is going-on
on its other side?

4

Let's Ponder!

What is the Mystery of the Creation?

(English Version of the original Hindi Book-

"Jara Sochen! Kya Hai Srajan ka Rahasya")

A Bold Analysis of
The Laws That Govern the Universe

By- SATYA PRAKASH VERMA
"SATYENDRA"

Let's Ponder!
What is The Mystery of the Creation?

Email - vermasatyendra@yaoo.com

Cover Design: Aditi Verma

Copyright ©: Satya Prakash Verma (2013 & 2019) - Second Edition

ISBN: 979-8-9862057-0-0

To
My Late Parents…

Janki Rani and Dr. Shivnarayan Verma

"From Darkness to Light
Kindly lead us, Oh Lord!"

"Oh God, kindly give us the power to realize the truth, and accept it as well."

Table of Contents

Introduction

In this book, the author has expressed his personal view on the creation of the universe. First, he has described the overview of some of the eminent science theories such as **The Theory of Electromagnetic Radiation, Relativity, Quantum Mechanics, the Standard Model of Particle Physics, String-theory, etc.** While doing so, he has empirically pointed out that these theories are not perfect; they contain a few loose ends.

The author, not limiting the book to the above extent only, has also expressed an entirely different viewpoint on the deep mysteries of the universe, namely- **the Black-Holes, the Dark-Matter, the Dark-Energy, and the origination of life, etc.**

Moreover, apart from the subject matter described above, this book also includes a lot of uncommon but valuable information that is not readily available to the readers from a single source, which adds up to the usefulness of this book.

The theories mentioned above are the most revered theories in the entire world, or rather, these theories are considered the Bible of science. As such, the specialists and the staunch followers of science will probably deem the author's act of fault-finding in these theories highly unwarranted, or rather bold, for challenging the great scientists who devised these theories. However, out of various logics put forward by the author in support of his viewpoint, the pragmatic and verifiable circumstances that flout these theories seem reasonable and hard to refute, at least a few of them, if not all.

Similarly, the author's effort to present his own perspective; a much different view from the conventional one, on some of the unresolved mysteries of the universe, would be considered an unauthorized, rather than a presumptuous effort, because he has crossed the unlaid limit for the non-specialists; even then the logic put forth by the author is based on empiric examples and is worth verification.

It is a bitter fact that though the theories referred to above are considered immaculate, they have no answer to some of the deep mysteries. Consequently, scientists worldwide are putting up their best efforts to devise a comprehensive theory. Even a trivial-looking point, if not considered earlier, may help to accomplish this goal; therefore, the tests/experiments proposed by the author, if conducted, may provide some clue.

At this stage, it is not possible to foresee what would be the outcome of the effort, if at all an effort is made in the future, to verify the experiments proposed by the author; however, if the outcome of such an effort matches with the results foreseen by the author, then a new avenue might be opened-up for developing a new theory, who knows? This book would become a threshold to the new-era science in such a case.

Preface to this Edition

Although copyright to the original Hindi manuscript was obtained from the U.S. Copyright Office in early 2013, the English version of this book* couldn't be published before March 2015. The said publication being a hasty and impromptu effort, I couldn't include all of my ideas and logic in that edition. Later, looking at the complexity of the subject matter, I included some additional details in the original manuscript to justify my viewpoint in a better way; with these additions, the size of the book has grown almost double. Even though the copyright to the modified manuscript was obtained in 2019, I could not publish it earlier. Anyhow, the revised edition with a new name is now in your hands.

This book mainly comprises three parts, an introductory part, the second part gives a brief overview of some of the **modern scientific theories, which are supposed to govern the functioning of the entire universe;** this part, apart from depicting a mere summary of these theories, boldly portrays some of the probable loose ends that exist in these theories, namely the **Theory of Electromagnetic Radiation, the Theory of Relativity, Quantum Mechanics, String Theory, Theory of Gravity, the Big-Bang Theory, and the Standard Model of Particle Physics, etc.** The third part discloses an altogether new perspective on the creation of the galaxies, stars, and their planets; moreover, this part sheds light on some unresolved mysteries such as **Blackholes, Dark-Matter, and Dark-Energy.** Furthermore, this book also suggests different ways by which the correctness of the **prospective loopholes,** which might exist in the aforementioned theories, can be verified.

Depending on their taste, most readers prefer to read famous bestsellers written by renowned writers and illustrious personalities. Nobody likes to read books on complex and off-the-track subjects written by an unknown person. Anyhow, I believe that the curious type of readers interested in the branch of science that governs the functioning of the Universe will find this book interesting and valuable. Besides, it is a knowledge bank too. The book will also introduce them to a new and different line of thinking.

This book is a discourse, or rather an inquisitive exposition that mainly promulgates the alternate perspective of the aforementioned scientific theories; besides, it emphasizes the need to identify and tie up the prospective loose-ends

---

* That edition of the book was published under the name **"Mysteries of the Universe-Unveiled."**

that might exist in these theories. To catch up with the spirit of this book, I would request the readers "to forgo their preconceptions, rather their allegiance and the incessant faith toward the existing theories and the originators of these theories," because, with a biased mind, they won't put up any effort to analyze and understand the logic put forth in this book.

I am sure that scientists and experts of physics would not agree with me because of the **bitter fact "It is tough for anybody to appreciate any thought that is against their firm belief, rather their indoctrination."** Whatsoever be the views of the scientists, I firmly believe that in the near or distant future, most of my presumptions would be found correct. Such an eventuality might result in revolutionary changes in mainstream science. I, because of this belief, earnestly desire that all the future scientists who are presently pursuing their studies must read this book, particularly the logic and the reasoning given therein, so that in the future, they might make up their minds to verify the truth and remove the undetected misconceptions (if any) that might exist in the present-day scientific theories.

However, this book not being a textbook prescribed by any of the universities or other educational institutes, I would earnestly advise the present-day students, especially that of physics, that **to do well in examinations, they should, at present, concentrate on the prescribed textbooks only, otherwise, they may not fare well in their exams.** However, after they secure a good job and settle down in their careers, the young scientists may, if they feel so, take up scrutiny and verification of, at least, some of my ideas. If anybody in the future does take up the work of confirmation of my predictions, then I would like that they shall do so without any bias toward the existing theories; they shall aim to establish the truth, not merely to discard somehow the ideas that challenge the correctness of the current scientific theories.

In the last, I would like to make it clear that though this book points out a few probable flaws in some of our prominent scientific theories, its aim is not merely to point out the faults; its motive is to promote a better understanding of these theories from all angles and also to emphasize the need for a thorough review of all such theories.

- Satya Prakash Verma

The Background of this Book and its Motive

For almost the last sixty years or so, I have, from time to time, pondered on various subjects with an entirely different and unusual angle, which is much different from that of the rest of the world. However, I never thought of writing down my ideas all through these years because I never considered myself an author or a scientist. I had to pay the price of this lethargy and indifference of mine; most of my thoughts and ideas have been lost forever.

Now, when I have outlived my life by more than 15 years (at present, the average age of Indians is about 64 years), I realized that it is vital to protect my remaining thoughts from burning to ashes along with me, thus letting my entire line of thinking to vanish forever. Now, I have decided not to bear the burden of some bizarre, queer, or puzzling thoughts myself alone; instead, I will put this burden on those who would willingly accept the same. The aforesaid decision of mine, to share my thoughts with others, has taken the shape of this book, which is on Cosmology and Astrophysics. *Although numerous books have been written to date on these subjects, this book is much different from all of them, so unusual that the readers would probably not be able to digest and accept it. Rather they would reject the ideas expressed in this book. Anyhow, had I not had anything new to write on this subject, I wouldn't have written this book. In this book, I have, apart from revealing a different and unique angle on the said subject, questioned the feasibility of the existing theories. I hope, at least a few, out of all the prospective readers, would examine my puzzling thoughts and try to find out their answers.*

I earnestly desire that the renowned specialists on this subject read this book because the questions and logic are specially meant for them. I wish that this sect of probable reads shall carefully study this book and seriously analyze those logics, which put-up questions on those theories. However, I don't have many expectations from them. I believe that instead of trying to give an appropriate clarification to the doubts raised in this book, they will outrightly reject these doubts because, presumably, they have such a firm belief in the so-called Mainstream Science that they hate to hear anything against these theories. Because of their bias, they are guaranteed to reject these ideas, which are much different from the conventions prevailing in the present era. This book is, therefore, specially written for the curious but non-specialist readers who, despite having no bias of any kind toward the complicated scientific theories, might avert racking their brains to understand these theories or critically analyze them; they simply take it for granted that these theories are absolutely correct and flawless. As against this general idea, I have, in this book, analyzed some of the well-established theories by using simple logic and easy to understand arguments to enable the prospective readers to examine these theories side by side while reading this book, and thereby, according to their wisdom and viewpoint, find out **"what is right and acceptable."** The main objective of this book is to induce the readers to develop a habit of *"not to accept even complicated-looking matters without analyzing and understanding such matters properly and thoroughly."* *I have complete faith that various flaws and controversies existing in the present-day scientific theories will become crystal clear to all those readers who will read this book carefully.*

Here, I would like to clarify that I'm not an expert in the subject matter; higher physics, higher math, or cosmological science have never been my subjects. I do not have any deep and crystal-clear knowledge of Mainstream Science, and as such, writing such a book is, in fact, my bold and unauthorized effort. To compensate for my inadequacy disclosed above, I had to contemplate a lot over a long span of time on various aspects of the subject matter. *This book has purely resulted from the said contemplation of mine.*

The only justification I can give to write this book is **"some of the solutions deduced by the great scientists who devised these theories didn't seem feasible, at least to me."** I, therefore, want to spread general awareness of such problems among as many readers as I can. In other words, I want to spread this disapproval of mine among as many people as possible by putting my efforts out here for consumption by others. **If this dream of mine comes true, i.e., if I achieve success in arousing the curiosity of even a few readers about the potential anomalies in these theories, then this would be an outstanding achievement and a massive reward for me.**

To achieve the aforesaid goal, I have, through this book, raised a few questions on some of the unnoticed anomalies that have been bothering me for a long. I hope these questions might arouse the reader's interest in this subject and compel them, too, to contemplate these anomalies. **For me, this book is a tool not only for myself alone to search the truth, but it is also a means by which I aim to induce as many people as possible to do the same.** *I believe that the effort put up by such people, whether sought to prove my thoughts wrong or to verify the correctness of my opinions, would undoubtedly bring out the truth. However, if no such effort is made, the truth could never be unveiled.*

In this book, I have, by going beyond my limits, also suggested a few solutions to some of the enigmatic problems. Even though my suggestions are based on proven rules of fundamental physics, my solutions might not conform to prominent scientific theories. However, if the suggested solutions are rejected without verification, then this would not only be unjust to me, but such rejection might also hamper the growth and development of science. **I can only hope that my thinking might not be considered disapproving or reverse thinking or considered an unnecessary effort to "split the hair;" in fact, neither have I opposed any theory just for simply challenging without any purpose, nor have my thoughts gone aberrant in any way.** After lengthy deliberations at my level, I brought out **what seemed correct to me.** I would request those readers who disagree with me to minutely evaluate various logics presented in this book before forming an opinion on such points.

There is a strong possibility that no one who is well-versed in science would, out of his bias, agree with my assertion that modern scientific theories may have some shortcomings. Foreseeing this possibility, I have, in chapter-1 of this book, described the circumstances under which errors are prone to creep into any newly formed theory unknowingly. Under such circumstances, it becomes almost impossible to identify when and where the error was committed. Therefore, *I have not hesitated to put questions on some of the theories of the present era without caring how great the scientists who put forth these theories were or how low my status is compared to them.* I have only questioned a theory that didn't seem proper, correct, or realistic to me. Raising questions is one of the easiest things in the world, isn't it? Readers would agree that *questions with a definite aim are the best tools for deducing correct answers, and science always seeks to answer each question.*

Very few of the readers might agree with the points raised in this book, whereas most of them, out of their bias, will disagree. *I feel if any* **probable error** *that comes into the notice of somebody is not questioned immediately, then such an error, if it actually exists, could never be eliminated.* **My intention is not to condemn anybody by pointing out their probable mistakes; I aim to assist the scientists, if they feel so, in identifying and eliminating all the possible errors that might have crept into our theories.** *In this world of uncertainties,* **under the circumstances described in chapter-1 of this book, anybody is likely to commit errors, may he be an eminent scientist or a layman.**

I want to clarify that English is not my first language; this book was written originally in *"Hindi," the unofficial National Language of India.* Since I aim to reach as many readers as possible, *I have re-written this book in English.* However, I first think in Hindi and then translate the same in English; therefore, the language of this book might appear a little odd to the readers because the sentences constructed in Hindi differ vastly from those conceived in English. Consequently, I would request the readers not to judge this book simply by the language; instead, evaluate my work by the ideas and logic brought-out in this book and the alternate angle with which I have viewed various theories.

While writing down my ideas, I realized how difficult is the work of writing; I didn't have to put much effort to conceive the basic concept of this book, I didn't have to toil to point out the perspective anomalies in different theories and develop various logics on distinct topics, to support my point. However, not being a ***professional writer*** *or an expert on the* ***English Language****, I had to put up comparatively lot more effort to express my ideas; while writing down my thoughts, I had to repeatedly cudgel my brain to select proper words and correctly use them; I had to devote a lot of my time to this purpose. In this effort, I realized that though it is pretty easy to conceive an idea, it is far more demanding to express them in words in a convincing manner and to convey the exact meaning behind those ideas; I realized how difficult and demanding is the profession of Writing;* ***hats off to all the Writers and the Poets!***

I have though completed this book, but not being an expert writer, I am not sure whether I have presented my thoughts clearly and adequately; whether I have succeeded to convey my thoughts precisely and articulately, I also don't know to what *extent I would be able to achieve my aim to intrigue, i.e., to arouse the curiosity of at least a few readers.* I believe that despite my limitations described earlier, I have completed my job to the best of my abilities; now your (reader's) turn has begun; you, the readers, have to decide the fate of this book. I am sure that if the book has any substance, it will establish its place sooner or later; otherwise, it will be lost in the darkness of anonymity. I am afraid that in case scientists/readers would not pay any attention to analyze various arguments brought out in this book, or they would negate these arguments without even trying to understand and analyze them, then also this book might meet the same fate, it will be doomed to oblivion; either the book itself or the future will decide the fate of it. *I feel that this book is like a bomb that can shatter at least a few misconceptions existing in the present-day theories and bring revolutionary changes in the world of science, or instead change it completely. However, this bomb is bound to fail to click or explode because masses always believe in miraculous solutions; their faith is so firm in the renowned personalities who deduced such solutions that there is no place for any argument against these theories. Accordingly, no one would believe an unknown person, like myself, who is trying to shatter these ostensible miracles. By nature, humankind is an easy-going critter; he always avoids the hard way or to go against his faith. Accordingly, nobody would be willing to accept the alternate ideas brought out in this book that contradicts the present-day theories. Therefore, I would advise, rather request the curious readers who are really interested in understanding the basic idea of the scientific theories covered in this book shall read this book with patience and without any bias towards these theories, which are considered immaculate and irrefutable. I am confident that an unbiased and rational analysis of the logic brought out in this book would enable the readers to understand the substance of this book. In case they do so, then and only then, they would appreciate the spirit of this book and enjoy it; such people, if convinced by the logic and reasoning put forth to support my predictions, will start to view these scientific theories from an altogether different angle.*

— Satya Prakash Verma

"Satyendra"

PART – 1

The Evolution of

Science

&

Its Fundamental Laws

1: The Journey of Science

Science and primitive man both evolved and flourished side by side; the same will continue in the future too. The monkey-like ancestors of human beings, such as "Sahelanthropus," "Orrorin," "Ardipithecus Kadabba," "Pan Prior," and some other similar species, came into existence as early as six to eight million years in the past. Species named "Australopithecus" having mixed Apes and Humankind features did exist about 4.4 million years ago or even earlier. As time passed, the above species became extinct or evolved into different kinds of primitive manlike creatures due to natural selection and transmutation. The existing human species were supposed to have evolved much later, about two to three hundred thousand years in the past. Of late, 2.6 to 3.4 million-year-old stone tools have been discovered; also, 300-thousand years old hearth has been found in Israel. These discoveries indicate that the ancestors of the human species or other similar species might have evolved much earlier than estimated. In the beginning, humankind lived like other wild animals; the only difference between them and other animals was that humankind had much bigger brains. The primitive man was far more curious; he used to observe everything with great curiosity; he also tried to understand almost everything that happened around him. This extra curiosity and his lust to know the "unknown" separated humankind very slowly from the rest of the animals.

A primitive man, according to the limited capacity of his not-so-developed mind, used to try to understand all the strange-looking events that did happen all around him; he also tried to solve different problems faced by him in day-to-day life. Continuous pondering over these problems resulted in the gradual development of his mind. His knowledge, as well as the ambit of his curiosity, both kept on increasing, though very slowly. From thence, perhaps, some bizarre, strange, and curious questions might have started to bother him, such as "What difference exists between live and dead bodies, what is life, what is death; how the death could be skipped, etc.?" Probably, some other problems that might have bothered him were "What the stellar objects like the Sun, Moon, and stars, are? Where they come from, and where do they disappear.?" Such complicated questions have been bothering the human race since eternal times; even in the present era, similar questions still bother him with the same severity and intensity.

How ignorant was the early man? How limited, parochial, or narrow was the ambit of his understanding and imagination, how baseless conclusions he would have deduced, etc., can be understood by the following story that I read in my childhood:

Once, in the long past, there lived a man in a small village in India who was known as "Lal-Bujhakkad." In the Hindi language, the word "Lal" means Red; however, in this context, it means "The best." Another Hindi word, "Boojh," means "Perception or Understanding." The term "Bujhakkad," which is originated from the Hindi word "Boojh," means the one who has the best Understanding. Accordingly, the title 'Lal-Bujhakkad" stands for "the most knowledgeable and wise man." Whenever the villagers failed to understand anything, they used to take the advice of Mr. Lal-Bujhakkad. One night, while passing through the village, an elephant left behind big round footmarks at various places. Since the villagers had never before seen an elephant, nobody could guess "what those round marks were?" Therefore, all the villagers approached Mr. Lal-Bujhakkad, who exclaimed, "Fools! Nobody except Lal-Bujhakkad knows everything; it was an antelope which, having tied big stone disks to his feet, jumped through the village." Everyone was convinced by this wise (?) proposition of the so-called wise man.

People like **Mr. Lal-Bujhakkad** are found everywhere, even in the present era too; the masses consider such people very wise; they have blind faith in them. Commoners believe that whatever these wise men say is absolutely correct without verifying such people's ideas. **The mentality of the common people and the way they think has not at all changed over all these years.** In fact, **a Lal-Bujhakkad is hidden within every man, who makes baseless and easiest-looking speculations to solve even the most complicated problem.**

Almost since the last 100-thousand years or so, various tribesmen, depending upon the limited capacity of their primitive minds, have fabricated some weird stories that our world was created according to the wishes of some Devine-self; such propositions were probably the easiest and the most acceptable solutions at that point in times, even in the present era too. Commoners always seek the easiest solutions for which they don't need to exert their brains by even a bit; this explains why miraculous solutions always attract them. However, such miraculous beliefs couldn't deter any curious people from traveling the Far East and Far West, searching for the homes of the sun, moon, stars, etc. Despite such unfruitful efforts, mankind never accepted defeat or gave up its efforts; his curiosity persistently compelled him to think over again and again and continue his efforts. The consistent racking of the brain resulted in the gradual enhancement of the ambit of his quest. In this effort, mankind established the relation between some constellations of stars and the cycle of different seasons; later on, such observations helped them in the cultivation and harvesting of various crops.

The never-ending necessities of human beings always got preference over his unquenchable curiosities. Even if we fail to understand any complicated and strange-looking phenomenon or fail to find the answer to any quaint question, our day-to-day works go on in the usual way at our own normal pace. The everlasting issues of "hunger" and "security" persistently bother humankind. The fulfillment of these necessities always gets priority over all other things. Mankind, according to his needs, made weapons, started to rear cattle, started cultivation, and built huts. Very soon, the hutments developed into villages and then into cities. His necessities also persisted in growing with such changes. Despite the intensity of those necessities, his lust to explore the unknown persisted in bothering at least a handful of curious people. The degree of the complicacy of those eternal questions, which existed in prehistoric times, couldn't yet be diluted a bit. The commoners, though, were equally bewildered by such enigmatic problems, were not willing to put up any effort to solve these problems; merely a few curious people, who were considered eccentric, obsessed, or even crazy; and who were not concerned with any worldly problem, continued their efforts to find the answer to such enigmas. On the other hand, depending on the capacity of their primitive minds, the influential people put forth different speculative solutions; however, those solutions were based on immature contemplations and limited observations, etc. During each era, such solutions would have quenched the curiosity of the masses only temporarily. However, after a short period of time, the shortcomings of such solutions would have come to the notice of the masses. Such observations of the groups would have led to new questions. Accordingly, new solutions would have been thought of repeatedly; the old speculations would have been modified time and again or even abandoned.

Right from the very beginning, only the witch-doctors, chiefs of the tribe, the elders who were considered wisest among all, or any other influential man having a superior position in the society enjoyed the authority to give speculative explanations of any natural phenomenon and foist their ideas on the commoners. The speculations of such influential people must have been considered sage and knowledgeable thoughts, which would have been transferred from one generation to the next. Only another powerful man or witch doctor might dare to give another explanation of such phenomena. Any attempt by a commoner to do so was sure to attract the death penalty. The self-pride and arrogance of the influential people occupying top positions in society and their "mental inertia," i.e., the dogged belief in their conceptions, constantly thwarted the development of any new thought. Such influential people never allow any new thought to flourish other than the prevailing ones.

While looking back into history, we find that **Socrates** was forced to take poison; the discovery published by **Copernicus**, which was *published* just before his death, was ignored; *Giordano (Filippo) Bruno* was burned alive in the year 1600 because he was probably the first man who dared to proclaim that all the stars are distant suns surrounded by their own planets. A little while after that, **Galileo** was put under house arrest for **discovering that the Earth goes around the Sun;** *the death penalty awarded to him for this discovery was revoked only on the condition that he shall agree publicly that the sun and all the stars orbit the earth.*

Many times, in the past, the obdurateness of the influential people resulted in the thriving of wrong concepts that lasted for several generations. As a result, the journey of science was diverted in a negative direction several times. Such obstructions, though, slowed down the rate of progress of science but couldn't stop its progress forever. Even under such circumstances, the ambit of knowledge persisted in growing gradually, though at a much slower rate. Despite such obstacles, the knowledge of humankind continued to grow; it was enriched step by step, i.e., the knowledge gathered by the earlier generations was refined and further improved and enriched by the new generations by way of some additions and also by the elimination of some of the miss-concepts. A series of such improvements and new misconceptions has continued since the beginning; the same is going on even in the present era, too.

The preliminary scientific concepts are always based on imagination, guesswork, hypotheses, etc. Later on, these concepts are tested on the criterion of truth. However, complete truth could never be discovered in the one go; usually, truth is found in several small steps; each of such part-truths could be discovered after a long gap of time. *It is ironic that though every part-truth is a truth in itself, the same is not a comprehensive truth; the same may differ from the absolute truth.* This fact could be understood from the example given here: *"at times, when a rat couldn't find a way to escape, it launches a counterattack on the cat."* If, based on this partial truth, one concludes that *"rats are capable of trouncing cats,"* *then it will be a wrong conclusion based on an incomplete truth. Based on this reason, sometimes, even an eminent scientist may unknowingly deduce wrong conclusions. The fact "an obvious, but incomplete truth may differ from the absolute truth" is evident from the example given below:*

An ancient story may clarify the above point: Once, some blind students were asked to touch and feel what an elephant is like? To have a real feel, different students touched or groped different body parts of the elephant. Based on personal experience, everybody developed a concept that was different from that of others. The student, who touched the leg, speculated that the elephant looked like a column. The one, who touched the tail, thought that the elephant was like a rope. Likewise, another student conceptualized that elephants are like walls, yet another student thought elephants are like pythons. **All of the above concepts, having been based on incomplete and part-truths, were totally wrong.**

Even though **the story of "Blind Students and the Elephant" is merely a story, the same has been repeated several times in the history of humankind-right from the primordial times to the present date; in fact, this is the way science has gradually grown on its journey of evolution.** *Scientists face similar situations on many occasions; they never get complete information before devising any theory. Instead, they discover part-truths in several small steps, each of which is discovered after a long span of time periods, which is* **analogous to a concept developed by a blind man who, at a time, touched only one body part of the elephant.** *Scientists can, therefore, consider only one aspect of a problem at a time; they encounter other aspects of the same problem at a much later point in time.* **At times, the incomplete truth, so discovered because of an incomplete input, might lead to misconceptions, falsified results, and blind-faith, etc.; all of them spread like contagious diseases. The blind faith so developed, whether it may be in the field of religion or science, always suppresses the truth.** *Sometimes, such misconceptions, conceived by some renowned personalities, are even considered very brilliant ideas and valuable achievements. As a result, the heritage of falsified knowledge has been transferred several times in the past to at least the next 3-4 generations.* This becomes possible because the common person blindly follows the renowned people who are considered extremely wise; usually, no one even bothers to verify the truth. Such a propensity of humankind is one of his greatest misfortune. **Misjudging any fact, or considering such misconceptions to be valuable discoveries, might cause science to divagate from its path of finding out the absolute truth;** *a very long and valuable time might also be lost in the elimination of such misconceptions.*

New and more comprehensive information may be generated frequently from different sources in the present era. As a result, the new findings might replace the old ones. However, the bank of knowledge can never fill up completely; of course, it keeps growing bit by bit. Accordingly, the old conclusions are modified in quick successions. *Whenever any new mystery is unveiled, it points to another and yet another mystery. The face of truth is hidden behind infinite layers of veils; if one tries to unveil the truth, then new veils come up one after another;* **the truth, therefore, remains as mysterious and elusive as it was before.** Under this situation, further questions come up one after another; *humankind's curiosity could never be quenched. Instead, the same is enhanced on every such occasion.* This growing curiosity helps science to keep progressing gradually, in various steps. Analogous to the phrase: **"Necessity is the mother of inventions,"** it is also true: **"Curiosity is the mother of science."** Till the time the curiosity of humankind remains unquenched, the onward march of science will continue.

Until humankind was living the life of a wild beast, the pace of development of science was very-very slow. After the end of the last Ice-age, when the life of human-being became a little easier, he might have gotten more time and chances to explore the **unknown.** As a result, new ways to improve agriculture and better construction methods were invented; new and improved weapons were also made. Some of the oldest civilizations, such as **"Egyptian," "Maya,"** and **"Inca,"** etc., made various discoveries; however, the books of the Maya and Inca people were declared *Satanic* and destroyed by their conquerors. Probably full details of the discoveries made by the Chinese scholars during the prehistoric ages are also not available. The ancient Aryans are supposed to have migrated to India sometime around three thousand years BC. The Aryans probably migrated from the Middle or North-East Europe, where after the end of the last ice age, the swamps, formed by the melting ice facilitated abundant natural springing-up of the legendary "Vedic Soma plant." The Aryans made remarkable discoveries in mathematics and cosmology; credit for inventing the concept of "zero" goes to them. Value of "Pi" (p) and "Pythagoras theorem," etc., were also known to them since the unknown era. The concept of **atoms** was also developed by them during the **"Rig-Veda Era."** Thousands of years before Christ, the Vedic Sage **Augustya,** in his book "Augustya-Samhita," described a method for constructing the Electric-Cell. An Aryan scholar **Vishnugupt** was the first astronomer to describe the property of the earth *to attract everything**. Later on, one of the Aryan scholars, **Aryabhata**, deduced that the Earth spins on its axis based on the westward movement of the stars. He also accurately calculated the speed of the Earth's spin, the time period taken by the earth to go around the sun, the distance between the sun and the earth, etc. He also very accurately calculated the earth's circumference, the time of eclipses of the moon, and the sun. Besides the above, the ancient Aryans also made various other discoveries. However, invasions by various outsiders resulted in the loss of most of their discoveries, especially after the destruction of the **"Nalanda"** and **"Taksh-Shila"** universities of ancient India.

The thoughts and the books of ancient Greek scholars, such as **Aristotle, Socrates, Pluto, Ptolemy, Archimedes,** etc., are still safe and well preserved, which could be considered authentic. Before the era of the above scholars, it was believed

---

* The Written reference of Gravitation was first quoted in the Atharva-Ved, under Prashn-Upanishad (1st millennium BCE).

"The Earth is like a flat plate or disk." Several weird beliefs related to our Earth prevailed in that era; according to some of these beliefs, it was thought that our Earth is poised either over the head of a giant snake or on the back of a tortoise. In some other countries, it was believed that the Earth was held up by a titanic giant, *Atlas,* over his shoulders. Such beliefs kept lingering on for very-very long periods of time; these beliefs were also accepted and adopted by various religions. During that era, it was also believed that the sun, the moon, and all the stars, etc., revolve around the earth, which is the center of the entire universe. Around 300 B.C., Greek philosopher **Aristarchus** predicted that the Earth goes around the Sun, but nobody took him seriously. For the next one and a half to two thousand years, the ancient belief remained unchanged. In the year 1514, **Nicolas Copernicus** said that the Sun does not orbit the earth. Instead, the earth goes around the Sun. However, for a period of further 100 years, nobody believed him. After the invention of the telescope in 1609, **Galileo Galilei** proved that the earth and other planets orbit the sun. *He was, unfortunately, punished for the aforesaid discovery instead of the grant of any reward or recognition.*

We thus see that later on, most of the predictions, not all, made by the prehistoric philosophers were found wrong. Based on this fact, we shall not infer that their wisdom and brains were underdeveloped or inferior. In fact, instead of their brain, their means were inferior; those philosophers were, indeed, owners of very sharp minds, very keen wit, and were the wisest thinkers of that era. No database of the knowledge or details of the observations made by their ancestors, etc., were available to them. They also didn't possess the modern facilities or sensitive and sophisticated instruments that the present-day inventors enjoy; there was nothing to guide them or shape the line of their thinking. They depended only on their imaginations and deliberations in the absence of any clue, guidelines, or systematic strategy. *Such unsubstantiated visions are always susceptible to aberration, especially if such speculations are based on wrong inputs.* This point could be well understood by the example, *"In case someone's calculator malfunctions, or else he has pressed the wrong button, then he is certainly going to get the wrong result. Likewise, if an effort is made to solve a problem without identifying its root cause, then the possibility increases manifold that he will get the wrong solution."* This could be best understood by the story of two wise professors. I read this story in my childhood days; the same is narrated below in brief:

Once, while walking down a street, two very learned professors saw a ball made of some alloy that was unknown to them. One of them, out of curiosity, touched the ball and was astonished to note that despite the bright sunlight and smoldering heat, the ball was much cooler at the top. The other professor also examined the ball, and he was also astonished to find that it was much warmer at the bottom. **Both the professors, who knew everything except the root cause of this problem,** instantaneously formulated two different theories and started to discuss their theories about the abnormal behavior of that unknown material. Just at that moment, a laborer came there and turned that ball upside down. When the professors inquired about the cause, the laborer explained: *"He rotates the ball, from time to time, so that it might not get overheated from a single side."* **Clearly, it is not possible to solve any problem without establishing and properly analyzing its root cause; any solution that is based on the wrong inputs is sure to go wrong. Similarly, when a doctor makes a wrong diagnosis, then he would not be able to find the true cure of the disease; he surely would give the wrong treatment. These examples clearly tell that any plausible theory that is based on incomplete facts and without establishing the root cause of the problem is likely to flout in the future.**

×　　　　　　　　×　　　　　　　　×

Indeed, some of the predictions made by our ancestors were afterward found erroneous; however, this doesn't mean that they were unwise people having underdeveloped brains. **Such mistakes were not resulted due to the limitations of their brains; instead, they resulted due to the limitations of their means.** In the absence of any database, **they were not able to establish the root causes of the problems they encountered-with.** No one is supposed to establish the root cause of any complicated issue in his first attempt. **The line of their thinking was probably, obsessed with the wrong speculations made by their ancestors. Even a wise of the wisest man, if obsessed with a misconception and prejudice, is bound to commit mistakes.** Since our ancestors were not blessed by proper means or rich heritage of knowledge, the credit for their achievements goes purely to their **wisdom, their efforts, and their obsession to find out the truth.** *It is the initiative that is important; anybody can commit mistakes, but very few people initiate action to discover the unknown.* The efforts made by our ancestors roused the curiosity in the minds of the next generations and inspired them to use their brains; they also blessed the next generations with the heritage of rich wealth of hard-earned knowledge. They were pioneers in this field; we are only following in their footsteps. We are highly indebted to them for today's achievements of humanity.

The brilliance and wisdom of our ancestors could be evaluated from the fact that, based on very ordinary observations and minimal means, they made some very remarkable discoveries. A discovery made by the Greek scholar and mathematician *Eratosthenes* is an excellent example of this fact. About 240 years B.C., he observed that the wooden poles installed vertically in the city of "Swenet" didn't cast any shadow during the summer solstice, i.e., on the local noon of June 21-22, when the sun reaches its highest position in the sky, i.e., directly above our heads. This city, located on the Tropic of Cancer, is nowadays known as "Aswan." On the other hand, similar poles installed in the city of "Alexandria," which is located about 800 kilometers north, did cast a shadow on the same day. Based on this very ordinary-looking fact, *Eratosthenes* proved that the earth is spherical in shape and is tilted on its axis at an angle of **27½°.** Based on the same fact, he didn't only calculate the circumference of the Earth, he also very accurately calculated the distance between the earth and the sun. *Even in the present era, how many people would make similar discoveries or even take notice of such an ordinary fact?*

The above-quoted example brings out the fact "All the work of the ancient scholars was not wrong; Science, based on the discoveries made by them, has progressed step by step." However, they, at the same time, also conceived numerous misconceptions. Probably, their minds were prejudiced by the old school of thought. Any misconception or prejudice, if infused in anybody's mind, is capable of influencing and shaping the thinking of the next several generations; it also thwarts the springing-up of new ideas for a long time. Typically, most people jump to the easiest solution without properly analyzing a complicated matter or finding the root cause of the problem, the reason being such solutions match with their preexisting conceptions or with the thoughts of the renowned personalities. Whenever such easier solutions are selected and accepted by the masses, then thinking of mankind becomes stagnant, it keeps revolving around a wrong point. Analogous to the phrase *"One has to tell hundreds of lies to hide even a single lie,"* it is also true *"If a misconception is once conceived by someone, then hundreds of misleading pleas may come up in his mind in support of that wrong concept."* In that case, the result obtained from any experiment would be interpreted and explained in such a way (by ignoring or hiding some facts) that the correctness of the prevailing conceptions/theories can be proved. *It is a human tendency, whenever anybody is bent upon proving something, then his mind begins to think only on such lines that would enable him to achieve his goal; the logic that would support his contentions would alone sparkle in his mind; he won't think in any other line.* It is the nature of the man that he can't realize his mistakes, and if at all he realizes any mistake, then he won't accept it easily; he would persistently try at his level best to prove that he was correct. Such a doggedness of mankind results in the stagnation of the progress of science for a short time only, not forever. *Under such circumstances, the direction of anybody's thinking may stray, be he an ordinary man or a specialist; if someone is wearing colored glasses, then he won't be able to see the true colors; this is not the fault of his eyes, colored glasses are responsible for the same.* And therefore, *faulty preconceptions, incomplete feedbacks, and non-establishing of the root cause of any problem, etc., are to be blamed for anybody's mistake, not any particular man in person.*

At times it also may so happen **"In case someone's mind is completely engaged in solving a complicated problem, then he considers only big-looking (complicated) and difficult solutions, though easy solutions might be available readily, the same might either fail to attract his attention or such solutions might appear worthless to him."** The following joke, which I read during my early childhood, explains this fact explicitly.

Once, the cat of a great scientist gave birth to a few kittens. When the kittens grew up, the scientist called a carpenter to make two holes in the door, the big one for the cat, and the smaller one, for the kittens. The carpenter pointed out that only one opening would serve the purpose, but the scientist didn't believe him. However, when the carpenter made a single hole in the door, then the cat, along with her kittens, easily passed through that opening. *This can be inferred from this joke* **"At times, almost everybody fails to realize that the easier solutions may also resolve enigmatic problems." Besides, the roots of some prejudices may, sometimes, occupy our brains so deeply that we normally select only those possibilities, out of numerous others, which match our preconceptions.** However, such a situation doesn't persist forever; at last, the latest discoveries or new thoughts, if correct, are bound to be accepted. Preconceptions may deter the progress of science for some time only, not forever.

Although we are now backed by very sophisticated equipment and very vast knowledge, the direction of our thinking is still based on scientific theories that are only 100 to 150 years old. On the other hand, the universe is as old as several billions of years. In fact, a theory can be said to be a good theory when it solves all the problems without fail; whenever any theory fails to provide a viable solution to even a trivial-looking problem, then such a problem shall not be ignored because of its trivialness; instead, the theory shall be reviewed and modified, if required, to solve such a problem too. However, most of the cosmological phenomena need several millions of years to complete, therefore, we have not yet gotten sufficient time to verify the truthfulness of our existing theories; much longer observations are probably, required to verify these theories.

As early as 5-6 hundred years B.C., our ancestors made several remarkable discoveries, and at the same time, they also developed some misconceptions; the roots of such misconceptions are still seeded very deeply in our minds. Our history is a witness to the fact that "The ideology that was once considered the extremity of wisdom, was, after some time, replaced by new ideas and new thoughts." Science has always made its progress in this manner only; at times, its onward march had to halt at several places on its way; however, its voyage kept on cruising, though in small steps. It has happened several times in the past, and, therefore, one shall not wonder if history repeats its course of action sometimes in the future too. The journey of science is still in its infancy; we still have a long way to go; *who knows what will happen in the future? One shall not wonder if the journey of science takes a new turn, it might adopt an altogether new direction in the course of its further development; who knows?*

2: A Broad Introduction To -
The Most Basic Laws of Science

Our universe is so vast and such a wondrous place that it always astonishes the scientists and the common man equally. After the 15th to 16th centuries, i.e., after the invention of the telescope, thousands of people are gazing at the sky, day and night. As a result, discoveries are made almost every day. To understand how big our universe is, it is, first necessary to understand *"From how far the light of the distant stars is reaching us and how much time it takes in this process?"* The rays of light travel at an unbelievable speed of ***299,792.458 kilometers*** in one second; in one hour, it travels one billion and eighty million kilometers; at this speed, light travels about 9.5 trillion kilometers in one year; this distance is called one light-year. Light, despite traveling at such a high speed, takes an unbelievably long time of more than 13.2 billion years to reach us from the furthest stars. The universe is probably spread over a far-bigger three-dimensional region. The effort to imagine such a vast distance, in kilometers or miles, may bewilder us; it is not possible to even imagine such a vast distance. The following facts may give a rough idea of the vastness of the universe, which is spread spherically over a radius of 13.2 billion light-years; our own galaxy, the Milky Way (the congregation of billions of stars, of which the solar system is an ordinary member), is spread over an area measuring only 1 to 1¼ hundred-thousand light-years, our nearest star is about 4 light-years apart from us, the light coming from the Sun takes only 8.3 minutes to reach us, light takes a time of about 5 seconds to travel across the diameter of the Sun, the Moon is only 1.27 light-seconds apart from us, and the Sun is about 1,300,000 times bigger than the Earth.

On the one hand, the universe is so vast, and on the other hand, the sub-atomic particles are so tiny that it is almost impossible to see them. When the magnified image of an *atom* (magnified over several hundred thousand times) is viewed, it looks like a tiny shimmering point. As of today, it is not possible to see the inner construction of the atom. Size-wise, the Hydrogen-atom is almost 10 million[th] of one millimeter; the nucleus of an atom is comparatively much smaller; it may be about 40 thousandths to 100 thousandths of the atom. The nucleus of an atom is made up of different kinds of sub-atomic particles that are even much tinier; they are so small that their size cannot even be imagined. Several kinds of such tiny particles are found within the atom, out of which some particles have very short lives; they decay within one billionth part of a second. Within these two extremities of the ***distances and time intervals*** discussed above, our scientists, to unveil the mysteries of nature, have to keep a vigilant eye around the year on the entire universe and almost all the natural phenomena happening at the sub-atomic level.

The aforementioned extremities might be beyond the imagination of a common man; therefore, at least some knowledge of the basic science seems necessary to broadly understand different achievements of modern science, as well as that of the working of the universe. This book is totally based on the most-fundamental knowledge of science; with the help of some very simple and fundamental laws of science, this book points out some possible shortcomings in some of the well-established scientific theories of the modern era. To understand my viewpoint, some of the readers might have to abreast themselves with the most basic lessons of science that they might have learned during their early student life.

No! Please don't panic by the name of science; no intricate mathematical equations or cumbersome formulae are included in this book. The information given below can be considered general knowledge or an introduction to fundamental science. Most readers might still remember all this stuff; all those who understand these terms may directly jump to the next chapter without wasting their time. However, the readers who have never learned even elementary science or who have forgotten the earlier lessons shall read these definitions whole-heartedly. They will understand these terms without any difficulty; all this stuff is very simple and easy to understand. If someone finds it difficult to understand the forthcoming chapters, he may repeat these definitions once again; everything would probably, become clear. ***Readers might have to concentrate a little on the subject matter because this book also brings out the alternate side of science, which is a little different from the prevailing ideology.*** Therefore, readers are requested not to read this book with any prejudice or disbelief. Although this book is very simple and easy to understand, readers are requested not to read this book with indifference or in a haste, i.e., without paying proper attention to various logic(s), otherwise, they won't be able to catch up with the spirit of this book.

The most fundamental scientific terms and their definitions are given below:

Force

Everybody must have heard that somebody is very strong, or a storm was very powerful, etc. Merely, by looking at a person, or a storm, it is not possible to estimate the exact strength of either of them, this could only be done by the fact that how much load a person can lift, or how destructive the storm was, i.e., how much force any person or a natural phenomenon can exert. It is not possible to see or measure the *force* directly; *"force"* is a quality, capacity, or influence, which can either move an object or at least produce a tendency in that object to move.

Work

The activity involving the physical displacement of an object is known as work. When any object is displaced from its position, even a little bit, then and only then, work is performed; no work can be done without applying a force. If an object couldn't be moved by applying a force, then no work would be performed; only a tendency to move that object in the direction of the applied force would be produced. It is not at all possible to move an object in the direction *opposite of the applied force*; this means **"It is practically not possible to perform negative work."** *The amount of work done is measured by multiplying the mass of the object and the distance it moved.*

Energy

Energy is an indirect capacity of an object, person, or machine, by which it can perform some work or at least produce a tendency to perform some work. In other words, energy is that indirect property or capacity of an object by which it exerts a force on other objects. The energy of an object cannot be measured directly; it can be measured by measuring "How much work it can perform."

Energy can neither be produced nor destroyed; it could only be transformed from one form of energy to another form; for example, vehicles are driven by converting heat energy into mechanical energy.

Power

In the colloquial language, it is hard to differentiate between *power* and *energy;* any one of these two words can probably be used for the same purpose. However, in the scientific sense, power means "How much work any object, natural phenomenon or a person can perform in a given span of time." In other words: at what rate do they expend or consume *energy,* or simply at what rate they can do the work. Machines, in a given span of time, can perform many times more work in comparison to any man or animal. Machines are, therefore, considered more powerful because they can perform more work than living beings in much smaller time periods.

Inertia

The property of matter by which it tends to continue or maintain its existing state of rest/uniform motion or its ability to resist any change in any one of the above two states is called **"inertia."** Because of its inertia, any piece of matter or object continues to remain in its existing position of rest until any force is not applied to displace it. Likewise, a moving object will continue to move at a steady, unchanged speed and unchanged direction until it is not stopped by applying a force. For example, brakes have to be necessarily applied to stop a moving vehicle.

Since the heavier objects offer more resistance to any change in their state of rest or motion, this resistance is directly proportional to their respective masses; the mass of an object is also considered its inertia. Accordingly, greater force is required to induce motion in the heavier objects, that is, to break their inertia. Likewise, greater force, proportionate to the mass of different moving bodies, is required to stop them. On the Earth, all the moving objects come to a halt due to friction, whereas, in the absence of any sort of friction in the interstellar space, the objects like satellites, spaceships, etc.; persist in keep moving continually due to their inertia; however, their speeds, or the direction of their motion, can't be changed without applying a force.

Speed and Velocity

Speed means *the rate of distance traveled by a moving object per unit time-period, i.e., how much distance an object, person, or energy wave such as light (photon), etc., travels in a unit time period, i.e., within the duration of one second, one minute, or one hour, etc.* On the other hand, the *"Velocity"* of an object is its **speed in any particular direction.** The difference between **speed** and **velocity** can be understood by the example of a car that is negotiating the curvature of the road at a steady speed. Even though the car maintains an unchanged *speed*, its *velocity* would still undergo continuous change due to continuous change in the direction of its motion.

In case the force, which causes an object to move, is removed after imparting a certain speed to that object, even then, that object, due to its inertia, will continue to move at a steady speed and in the same direction, subject to the condition that no force causes its speed to accelerate or retard. The bullet fired from a gun is the best example of this fact; the bullet, after leaving the gun, would continue to move if not stopped by the friction of air and the gravitational force of the Earth.

Acceleration

In case the force, which causes an object to move, is not withdrawn, then the speed, or rather the *velocity* of such an object, would continue to increase under the influence of that force. This increasing speed, rather than the rate of increase in speed, is called the *"Acceleration."* For example, the velocity of the falling objects continues to increase because the gravitational force, which continues to act on such falling objects, causes their velocity to increase gradually and persistently.

The *rate of* *change* *in* *the* *direction* *of* *the velocity* of any object is also known as **acceleration.**

Momentum

In the language of physics, **momentum** represents energy stored in any moving object; such stored energy is proportional to the product of its mass and velocity. Everybody might have noticed that heavier hammers are more effective while driving nails into the wall. This suggests that any moving body possesses some energy stored within it; the heavier the body is or the faster is the rate at which the same is moving, the higher would be its stored energy. Such energy, stored in moving masses, is called momentum, which is proportional to the mass of the object and the driving force that causes the object to move.

Angular Momentum

Similar to the property of the momentum of the moving bodies, energy is also stored in the rotating objects; this energy is also proportional to the mass of rotating bodies and their rotational speeds. Angular momentum is an important property of rotating bodies because it resists any change in the direction of the rotational axis of such rotating objects, or in other words, it gives stability to the axis of rotation of such rotating objects. Bikes or motorcycles, etc., have become an inseparable part of our day-to-day lives; the angular momentum of their rotating wheels and the engines provides balance to such moving bikes.

The Centrifugal Force

Everybody might have observed, "When an object tied to a rope is swung in a circular path by rotating the rope, then some tension is produced in the rope which depends on the mass of the rotating object and its rotational speed." The heavier would be the object or faster the rope is swung, the higher would be the tension produced in the rope. If the rope is released, then the object flies off in a straight path and falls at a far-off distance. The force applied by the hand acts on such an object through the rope, which tries to throw it away. As a result, a velocity in a straight line and perpendicular to the rope is produced in that object; the object, because of its inertia, tends to move in a straight line; on the other hand, the rope pulls it back, it doesn't allow that object to go anywhere. Accordingly, the direction of its velocity keeps on changing continuously; the object tries to move in a straight path, but the rope forces it to go around in a circular path. This produces tension in the rope, which increases with the increase in the rotational speed. The force that causes such an object (moving in a circular path) to move away from the center is called the **"Centrifugal Force."** This tension produced in the rope produces another force in the rope, which acts in the opposite direction, i.e., toward the center. Such a force is called the **"Centripetal Force."** In fact, the centrifugal force produced in this way is resulted due to inertia; the object tries to move in a straight path, but the centripetal force acting toward the center causes it to move in a circular path.

Matter and its construction

Almost everyone might be conversant with the fact that matter is made up of very tiny particles such as molecules and atoms. The matter is normally found in three forms, that is, elements, compounds, and mixtures. The elements consist of innumerable atoms of the same kind. The molecules of different compounds are made up of two or more kinds of atoms that are combined chemically due to which an entirely different substance is formed; properties of the compounds so formed differ from the parent elements; the molecules of different compounds or elements in a *Mixture* could be mixed in any ratio without being coalesced chemically. As a result, each ingredient of a mixture retains its original properties.

Almost everybody knows that atoms are not indivisible; they are also made of much tinier particles. It is believed that the construction of the atom is somewhat similar to our solar system; the atoms have heavy cores known as "Nucleus," which are made of positively charged particles "Protons" and electrically neutral particles known as "Neutrons." Almost the whole mass or weight of the atom is centered in its nucleus. Similar to the solar system, "Electrons" (which are negatively charged particles that are supposed to be almost weightless) revolve around the nucleus. The charge of the electrons and that of the protons are equal in value but are opposite in nature. Different elements are made up of different numbers of electrons and protons; however, the total number of electrons in any normal atom is exactly equal to the number of the protons; as a result, the electrical charges of these particles neutralize the effect of each other, but they don't annihilate each other; as a result, the atoms of every element are normally found in neutral or charge-less conditions. In short, the matter is made up of different kinds of particles and different kinds of electrical charges; this means that *"Energy is intrinsic to the matter"* in the form of electric charges. The matter, which may appear to be a still, motionless, and homogeneous solid body, is, in fact, a collection of particles that are not at all still or motionless; the electrons within their atoms move perpetually in a violent flurry or a swirling motion. However, this fact couldn't be noticed easily.

Atoms of most of the elements are normally not found alone; they normally combine either with other atoms either of their own kind or with the atoms or groups of atoms of different elements. Such combinations produce molecules of either the same element or of a compound having entirely different properties. Molecules or atoms of any matter, under normal conditions, could never come very close to each other; they always maintain a minimum distance between them. They may sometimes collide, but immediately after such a collision, they get pushed away abruptly. This means that under the normal condition, they can't merge to form a new element; however, deep within the cores of the stars, where infinite pressure and temperature exist, two atoms of Hydrogen get merge into each other to form one atom of Helium. However, even during such a merger, the positive charge remains in the nucleus, whereas the negative charge continues to orbit the newly formed nucleus. It is believed that matter and energy, both, could neither be created nor destroyed.

The plasma state of the matter

Everybody knows that at different temperatures, water has three different states; these states are- ice, water, and steam. Not only water, almost all the elements, and many other substances have all of the above three states. If heated to a proper temperature, steel and all other metals first melt to become liquid, then they are transformed into the gaseous state. Likewise, when any gas is cooled down, then it first liquefies and thereafter solidifies. The entire universe is made up of matter that is found in different states.

Apart from the aforesaid three states, matter also has a fourth state too, the plasma state. At very high temperatures, electrons start to break out from the molecules or atoms. If an electron is liberated from any atom, then such an atom becomes positively charged; on the other hand, if one extra electron is added-up to it, then such an atom becomes negatively charged. Such charged atoms or molecules are called "Ions," and this process is called "Ionization." The ionized gas, when heated intensively, turns into a plasma state of matter. The electric bolt, an electrical spark, a fire flame, etc., are some of the examples of the plasma state of matter.

Besides high temperature, ionization could also be achieved by very high voltages and strong radiation, etc. The glittering flash of light seen during a thunderbolt is caused due to ionization of air molecules. Deep inside the space, where the temperature drops to almost absolute zero, or very near to it, that is, at near -273.15°C or 0°K, the clouds of hydrogen gas found in the interstellar space also get ionized due to very powerful radiations, such as- gamma-rays, X-rays, cosmic rays or ultraviolet rays, etc.

In the plasma state, the properties of matter undergo a lot of changes; matter becomes a good conductor of electricity; besides such a change, the ionized matter is also affected by the magnetic field. The matter in the plasma state, because of its magnetic properties, could be controlled and confined at the desired place by applying a magnetic field; apart from gases, the Stardust, which is available in abundance in space, is also found in the ionized condition.

Mass or the quantity of matter

In scientific terms, "Mass" means the quantity of matter contained in any object or body. Most people are not able to differentiate between the mass and weight of an object; they think that both mass and weight are the same entity; the unit of both of them is also the same, that is, "Kilogram." whereas both are much different from each other. The weight of any object is, actually, the force by which the earth pulls it towards its center of gravity. In interstellar space, where the gravitational force is almost zero, all objects become almost weightless. If the weight of any object is measured at different planets by the same spring balance, then the same object would register different weights on different planets, but its mass would remain unchanged. It is clear from this example that "Different planets exert different amounts of the gravitational pull on the same mass," due to which different readings of the weight of the same object were recorded at different planets. It could thus be concluded that the weight and the mass differ vastly from each other.

The difference between the mass (quantity of matter) and weight can be understood in another way too. In case one kilogram of corded or puffed cotton is compressed and compacted to the maximum extent, then, despite the reduction in its volume, its weight would not reduce at all because the quantity of cotton would remain unaltered; however, its compactness, or density, that is, the quantity of matter per unit volume would be increased. This means that the more is the density, the more would be the mass per unit volume. *It could thus be concluded that the mass of a body, that is, the quantity of matter stored within an object, depends on the total number of atoms, more accurately, on the total number of protons, neutrons, electrons, etcetera, contained in that object.*

Einstein envisaged that energy and matter could be interchanged into each other. Accordingly, mass is also defined as the total **energy** of all the subatomic particles contained in any piece of matter. Therefore, the mass is also expressed in the term **"Electron-volt,"** which is the unit of **energy ("Electron-volt"** is the energy gained by an electron from an electric field of one volt.). Accordingly, scientists believe that the mass of the subatomic particles represents their energies; that's why both of them have the same unit of measurement. Scientists also believe that a force-carrying particle named "Higgs-Boson," bestows energy or mass to different matter particles. A brief overview of this belief is given in the next chapter (chapter-3) of this book.

The inertia of an object, that is, its property to resist any change in its velocity or its state of rest, depends directly on its mass. In fact, the inertia of any object is produced due to its mass or quantity of matter contained in that object. In other words, the more mass an object would possess, the greater would be its inertia; the inverse of this is also true; the lesser is the mass, the lesser would be its inertia. Perhaps, this is the reason why scientists consider both of them the same entity.

Magnetism and The Ferro Magnets

Almost everyone must be conversant with the property of "magnetism;" everybody must have observed the magnets either attracting pieces of iron or repelling other magnets. Everybody might also be conversant with the magnetic effect produced by the electric current that flows in a closed circuit. Analogous to the electric current, electrons revolving around the nuclei of different atoms create their own magnetic fields. Normally, the electrons orbiting the nucleus of an atom form pairs; the electrons revolving in such pairs cancel the magnetic fields created by each other. However, every electron does not form such pairs; the substances in which atoms, or molecules, comprise some unpaired electrons exhibit magnetic properties. For example- Manganese, Cobalt, Iron, etc. The magnets made of such materials are known as "Ferro-magnets."

Besides ferromagnetic materials, some other materials also exhibit magnetic properties, though in different ways. A brief description of the different types of magnetism is given below.

Ferri-Magnetism and Antiferromagnetism

The magnetism of both the above varieties, though, is similar in nature; both of these kinds of magnetism have different properties. *Ferrimagnetic materials* comprise two different magnetic substances; in this type of magnet, the magnetic field created by the electrons revolving within the atoms of one of the substances aligns with the external magnetic field, whereas the magnetic field produced by the electrons of the other substance, opposes the same. In other words, the magnetic fields created by the neighboring electrons of both of these different types of magnetic materials point in opposite directions. However, in the *Ferrimagnetic materials*, the field strength in one of the directions is comparatively stronger, whereas in the *Antiferromagnetic materials*, the magnetic fields created in both the directions being equal in strength cancel the effect of each other. Therefore, the magnetic property of such a substance can't be detected easily.

Para Magnetism

Any magnet, under normal temperature, attracts a piece of iron or exerts a repulsive force on the other magnets. However, when magnets are heated up to 800°C or above, they normally lose their magnetic property. Weak magnetic fields normally fail to affect any magnetic material at such a high temperature. Only a very strong magnetic force can attract different magnetic materials at such a high temperature. This kind of magnetism is called "Paramagnetism."

Diamagnetism

Almost all the nonmagnetic and carbonaceous materials possess the property of diamagnetism. Contrary to all other sorts of magnetic materials, magnetic fields never attract diamagnetic materials; magnets always repel them. Such a repulsive force is so weak that it couldn't even be noticed. However, at very low temperatures that normally exist in interstellar space, magnetic fields exert a potent repulsive force on all sorts of diamagnetic materials.

The property of magnetism is even found in some of the gases also; the molecules of Carbon-di-oxide gas, due to diamagnetic properties, rotate in the direction perpendicular to the applied magnetic field. Oxygen and Lithium both have paramagnetic properties, and the water molecules possess diamagnetic properties. The solutions of some chemicals made with water or alcohol also react to the magnetic forces. Almost all carbonic substances exhibit diamagnetic properties.

Superconductivity

Almost everyone knows that the electrical resistance of different good conductors of electricity increases at high temperatures, whereas their resistance decreases at low temperatures; in other words, conductivity increases at lower temperatures. Electric current is, in fact, a flow of free electrons through any conductor. The speed of electrons orbiting within the atoms of any conductor increases at elevated temperatures; as a result, the flow of free electrons that are flowing through the conductor is obstructed; which, in turn, increases the resistance of that conductor. The opposite of this is also true; the resistance of a conductor decreases at low temperatures.

Scientists have observed that when some alloys or nonmetallic substances are cooled to almost absolute zero (0°K), then their conductivity increases to such an extent that once the electric current is established in a circuit, then such current continues to flow even when the source of the current is removed. Such extreme conductivity is called the *"Superconductivity."* Such a low temperature that produces superconductivity doesn't naturally exist on the earth. However, this (low temperature) is one of the most common features in interstellar space. In Octomber'2020, Rochester University, New York's scientists have developed a compound of Hydrogen, Carbon, and Sulfur, which exhibits superconductivity even at 15°C, at normal pressure.

Superconductivity and Magnetism

Generally, super-conductive materials don't exhibit magnetic properties at normal temperatures; magnets neither attract nor repel them. When a superconductive material, having an especial composition such as *"Yttrium-Barium-Copper-Oxide" alloy* ($YBa_2Cu_3O_7$), which comes under the type 2 category of the superconductors, is cooled below a critical temperature by dipping it into liquid Nitrogen, then it exhibits a very strong diamagnetic property. In case a very strong permanent magnet is placed over the superconductor, thereafter, if that superconductor is cooled by liquid Nitrogen, then as soon as the temperature of the said superconductor drops below a critical temperature, the permanent magnet placed on the said superconductor suddenly levitates in the air, the reason being as soon as the temperature of the superconductor falls below a critical limit, the magnetic field of the permanent magnet induces the free electrons existing within the said superconductor, to flow in a circular path, in an unrestricted manner. As a result, a magnetic field of the same polarity and strength as that of the said permanent magnet is induced in the said superconductor. In other words, a *"mirror image"* of the permanent magnet is created in the superconductor; this mirror image repels the said permanent magnet and thereby causes it to levitate freely in the air.

The Effect of the Magnetic field on the Non-magnetic Materials

Apparently, a magnetic field doesn't attract non-magnetic materials. However, almost all the non-magnetic materials are, in fact, diamagnetic by nature; therefore, all magnets repel them with such a feeble force that it is very hard to take any notice of the same. This picture completely changes in very strong magnetic fields, which repel all the carbonaceous materials such as wood, fruits, and even living beings such as mice and frogs, etc., with such a strong force that they are forced to levitate against gravitation. Both gravitational and magnetic forces act simultaneously on such materials side by side; however, the effect of only the stronger force could be noticed. This fact suggests that in interstellar space, where the temperature remains closer to absolute zero, all the natural phenomena might be governed by the combined effect of these two forces.

The London Moment

Sometimes, during the period 1930 to 1934, scientists observed that the spinning superconductors generate a magnetic field, the polarity of such a magnetic field exactly lines-up with its axis of rotation. This phenomenon is called the **"London-moment."** Very low temperature to support superconductivity is a common feature in interstellar space, and such a combination of **superconductivity, magnetic force, and gravitation, which rule the interstellar space, may play an important role in almost all cosmic events.** It is also believed that rotating mass may induce a weak gravitational field; however, to the best of my knowledge, such a phenomenon, beyond any doubt, could not yet be detected.

Accretion

In general, "Accretion" means the growth or accumulation of wealth or people in one place. In cosmology, accretion means a gathering of particles at any place, which results in the gradual growth of additional layers of matter. In the years 2003-04, astronauts Donald R. Pettit and his colleague Stanley G. Love, during their stay on the space station, performed an important experiment in almost zero-gravity conditions. They shook a transparent plastic bag containing some granules of grounded coffee, salt, sugar, etc. After a few minutes, they observed, "The granules contained in the said bag started sticking together to form small clumps." *It is believed that the formation of such clumps was caused by the static electricity that was generated due to the mutual rubbing of those granules.* This property of matter is of great importance in astrophysics. Scientists believe that during the process of formation of the stars and planets, etc., (Chapter-9), this property, as a first step, helped matter particles to congregate and clump into compact clusters. With time, such clusters persisted in growing gradually; and in a period of the next few million years, they formed compact cores of stars and planets, etc.

At the age of 13-14 years, I dropped some small pieces of paper in a pot containing water; thereafter, I churned the water to form a whirl. After about a short period of about half an hour or so, I observed that the pieces of paper had collected together in several small clumps; most of the pieces did collect at the bank of the container. I thought that this was caused by the surface tension of the water; however, such grouping of paper pieces might have been caused due to the property of accretion. I am not sure about this phenomenon; however, if accretion can cause small objects floating on the water surface to congregate, then static electricity could have no role in this phenomenon. It seems to me that some sort of very feeble force of attraction, which becomes effective in the absence of any other force, might be behind such congregation; this idea needs verification.

The Fourth Dimension

Almost everybody might be conversant with the concept of the three dimensions, i.e., the length, breadth, and height. The maps can show only two out of these three dimensions, that is, length and breadth only; the third dimension is necessarily required to determine the height of any object. However, at times, some of the readers might have noticed that at a particular time, a particular article was kept at a particular place; however, after a lapse of a few hours or so, some other article is seen at the same place. This fact reveals that though the place remains the same, different articles might be kept at that particular place at different points in time. In case we can go back in time, then we can find an article, which was kept at that place at any particular point in time. This fact indicates that the dimension of time is also required to locate that article. However, the meaning of the fourth dimension of "time" is not limited to this extent only; in this continuously changing, rather than, expanding universe, the fourth dimension, which is the combination or continuum of "space (distances) and time," is essentially required to establish the locations of different celestial bodies. This continuum of space-time is also said to expand persistently with almost the speed of light. The system of four dimensions determines the location of any celestial object at any particular point in time; this system includes the location of any particular event, as well as the time when that event did actually happen.

The vastness of the universe is beyond imagination; accordingly, a new unit of "light-year" is required to measure very large distances existing within the ambit of the universe. The light rays travel at a speed of almost 300,000 Km per second. Despite such a high speed, the rays that were emitted from the furthest stars, about 13 to 14 billion years in the past, are reaching us at present. It is not sure whether such distant stars do still exist; if they do exist, then, by now, they would have moved far-away from the place where they presently appear to exist. In fact, in the present, we are watching the past of all the stars; no means are available to determine their present locations. In case a star, from which light takes more than 10 billion years to reach us, dies today or ceases to shine, even then, we would continue to see the same for the next 10 billion years or so. If the human species somehow survive for the next 10 billion years, only then our future generations would be able to see it dying. Likewise, if a star takes birth today at a location that is about 10 billion light-years away from us, then the living beings on the Earth would see it taking birth, only after 10 billion years from hence. This simply means that the star that has died today will continue to exist for us for the next 10 billion years or so; similarly, the star, which has already taken birth today, wouldn't have any existence for us for the next few billions of years. This weird phenomenon is the magic of the fourth dimension of time. What deems to be the existing time of different stars, i.e., their "present," is, in fact, their past! The far-away, we look back in the sky, that far back in the time we reach out; distances and time, thus appear to have merged and mingled into each other.

Any place on the Earth can be located by the three-dimensional coordinate system because all such places are stationary in the reference frame of the Earth. This is not possible in outer space, where the location of everything is continuously changing with time. All the planets and all the stars though are moving away from their locations, most of the distant stars appear to be stationary! How is it possible? Our sun, along with all of its planets, is orbiting the galactic center (center of the Milky-Way galaxy) at an amazing speed of about 250 kilometers per second. Our Sun, moving at this speed, takes about 250 million years to complete its one orbit. Hundreds of billions of stars contained in the Milky Way galaxy, including our sun, orbit the galactic center with almost equal speeds so that the distances between different stars remain ostensibly unchanged. This is the reason that the Milky Way always retains its spiral shape. Similarly, different stars seem to retain their same relative locations over very long time periods. At least, no change in their respective locations could be noticed during the life period of a man. As a result, different groups of stars appear to retain different shapes; some of such groups of stars or the constellations of stars can be seen by the naked eyes. However, the locations of different celestial bodies always keep changing in this constantly changing universe. Their locations are, therefore, determined concerning the line joining our Sun and the nearest star named "Alfa Century." The locations of different Stars, so determined, denote the location of their past (not the present ones). Although all the stars have moved to their present time, their exact locations could neither be seen nor determined.

Negative Numbers and the Higher Science

Probably, the concept of negative numbers was developed due to borrowing and lending money or other articles. The mathematics of the negative numbers is, up to a certain extent, similar to the normal or positive numbers. The repetitive addition of the same number is called multiplication, and likewise, the process of dividing something repeatedly into equal parts is called Division. However, the multiplication and division of the negative numbers differ from the positive numbers. The negative numbers are opposite to their positive counterparts; therefore, the multiplication and division of the negative numbers is a process reverse in nature that applies to the positive numbers. If a negative number is reversed once, then it becomes positive, and if reversed yet another time, it again becomes negative. Analogous to the above phenomenon, if two negative numbers are multiplied once, then their product becomes positive. However, on multiplying them by yet another time, their product once again becomes negative; the process of division of the negative numbers is very much similar to their multiplication. *Contrarily, the negativity of the charge of an electron in any of its multiples, whether in the odd or even number, always remains unchanged. This fact indicates that Nature probably, doesn't necessarily follow man-made rules;*

it might have a different kind of mathematics of its own. In this context, please also see the next chapter (Chapter-3), "Development of the Atomic Theory."

Based on the above fact, one should always keep in mind that everything that we find in *Nature* is real; nothing is imaginary, nothing has a value lesser than zero (negative); this fact indicates that negative numbers are not natural or real numbers. For example, an eruption of a volcano can never be reversed; similarly, the rays of light, if not reflected, can never come back to their source. Therefore, only the real things, which do physically exist, could only be borrowed or given back in the actual life; imaginary or nonexistent things couldn't be given or taken back. It is not, at all, possible that while in a necessity, anyone may take out some money from his empty wallet and at a later point in time, when he puts back some money in the same wallet, then an equal amount of money, borrowed earlier, may go back in the wallet, and vanish automatically. However, numerical problems based on similar imagination can be solved by mathematics. This simply means that at times we, by mathematics, might obtain such a solution to complicated problems that might not, at all, be practical or realistic.

In the above context, it is for contemplation that even though the product of two numbers having similar signs, i.e., both the numbers having positive signs or both having negative values, can never become negative; even then the imaginary and nonexistent numbers are in vogue in mathematics; the square root of the digit "minus one" ($\sqrt{-1}$) is denoted by an imaginary and a nonexistent number "*i*," which is a mathematical term. Although any such number does not exist in reality, even then, mathematicians and scientists have found a solution to this problem also. Math becomes still more complicated and difficult in higher science. Most of the equations might contain imaginary and nonexistent numbers. Besides the use of imaginary and nonexistent numbers, some more nonexistent entities such as negative energy (please see the next Chapter) as well as imaginary time, etc., are at times assumed in modern science to solve some of the complicated problems. This means that even as of today, we still have to depend on imagination and assumptions to solve such problems. Unlike the primitive ages, the imaginations of our scientists and physicists are not so baseless; however, assumptions are, after all, only assumptions; in case they are based on a wrong idea or wrong input, then such imagination is bound to go wrong. ***There is still some possibility that imaginary factors, things or quantities, etc., might not exist, at all, in Nature or in reality.*** This implies that if any of our assumptions go wrong during any stage, then it couldn't be detected until the results obtained based on any plausible theory differ vastly from the actual ones; the necessity of reviewing any such theory is felt only under such circumstances. However, at the scale of the universe, the completion of almost every celestial phenomenon takes an unimaginably long period of time; that's the reason why any such mistake might slip its immediate detection; even the entire lifetime of one generation might not be enough to detect and eliminate such mistakes.

PART – 2

The Eminent

Scientific Theories

-

Brief Overview

&

The Alternate angle thereon

3: Quantum Mechanics - The Science of Subatomic Particles

The universe comprises an infinite quantity of matter in different forms varying from very fine and minute particles to very big and massive stellar bodies. Matter, in all these forms, is spread all over in the vast interstellar space. Besides matter, space is also believed to be filled with infinite energy. This energy is not only stored in space; it propagates in all directions in the form of different kinds of radiation. As per the prevailing scientific theories, energy propagates in the form of waves, as well as in the form of very minute particles known as photons, i.e., energy has dual properties. It is also believed that at a far-remote point in time, the entire matter and, subsequently, the entire universe was created by an abrupt explosion of energy; before that, the universe was confined to an infinitesimally small point. Since modern science is based on very complex theories, it seems necessary for the readers of this book that, to understand these hypotheses, they shall acquire at least a very basic knowledge of science related to *energy and subatomic particles.*

The discipline of science related to the study of tiny particles and their behavior is known as *"Quantum Mechanics."* This is an entirely different branch of science that was developed sometime during the period 1920-26. The word *"quanta"* (plural of quantum) was first used in the year 1900 by physicist Max Planck. Earlier to this date, it was believed that the energy given off by any hot body depends directly on the frequency of the radiation; this simply means that at very high frequencies, the rate of emission of energy would increase manifold. In December 1900, Planck suggested that light cannot be emitted at an infinite rate; instead, it is emitted in certain packets of energy that are known as *"quanta"* of energy. The amount of energy contained in each of such quanta can be represented by the *"frequency of the waves emitted multiplied by a constant that is known as **Planck's Constant.***" *He also suggested that at higher frequencies, the emission of a single quantum of energy would require more energy. In other words, the increase in energy (frequency) would not result in the emission of uncontrolled radiation;* **this property of light (including all other sorts of energy-radiations) limits the rate of the energy radiation at higher frequencies.**

Quantum mechanics is based on the *"Uncertainty Principle;"* in quantum mechanics, no definite or specific solution to a complicated scientific problem is predicted; instead, out of several possible solutions, the most likely solution is selected. This method though is very effective to find out the solution to any intricate problem; I feel that this method has a probable disadvantage too; the solution so obtained by this method depends on only those solutions that were considered during the solution-finding process; therefore, the solutions, which couldn't be foreseen during this process, are eliminated right away. Besides, it is also possible that the solutions so eliminated without being considered might include the *true solution;* who knows?

Anyhow, the ***principle of uncertainty*** is of great importance in the world of very tiny particles; therefore, a very brief introduction of this principle is given below:

Uncertainty and Unpredictability

Sometimes, during the 18th-19th century, some scientists were trying to establish a theory that would allow the prediction of everything, including human behavior and even the future too. Such a prediction necessitated the exact determination of the present positions and speeds of different particles so that the future positions and speeds of those particles could be determined accurately. In order to fulfill this requirement, tiny particles had to be exposed to very powerful light. Soon scientists noticed that if particles are exposed to intense light, then their speeds start to increase. Thus, it became clear that the exact speed and exact position of any particle at any particular instant can't be measured. Therefore, the idea to formulate such a theory had to be abandoned. However, uncertainty in the positions and velocities of the tiny particles led to the formulation of the famous *"Uncertainty Principle."* According to this principle, *"**Uncertainty is an inescapable property of all the entities of the world; nothing in the world is exempted from this property. According to this principle, uncertainty in the velocities and positions of particles can never be lesser than a certain value.**"* In honor of Max Planck, this "certain value or limit" is named as *"**Planck's Constant.**"* This principle enjoys a very respectable status in particle physics and quantum mechanics; most of the complicated issues are solved by this principle. These days, the speed and position of the subatomic particles are not considered separately; instead, the value of the speed and exact location of the particles has been replaced by the combination of both of these entities, which is now known as the *"**quantum state**"* of the particles.

My perspective on the Uncertainty Principle

One may wonder, if at present, it is not possible to measure the exact positions and speeds of tiny particles separately, then would it not become possible in the future to take accurate measurements of both of these entities, or *whether the same is impossible for mother Nature too?* (Kindly see the Attosecond Technology in the same chapter, under the sub-title- "Visualizing the Atom and Electrons moving in its orbit,"). Many things, which are apparently uncertain, are several times repeated in nature, but it seems impossible to predict them in advance; for example: - *through which path the thunderbolt would pass?* This might seem uncertain to us; however, nature practically chooses the correct and the only possible path for this purpose. Of course, the next time, under the changed circumstances, it may choose any other path that would be the only possible and practicable path for that particular instant of time. Many such uncertain things often happen in nature, which, seemingly, are not governed by any rule; for example: - *when and how the clouds would form? Or, when a cracker or a bomb is exploded, then how many pieces would it be shattered in, and in what manner would its pieces scatter? And how would its smoke spread?* Humankind might not solve such problems by any rule; however, it is not at all difficult for nature, which instantaneously finds out the only possible solution. There being no means available to humankind to determine such events in advance, the same seems uncertain to him; however, nothing is uncertain for mother nature; it can instantly solve any such problem in its own ways.

If some dust particles are sprinkled in the blowing wind, then mankind would not be able to find out in what manner different dust particles would be blown away, and also, at what different places would they get deposited? Hundreds of data that are very difficult to collect would be necessarily required to predict a possible solution to this problem. Somehow, if the required data could be collected and fed in a supercomputer, even then, the direction and speed of the wind might change several times before pressing the "Enter" button. What I mean to say is man can predict such things to some extent only, not exactly. On the other hand, nature finds out the only possible solution instantaneously. Einstein had very rightly said, "God does not play dice." Accordingly, nature always carries out its work following the situations prevailing at any particular moment. *Therefore, it could be said that all the deeds of nature are* **"unpredictable"** *instead of being* **"uncertain."** Although natural phenomena may appear uncertain, they are, in fact, not, at least for Mother Nature. The entire universe, the entire lot of atoms and subatomic particles, etc., always follows the rules of nature. Since the circumstances prevailing at different time-instants and places keep changing persistently, nature accordingly keeps modifying these solutions to match with the changes that have taken place; it is, therefore, not possible for us to predict such solutions in advance.

The man-made rules are primarily based on *imaginations, presumptions, and observations made over limited periods of time;* the lengths of such observations are very-very small on the scale of the universe. These rules might give correct results within certain limits, not always. These rules, beyond such limits, might sometimes give erratic results; therefore, everything might seem uncertain. When it is not possible to deduce a direct and exact solution to any complicated problem, then scientists deduce probable solutions that are based on the uncertainty principle. Mankind is free to revise these rules if the necessity is felt. On the other hand, nature always functions according to its own rules, whether mankind has or hasn't set out any rule; or else, the rules framed by mankind differ from nature's rules; nature is not bound to abide by the man-made laws. This doesn't mean that whatever happens in the universe is always right and good, or nature will always deduce a similar solution under similar circumstances. At times, such circumstances may arise in which more than one solution might be possible. Under these circumstances, nature chooses the most practicable and straightforward solutions; however, such solutions cannot be changed or reversed - even if they are not the best solution (for example- the destruction resulting from an Earth-quack); nature does not have the option to change or discard its decisions. Such decisions, instead of uncertain, are **Nature's Choice**. Since the laws of nature always remain unchanged, therefore, whatever has happened in the past, or what would happen in the future, will happen according to the well-set laws of nature. Accordingly, if the laws of nature could be understood properly, then we might be able to predict the future-happenings to some extent, at least, subject to the exact prediction of the circumstances that might prevail in the future. Therefore, it seems to me that *nothing is uncertain in nature;* instead, the same is ***unpredictable***.

Development of the Atomic Theory

The ancient people believed that each and everything in the world is made up of five elements, which are: - the earth, water, air, fire, and sky (Aether). Contrary to the above belief, none of the above entities is an element; in fact, different kinds of matter, whose molecules comprise atoms of the same kind, are known as elements. The smallest particle of an element known as the atom possesses the properties of the same element. The idea of the atom was first developed in India about 600

BC, or even much earlier than that. A similar concept was also developed by the Greek philosophers about 500 BC. In the Greek language, the word *atom* means "indivisible."

In the modern era, almost everyone knows that even the atom is also made of subatomic particles; some of them possess an intrinsic electrical charge. The electric charge was probably known to mankind since the prehistoric ages; however, the Greek philosopher, named *Thales*, was the first man who, in 600 BC, mentioned in writing about the electrification of "Amber rod" by rubbing it with fur. Different names such as "electric," "electricity," "electron," etc., were derived from the Greek name Amber. The first machine to generate static electricity was made as early as 1660. Later, in the year 1733, it was assumed that the flow of two different kinds of liquids produced two different types of electricity. A little later, in the year 1750, American scientist Benjamin Franklin proved that electricity is of one kind only, not of two kinds. Subsequently, Michael Faraday also established in 1839 that though electricity is only of one kind, it has two different poles. These poles, whose properties are opposite of each other in nature, were named as positive and negative poles. About 217 years in the past from hence, that is, in the year 1805, John Dalton introduced the concept of atoms. Later on, the concept of two opposite poles, or opposite charges of electricity, was also adopted in the atomic theory. This concept of two opposite kinds of electric charges is still in vogue.

In 1897, physicist J. J. Thomson observed that the Cathode Rays are, in fact, a shower of extremely minute and almost mass-less tiny particles, each of which possesses a unit negative charge. He suggested that these particles are part of the atoms and must be coming from within the atoms. These particles were given the name "electron." A little time thereafter, that is, in 1909, some scientists observed that when positively charged "alpha" (a) particles are fired on extremely thin gold foil, then most of these particles get passed through it without much deflection. However, very few particles were found to deflect by very large amounts. Based on the aforesaid observation, Scientist Rutherford concluded that the atom must have a positively charged nucleus surrounded by electrons. He further predicted that the electrons orbit the nucleus due to their mutual attraction. Since atoms don't have any sort of electric charge, it was assumed that the electrical charges of the nucleus and that of the electrons orbiting it must be equal in value but opposite to each other in nature.

Scientists didn't have to think much while deciding the nature of the charges of the nucleus, as well as that of the electron, because the properties of these charges were determined much earlier. However, since the negative sign represents values lesser than zero, any ordinary man might think that in case some positive charge could be taken out of a neutral particle, then the same would develop a negative charge. Likewise, if double the quantity of positive charge is added to a particle, such as an electron having a unit negative charge, then it should become positively charged. Everybody is probably aware that when two dissimilar insulating materials are rubbed together, then some electrons from one of these materials are transferred to the other one. The material that gains extra electrons becomes negatively charged, whereas the other one that loses some electrons becomes positively charged. However, the charge of any subatomic particle can't be transferred to another subatomic particle by any means; neither it is possible to add an extra positive charge to the electrons to convert them into positively charged particles or even make them neutral, nor the protons or positrons can be converted into electrons by removing two units of positive charge from them. The electron, in the hydrogen atom, orbits the nucleus from a very-very small distance of 20 millionth part of a millimeter; however, the electric charge of any of these two particles is not able to jump from one particle to the other one, even from such a small distance. It could be inferred from the aforesaid facts that different kinds of charges are inseparable and intrinsic properties of the respective charged particles; in any case, their charges could neither be separated from them nor be converted into the charge of opposite nature, that is, the values of their respective charges can neither be increased nor be decreased. *Both kinds of these charges exert similar and equal attractive forces on all the neutral substances. These two charges can merely nullify the effect of each other; they normally do not annihilate the electric charge of the opposite kind of particle or annihilate such a particle itself; they always retain their independent and unique charges, as well as their separate identities.*

In the above perspective, it doesn't seem justified to say that charge of either the nucleus or electron is really negative in nature because the charge of either of them is neither a deficiency of any other kind of charge nor the value of any of these charges is lesser than zero; therefore, both of these charges are real and inseparable properties of the respective particles; despite the fact that these charges are opposite in nature to each other, they have independent, and separate existences. In other words, both kinds of these charges are two different, independent, and real properties that certainly have definite values above zero. The aforesaid perception might appear weird and impossible because if the charges of both of these particles are equal in magnitude and opposite in nature, then how is it possible that both of them have values above zero, that is, despite *being opposite in nature, they point in the same direction?* Although this seems impossible to us, even then this might be possible in Nature; for example: If a man travels northward from any point located on the Earth's surface, and at the same time another person travels southward relative to the same point, then it cannot be said that one of them would travel in a negative direction, i.e., he would travel through a negative distance; the distances in both the directions are though the opposite of each other, they would be positive concerning the starting point. Similarly, the distance measured from the center of a circle or a sphere would always be positive in whatsoever direction the same is measured. If the distance is measured backward from any point situated on the periphery of a circle, i.e., toward the center, then we can never go in a negative direction; as soon as we cross its center, we will again move in the positive direction relative to the center. If we dig a hole on Earth's surface, even then, we will be at a positive distance from Earth's center. Likewise, the two opposite walls of a balloon can't be said to be positive and negative regarding each other. The pole strength of any magnet (irrespective of the fact that they exert force in opposite directions of each other) is not considered positive or negative. *It, thus, seems to me that the electrical charge of the electron shouldn't be treated as negative in nature because the same is neither the deficiency of positive charge nor its value is lesser than zero; it too exerts a positive force on all the neutral objects.*

The atomic theories that were made during the last 100-125 years must have been based on the results obtained from different experiments and complicated mathematical calculations; therefore, there seems not even the slightest chance of any error being crept into these theories. We all know that modern science depends on highly complex mathematics. *This fact gives rise to the possibility that if an equation is formulated depending on the negative value of the charge of the electron, then most likely, the solution to such an equation might contain a negative, imaginary, or non-existent number; however, I feel that all the physical and material things found in Nature are real, not imaginary.* For example, though mirror images, dreams, thoughts, etc., are nonphysical or nonmaterial entities, even then, they indeed do exist either in the form of light rays or electrical signals. All real/physical things have values above zero; they never have negative values, i.e., lesser than zero. Therefore, results obtained in imaginary or nonexistent numbers might be misleading, and in turn, they might make things even more complicated. However, it is up to the scientists, not an ordinary person *like me*, to decide whether the existing system of negative charge or imaginary and nonexistent numbers, etc., shall be reviewed, or the same shall be continued as it is. I, however, feel that in case the charge of the electron is called negative, only for namesake, then it makes no difference whatsoever; however, this will be only a relative term to indicate that the nature of the charge of the electron is opposite of that of the proton or positron; this sign should not bear any relation to its absolute value or its vector quantity, because nature of the charge of the electron, doesn't change due to any mathematical operation like multiplication or division *(See Chapter-2, under section "Mathematics of the negative numbers & Higher Science").*

The Atom and its Construction

First of all, during 1902-1904, a few models of the structure of the atom were proposed. However, at that time, nobody had any idea that the atom possessed a nucleus too. A little later, from 1909 to 1910, the experiments performed by Rutherford did establish that atoms comprise a positively charged nucleus, which is surrounded by negatively charged electrons. Based on this conception, Rutherford, in 1911, proposed a model of an atom that was similar to the solar system, in which the electrons, instead of gravitational force, orbit the nucleus due to the force of attraction acting between these two particles. However, there was one limitation in this model; according to the theory of electromagnetic radiation that James Clerk Maxwell proposed in the 1980s, "Whenever any charged particle, such as an electron, is subjected to acceleration or change of direction, it shall lose some of its energy in the form of some kind of radiation." This theory implies that an electron orbiting the nucleus shall gradually lose its energy due to persistent change in its direction; as a result, it should spiral inward until it falls into the nucleus. Later on, scientist Niels Henrik David Bohr, in 1913, came up with a partial solution to this problem. He suggested that the electrons don't orbit the nucleus in any arbitrary orbit; instead, they orbit the nucleus from certain specified distances, which have certain fixed levels of energies. The electrons orbiting in the fixed energy levels won't lose any energy, and therefore, their energy levels would remain constant. The Rutherford-Bohr model of the atom was further refined sometime in 1925-26, that is, after the theory of "quantum mechanics" was formulated by different scientists.

Quantum mechanics predicts that subatomic particles have the properties of both waves as well as particles. The electrons in the form of the matter-waves can orbit the nucleus from such distances only, which possess a certain level of energy that would correspond to a whole number of such waves. The paths at which electrons orbit the nucleus are known as "orbitals," "shells," or "energy levels." It is believed that when an electron moves closer to the nucleus, it performs some work and loses some energy proportional to the work done. Based on the aforesaid idea, scientists believe that the electrons that orbit the nucleus from a greater distance (radius) possess more energy. Accordingly, if an electron gains some extra energy or absorbs some energy from any other source, then it would jump out from an inner orbital to the outer orbital. Similarly, if any electron loses some energy, then it would jump from an outer orbital into an inner orbital; the energy so released can be seen in the form of a flash of light, which can be noticed by very sensitive instruments.

Somehow, I feel that the above theory does not explain "Wherefrom the electron gets the energy to move continuously, or how the electron in the capacity of matter-wave moves in a circle around the nucleus because the waves normally originate from a point and propagate like the expanding circles; waves never revolve around a point." Moreover, the prediction **"Electrons don't lose energy while they move in the orbitals having a fixed level of energy"** *doesn't seem correct because the electrons, due to their inertia, can though persist in moving in a straight line, but while moving in an orbit, the electron necessarily consumes some energy regularly, for changing its direction continuously; since the electron never spirals into the nucleus, it must consume some energy from an inexhaustible energy-source (probably, from the attractive force acting between the nucleus and the electron) to keep moving in a stable orbit.*

Further, the concept, "*The mass of a particle represents its energy,*" *gives rise to another question* "*Do the electrons occupying higher energy levels possess higher masses because mass is the measure of energy?*" *However, the possibility under question seems totally impracticable. One more question that remains unanswered is- Relativity (please see Chapters- 5 & 6) envisages that Gravity is not a force; the planets, though they move in straight lines, appear to orbit the Sun because they follow the curvature of warped space. However, since rules shall always be the same, it is not clear how the force of attraction acting between electron and proton compels the electron to move around the nucleus in a circular path? Does the force of electrical attraction also result from the warping of space?*

Visualizing the Atom, and the Electrons

From 1900, or even earlier, scientists had been trying to understand the inner structure of the atom. In that era, it wasn't possible to see the atom; even then, in 1909, by the Oil Drop Experiment, different scientists succeeded in establishing the relation between the mass and the charge of the electrons. Almost during the same period, scientist Rutherford established that a cloud of negative charge envelops a tiny but dense core, which possesses a positive charge. During that era, the properties of subatomic particles were also studied by very simple equipment such as the cloud chamber; no other means or sophisticated equipment were available. Later, during 1925-26, after the formation of quantum theory, scientists developed a concept that electrons and other subatomic particles behave like waves. It was also predicted that even molecules also behave like waves. Therefore, the term "matter wave" was tossed by the scientists of that era.

Atoms were seen for the first time in 1956 by a very powerful "electron microscope." After enlarging by several hundred-thousand times, the image of the atom appeared like a bunch of very small shimmering points. In 1981, a single atom could be seen for the first time by "Scanning Emission Microscope." Even by that time, the inner structure of the atom couldn't be seen. In 2005-06, that is, about 15 to 16 years ago, glimpses of an electron could be seen by very advanced equipment such as the "Field Emission Microscope" and "Atomic Force Microscope." The image of the electron looked like a mist or fog that envelopes the nucleus from all around. The said envelope appeared comparatively much denser in the center; however, going away from the nucleus, the density of the said cloud gradually faded away or thinned out; the said envelope appeared very thin and defused at the outer edges. Such an image of the electron supported the idea that in reality, electrons are waves, or rather, they are "matter waves;" however, there seems another possibility that **this blurry, hazy, and fog-like appearance of the electrons might have resulted due to their very high speeds at which they orbit the nucleus.**

Later, in the year 2008, Swiss scientists exposed the atom with a pulse of a laser beam for a very-very small instant of one *"Attosecond,"* that is, a period of one-billionth-of-one-billionth second or $10^{-9} \times 10^{-9}$ second or 10^{-18} second. This small exposure not only enabled the scientists to see the electron, but it also enabled them to make a movie of the electron while the same was in motion on its orbital *(Hereinafter, this technology that enables us to see the electrons moving either in their orbits or in straight lines, and make a movie thereof, has been referred to as the "Attosecond Technology")*. In the above experiment, it was found that, in the *Hydrogen Atom,* the electron takes a time period of about 150 "Attoseconds," or 150×10^{-18} seconds, to complete one revolution around the nucleus. Diameter-wise, the Hydrogen-atom measures about 1.1 Angstrom (Angstrom means 10^{-10} meters or about ten-millionths part of a millimeter); accordingly, the electron orbiting the Hydrogen atom's nucleus travels at a speed of about 2300 kilometers per second. The electron, traveling at such a high speed, completes about 6.66×10^{15} revolutions around the Hydrogen-nucleus in one second; that is, the electron completes 6.66 thousand trillion ($6.66 \times 10^3 \times 10^{12}$) orbits or 6.66 million billion ($6.66 \times 10^6 \times 10^9$) orbits in a second. This fact clarifies that it is not possible to focus a camera on the electrons due to their tiny size and extremely high speed; *probably, this might be the reason why they appear like a blurry fog.*

In December 2011, MIT America exhibited a camera capable of shooting one trillion (10^{12}) photo-frames in a second, that is, *even faster than the speed of light.* This camera is not only capable of taking snapshots of the moving light beams; it can also make a movie of a light beam propagating on its way. In such a movie, light is seen to move like an expanding bubble, expanding at a steady speed without any fluctuation whatsoever; this fact puts the wave nature of light under question. If the Attosecond technology could be combined with the aforesaid camera, then we might also be able to make a movie of the inner structure of the atoms, that is, of the electrons moving in different orbitals within the atoms of different elements.

The atomic structure of the heavier elements having an atomic number around 100 or above must be very-very complicated. When compared with the force exerted by the nucleus of the Hydrogen atom, on the electron orbiting it, the nucleus of such an element, because of its much higher positive charge, shall exert about 50 times greater attractive force on the electron orbiting it in the innermost orbit. And at the same time, all the outer electrons must be pushing the innermost electron with an equally strong force toward the nucleus. It is very difficult to imagine the inner structure of such a complicated atom. If a movie of the inner structure of such a complicated atom could be studied by combining the Attosecond technology and the above-mentioned camera, then we might be able to see the realistic image of the inner structure of such an atom and study the same in a much better way. In that case, we might also be able to determine whether or not our present conception of the atomic structure matches fully with the actual atomic structure. *If at all, any difference between the theoretical and actual atomic structures is found, then all the work done in this field so far would require a thorough review. In that case, our existing theories might also need some modification.*

"Spin" of the Tiny Particles

Sometime during 1920-22, the duo of scientists **Otto Stern and Walther Gerlach** made a remarkable discovery. They found that when a beam of neutral particles, such as silver atoms, are made to travel through an inhomogeneous magnetic field, which is produced between two magnetic poles having uneven strengths, then some of the particles are deflected, either in the direction of the stronger pole or away from it, that is, some of the atoms were attracted by the stronger pole and some of them were repelled by the same pole. The scientist duo also noted that the stronger magnet caused the neutral particles to deflect by an equal amount in both directions. Based on this observation, they concluded that the tiny particles must have their own magnetic fields, which is similar to the field of a spinning bar magnet; that's why such particles have the capacity to maintain a fixed direction of their magnetic polarity. This could be the only reason that the stronger magnetic pole either attracted or repelled the neutral particles like the tiny rotating bar magnets.

The conclusion arrived at by the said scientist duo could be understood by the example of very simple toys such as the top or the gyroscope. These toys, while spinning, have the capacity to retain the direction of their axes of spin. This property of the spinning objects, to maintain the steady direction of their axis of rotation, is known as the "angular momentum." In analogy to this property, scientists concluded that the tiny particles too must have similar property by which they retain the direction of their magnetic fields. Later on, scientists further noticed that though the direction of the magnetic fields of these tiny particles could be changed, it is not possible to change the strength of their magnetic fields. Looking into the intrinsic electric charges of the particles like the electrons and protons, etc., scientists, at first, thought that their magnetic fields are generated by the spinning of these particles; as such, these particles maintain a stable direction of their magnetic fields. Based on this assumption, this property of the subatomic particles was given the name *"spin."* However, the scientists very soon realized that such a strong magnetic field could only be created when these particles were rotating at faster-than-light speeds. Such high speeds of rotation of the atoms having been considered impossible, scientists had to drop this idea, anyhow, even as of today, this property of the subatomic particles is known as "spin."

Scientists believe *"particle spin"* is a property intrinsic to subatomic particles. This property doesn't depend on their masses or electric charges. Even after a lapse of more than 90 years after the discovery of the aforesaid property, no one knows how and why this property of spin is produced in subatomic particles. However, according to the string theory* (which was formulated about 50 years after the discovery of this property), all types of elementary particles are made of vibrating strings. *If this notion is correct, then is it not possible that the magnetic fields of these particles are produced due to some repetitive periodic changes within the subatomic particles?*

Classification of the Particle-Spin

In case an object having a regular shape is rotated slowly, then after some time, it again looks similar to what it did appear in the beginning. Analogous to the above phenomenon, the property of "particle spin" tells us what they look like from different angles. Since it is not possible to see the subatomic particles, their similarity is judged by the similarity of their properties. These particles have different kinds of properties such as different types and amounts of electrical charges, masses, spins, energies, speeds, etc.; out of them, some of these properties, such as spin, charge, and speed, have definite directions. Out of all these properties, some properties are

---

* A brief outline of this theory is given in chapter-7.

variable in nature, and others have fixed and finite values. The system formed by the combination of all these properties of the subatomic particles decides how any particle would behave in a field of force. After the applied field of force is removed, this system, which is known as "Quantum State," comes back to its initial state. The symmetry of the subatomic particles is judged by their coming back into the same quantum state. The number of turns by which a particle is required to be rotated to maintain its symmetry determines the value of its spin. If turning a particle by 180° brings it back into its initial state, then after completing one full circle, that is, turning it by 360°, it would come back twice to its initial state. The particles of this category are said to have "spin 2." Likewise, if a particle, by turning it through one complete circle, comes back to its initial state only once, then it would be placed under the category "spin 1" type particle. If any particle looks the same only once after rotating it through 2 full circles, then after completing only one circle, the same would be halfway to return into its initial state. Such particles are known as "spin ½" particles. And, if the quantum state of a particle remains unaltered even after turning it by infinity times, then it is known as a "spin 0" particle.

The Shape of the Electron

The shape and size of the electron couldn't yet be determined. During the last 8 to 10 years, several photographs of electrons are said to have been taken by different laboratories/scientists who used different technologies; none of these photographs are alike or similar. Therefore, it is not possible to ascertain what exact shape the electrons do have? In some of the photographs, the electron looks like a spherical object having a few glowing spots of different colors, as if a bright light is beaming out from these spots. During the beginning of the year 2010, several photographs of the spin of the electron were said to have been taken by "Scanning Tunneling Microscope." Different photographs of the electron and its spin are available on the internet. In the photographs of the said spin, several glowing objects of conical shapes are seen. Researchers believe that these glowing cones represent the spin of the electron. Alternatively, these cones might represent the three-dimensional waves, which are formed by the spin of an electron. In case these photographs are authentic, then the apex of these cones might have some relation to the spin of the electron, or alternatively, this apex represents the peak value of its charge. If possible, efforts should be made to determine:

1. whether electrons at rest also exhibit such cones.

2. Whether luminescence of these cones remains unchanged or varies periodically.

3. Whether the locations and sizes of these cones, as well as the value of the charge of the electron, always remain constant, or these quantities fluctuate in a particular manner. &

4. whether the apex of these cones bears any relation with the direction of the magnetic field or, alternatively, with the value of the charge of the electrons.

The Internal Structure of the Atomic Nucleus

When the neutrons and the protons are made to collide head-on at very high energy, they break up into yet smaller particles that are much different in nature. This fact reveals that neutrons are composite particles made of even much smaller particles. These component particles are known as *"Quarks,"* which are the elementary constituents of the nucleus. As per the latest information, quarks exist only in six (6) varieties, which have different amounts of masses and charges. Different *kinds* of quarks, for the namesake, are called "flavors." All six types of quarks have *electric charges in fractional numbers*, not in whole numbers. Three of these flavors of quarks have positive charges; each one of these positively charged quarks has $2/3^{rd}$ of the unit positive charge. These quarks, in the ascending order of their masses, are respectively named as *"up,"* *"charm,"* and *"top"* quarks. Each of the remaining three types of quarks has $1/3^{rd}$ of the unit negative charge. The quarks of this category, in the ascending order of their masses, are respectively known as *"down,"* *"strange,"* and *"bottom"* quarks. Only two out of these six kinds of quarks, namely "up" and "down" quarks, are of stable nature, both of which possess the least mass. The remaining other types of quarks are unstable in nature and are found at very high energy that is produced either in the particle accelerators or while the cosmic rays or gamma (g) rays collide with an atomic nucleus. Quarks with higher masses decay very rapidly into the less-massive varieties of quarks. Scientists believe that a very compact and heavy substance known as **the strange matter** is made of **strange quarks** (Please refer to Chapter-9 for more details). A new type of quark was also discovered in December of 2011, which has been named *"Beauty quark;"* more information on this variety of quarks is not available.

Besides the electric charge, quarks also possess another kind of charge, which, for the namesake, is called the *"Color Charge;"* this kind of charge has, in fact, no relation with any color. The said **color charge** has three varieties, which are named after different colors, that is, *"red,"* *"blue,"* and *"green."* According to their color charges, each flavor or type of the quarks (up, down, etc.) has three different sub-types, one each in the above colors. Quarks of one color don't attract other quarks of the same color. However, they have a great affinity for the quarks of other colors, due to which they are never found alone; they are normally found in the combination of one of each of these three colors to form a so-called colorless particle. Sometimes quarks of one color combine with another quark of anticolor; as a result, an unstable pair of quarks, called *"meson,"* is produced; only three quarks of different colors can form a stable combination. Protons and neutrons are made by different combinations of quarks. The protons are made up of two "up" and one "down" type quarks; a total of their electric charges, therefore, becomes $\{(+2/3) \times 2\} + (-1/3) = 1$. Whereas neutron is made of one "up" quark and two "down" type quarks, the summed-up value of their total electric charge becomes mathematically zero, that is, $[\{(-1/3) \times 2\} + 2/3] = 0$.

The "Standard Model" of Particle Physics

Scientists believe that every function of the universe is carried out through a few basic building blocks, known as *fundamental particles,* and four *natural forces,* that is, the **"strong nuclear force" (strong interaction), "electromagnetic force," "weak nuclear force" (weak interaction) and "gravitational force." The matter is supposed to have been created by the first three forces and a few elementary particles; gravitational force is not supposed to play any role in the construction of the matter.** Scientists further believe that all the natural forces perform their work by exchanging different kinds of force-carrier particles, rather by emitting and absorbing such particles. The understanding of *"How the matter particles and three nuclear forces are related to each other"* is encapsulated in the **"Standard Model of particle physics,"** developed in the early 1970s. A broad introduction to various kinds of particles, excluding the mathematical framework of this theory, is given below.

To honor the scientist "Enrico Fermi," different kinds of matter-particles are known as *"Fermions."* Similarly, the Force-Carrier-Particles are known as *"Bosons"* to honor the Indian scientist *"Satyendra Nath Bose."* Fermions are further divided into two sub-verities; the particles that can't be split any further, such as *electrons* and *quarks,* are known as *elementary or* fundamental particles; on the other hand, the particles, which are made up of two or more elementary particles are called *composite particles* or *"Hadrons;"* protons, neutrons, and atomic nuclei, etc. come under this category. Hadrons are further divided into two subcategories, that is, *"Baryons"* and *"Mesons." Out of these, "Baryons" are stable or long-lived particles made of three quarks, whereas "Mesons" are short-lived pairs of quarks and antiquarks.*

The Elementary Fermions (matter particles) have two subcategories- *"Quarks" and "Leptons,"* the introduction to six types of **quarks** has already been given beforehand. The remaining verities of fermions, *which normally do not constitute the atomic nucleus*, are placed under the category of *"Leptons,"* *which* react with the weak interaction only; strong interaction doesn't affect them in any way. The Leptons also have six sub-varieties. Out of this lot, three leptons have a unit negative electric charge; *"electron"* is the lightest-one of them. The remaining two charged leptons are known as *"Muon"* and *"Tau,"* each of which has a unit negative charge; however, these two particles are several times heavier (more massive) than the electron; therefore, they are unstable and short-lived ones. The remaining three leptons are neutral particles having no electric charge. These leptons, according to the ascending values of their masses, are, respectively, known as *"Electron Neutrino," "Muon Neutrino,"* and *"Tau Neutrino."*

It could thus be seen from the above that in all, there are only 12 types of matter particles, which are known as *"fermions."* Each of these matter particles has one antiparticle, too; accordingly, there are in all 24 matter particles. Everything in the universe is created out of different combinations of these 24 particles. On the other hand, if the particles and antiparticles come into contact, then they will annihilate each other; the combination of matter and antimatter can destroy even everything in the world. It is understood that almost a similar description of the creation and destruction of the world is given in the *"Bhagavad Gita"* too, which was written as far back as several thousands of years BC; this seems incredible, rather than unbelievable.

All the matter particles that are classified under the category "fermions" possess a very special property, which, on the name of the scientist *"Wolfgang Joseph Pauli"* is known as *"Pauli's exclusion principle."* According to this principle, *"the identical fermions can never occupy a similar quantum state at the same time and same place. In case two fermions occupy a similar state at the same time and place, then, at least, one of their properties, such as spin, charge, or velocity, etc., shall differ from that of the other."* This is a very important property of all the matter particles because they acquire their **shape and stiffness** etc., due to this property only.

The standard theory predicts that the natural forces, such as "strong," "weak," and "electromagnetic" forces, interact with the subatomic particles by the exchange of very tiny and mass-less particles, which are known as *"force carrier particles" or "mediators of different forces."* Each force has its own force-carrying particle, or in other words, *"different forces act through different kinds of mediator (carrier) particles."* These force carriers, or mediators, are thought to be mass-less, but some of them have been found to have different amounts of masses, which decides their range of effectiveness; however, none of them has any sort of electric charge. To honor the Indian scientist, Mr. *"Satyendranath Bose,"* all the force-carrying particles are, for the namesake, called *"Gauge Bosons,"* or simply *"Bosons."* The value of spin, of all the Bosons of different forces, is in integer numbers, such as 0, 1 or 2, etc., that makes them different from the "Fermions," because the value of spin of all kinds of fermions is ½. All the bosons have a very special property that *they don't follow the Exclusion Principle.* **It is believed that by virtue of this property,** *there is no limit to the number in which the Bosons can be exchanged between different matter particles.*

The Bosons (force-carrying particles) are believed to be virtual particles. They couldn't be detected directly or seen like real photons, which are emitted by different hot or glowing objects. This could be clarified with the example of a magnetic field. The force acting between two magnetic poles is said to have resulted due to the exchange of virtual photons between them (not the light-carrying or real photons, which can be seen or detected directly). Although the emission of these particles can't be detected, even in a particle detector, the effect of this emission can be felt and measured conspicuously. Even though it is not possible to detect virtual particles, scientists have adequate proof of their existence.

Scientists believe that the force carriers are not real particles; instead, they result from the interaction of two or more energy fields. The real or virtual particles are thought to be waves; *scientists believe that an electron is a disturbance in the **electron field** and a photon is a disturbance in the **electromagnetic field;** such particles can be sent to other places in the form of either the beams of electrons, protons, or light (real photons); this is, however, not possible with the so-called force carrier particles.* On the one hand, the energy fields of the real particles never die off; therefore, they have long lives, on the other hand, the virtual particles don't have their own energy fields, when the energy fields of two or more Fermions interact with each other, then a disturbance is supposed to be produced in their fields, *this disturbance is considered virtual particles; after the cause of these disturbances is gone or removed,* such disturbances immediately die-off; they, therefore, don't have long lives.

A series of questions arise on the aforesaid concept- "How a real matter particle, in the form of a disturbance, is created in a field of force, and how this disturbance remains confined in an extremely small location having a subatomic size, that too for billions of years? Why does such a disturbance not spread out in that field of force? Why doesn't such a disturbance, analogous to water-ripples, die off after some time, or instead of remaining confined to a very small location, why doesn't the same permeate gradually throughout the entire field like all other waves? The disturbances in these fields shall calm down gradually or attenuate with time; accordingly, the matter-particles, too, shall decay gradually during their lifespan? How do the photons travel through the distances of billions of light-years? Why doesn't the disturbance that creates photons subside after some time? How the strength of these fields always remains constant."

No satisfactory clarification on the above points seems available anywhere.

"Higgs Boson" The Particle that Bestows Mass

Almost the whole mass of an atom is concentrated in its nucleus, i.e., within the neutrons and protons contained within its nucleus. The larger is the number of neutrons and protons comprising the nucleus, the more massive it becomes. The mass of the proton and neutron is, respectively, 1836 times and 1839 times greater in comparison to that of the electron. The Proton and Neutron are both made up of a combination of three quarks; however, the combined mass of these three quarks, when taken together, is much lesser than the mass of either the proton or the neutron. A question, therefore, arises *"Wherefrom this extra mass has produced?"* Alternatively, **Whether this increase in the mass has resulted due to the coalescing of the quarks with each other?"** Scientists believe that different quarks are held together by the exchange of force-carrying particles named "gluons." Gluons are not supposed to have any mass; however, protons and neutrons probably gain extra mass from the gluons' energies; scientists alone are considered capable of clarifying *this point.*

Before 1964, scientists believed that all the kinds of force carriers, namely *"bosons,"* were mass-less particles. However, contrary to this belief, it was later found that the force carriers of the *"weak interaction"* have high masses. Now, if the bosons were fundamentally mass-less, then how and wherefrom some of them have acquired high masses? In 1964, the solution to this problem was suggested by *Peter Higgs.* He suggested that an invisible energy field that permeates throughout the universe exists in a vacuum or interstellar space. The said field interacts with different particles in different ways. Whenever any matter-particle, or boson, enters or passes through this field, then the force-carrier particles of this field, which are known as *"Higgs Bosons,"* interact with such particles and thereby transfer different amounts of *energy,* i.e., *"mass,"* to different particles. This energy field is known as *"Higgs Field"* in honor of *Peter Higgs.* The amount of energy or mass transferred to any particle depends on the capacity of that particle to absorb mass, i.e., the mass gained by any particle is proportionate to the amount of energy absorbed by it. This is analogous to a swab of cotton, or any other porous matter, which, while passing through water, depending on its porosity, absorbs water in different quantities, thereby it becomes heavier. The weight gained by that material, due to the ingress of water, depends on its porosity, that is, the capacity of that material to absorb water. *It is clear from the above description that the "Higgs Field" doesn't create or generate mass, it simply transfers mass to different particles, and this gain of mass depends on the capacity of different particles to absorb mass in different quantities.*

Looking into the hypothetical properties of the said particle *"Higgs-Boson,"* that is, its capability to bestow mass upon various particles which may exist anywhere in the whole of the vast universe, renowned physicist **Leon Lederman** inadvertently called this particle the *"God Particle."* However, the said particle has no relation with God; it is merely a particle having some special properties, and of course, it too is one of the creations of the "God" or Nature.

Although the existence of "Higgs-Boson," was predicted as early as 1964, it had been eluding the scientists for a very long period of time. During December 2011, when probably the nuclei of Lead were collided head-on, at very high energy, a new kind of fundamental particle was observed, which had an extremely short lifespan. Properties of this newly discovered particle were found to bear some similarities with the hypothetical "Higgs Boson." Later on, scientists, on 14th March 2013, tentatively confirmed that the newly found particle is, in fact, the *"Higgs Boson."*

The fact that "The Higgs-Boson has an extremely short lifespan of about 1.56×10^{-22} seconds" attracts a series of questions. The prospective particle "Higgs-Boson" is produced at a very high energy of 126 GeV and above. Such high energy would have got produced a little after the big bang. Thereafter, within its extremely short lifespan, i.e., before decaying after completing its life, it should have bestowed mass upon the entire lot of the particles produced at that instant. This, however, seems impossible because the Higgs-Field, during such a short time period, could have spread spherically over a very short distance having a radius of 4.68×10^{-11} millimeters, which is even several thousand times smaller than an atom. The said particle would have vanished immediately after the energy level fell below the required level; accordingly, its energy field would also have died off immediately thereafter. In case the matter particles were created a little earlier, then the said Higgs-Field would not have reached out to them. However, the matter particles were probably created at a much lower energy of 1.88 MeV or so; therefore, the said Higgs-Field might have died off before the creation of matter particles. In that case, the mass couldn't be transferred to the newly created matter particles. And, in case the matter particles and Higgs-Boson were both created simultaneously, then the entire universe would have been squeezed into a zero-sized point because of the concentration of the entire mass in such a small place; however, it did never happen. Another point, which goes against this ideology, is "Since matter particles at the time of their

creation were moving at faster-than-light speeds, Higgs Field couldn't have reached out to them."
Further, the fact that the force carrier particles of weak-interaction are also supposed to possess
mass gives rise to another series of questions- those questions are:

1. How the force carrier particles of the weak nuclear force could absorb mass because they are not real particles, instead of real particles, they are merely disturbances in the energy fields of other particles.

2. *At very high energies, the electromagnetic force and the weak interaction become one and the same, this means that the Bosons of the E&M force might have acquired some mass, and simultaneously, the carriers of weak interaction would have lost some mass so that the energies of both become equal. However, particles gain mass at high energies instead of losing it; this is against the law. Moreover, how can the mass, once absorbed by any particle, desert it?*

3. *How can the matter in a wave's capacity acquire mass because the waves are not supposed to have any mass? All these facts and questions suggest that the concept of the Higgs-Boson shall be reviewed thoroughly and modified if felt necessary. In this context, please also refer to Chapter-17.*

The Direct Detection of the "Gluons" (The Pluto Experiment)

Earlier, when the electrons and the positrons were made to collide, then co-planar jets of the quark and the antiquark were produced. Later, during the middle of 1979, a similar experiment was again conducted at the "Pluto Collider," DESY Laboratories, Hamburg, Germany, with much higher energy. In this experiment, an extra jet of gluons was also ejected in the direction perpendicular to the co-planar jets of the quarks and the antiquarks. The result of this experiment, known as the "Pluto Experiments," puts a question on the conception that "The particles and their antiparticles, on coming in contact completely annihilate each other, and as a result, only pure energy is emitted in the form of photons." This series of experiments attract yet another question: "where did these quarks and gluons come from?" The possible answer is: "The energy of the collision had momentarily transformed the original particles into quarks or gluons." However, I see yet another possibility: "Quarks and gluons were produced as debris that was left over after annihilation of the electrons and their antiparticles." In the said experiment, the emission of jets of quarks and gluons gives rise to one more question that is: "Do the so-called elementary particles (Electrons and Positrons) are made of even much tinier particles (quarks), which were held together by gluons?" Such a possibility can't be ruled out completely because the smallest unit of the negative charge is $^1/_3{}^{rd}$ of that *of the charge of an electron, whereas the electron possesses a unitary charge, i.e., in a whole number.*

Scientists have also observed that when quarks are made to collide with each other at sufficiently high energies, they sometimes transform into antielectrons and sometimes into electrons; such a result is just the opposite of the result obtained at the Pluto collider. Such a result poses a big question *"Since the quarks possess fractional electric charges, their conversion into electrons or antielectrons seems impossible because the converted particles possess the charge in a whole number."* Had the electron been a composite particle made of quarks, then such a conversion might have been possible, but the electrons are considered elementary particles without any internal structure. This fact raises several questions on the convertibility of the quark into an electron or a positron; these questions are:

1. *"How a particle having a fractional charge can be converted into another particle having a charge in the whole number?"*

2. *"Where from the converted particle did get this extra charge, and how?"*

The fact that *"Quarks possess many times more mass in comparison to an electron or an antielectron"* poses yet another question *"If the quark was converted into electron/antielectron, then where and how its extra mass is disposed-off; do such collisions cause particles to lose energy instead of gaining more energy?"* *Conversion of quarks into electrons/anti-electrons poses* yet another question: *"Where the carriers of the strong nuclear force (gluons) that were associated with the quarks are disposed-off, and how?"* These unresolved questions point to a profound mystery that is yet to be explored.

The Matter in the form of the "Matter Wave"

Scientists have observed that sometimes the subatomic particles, atoms, and molecules too, behave like particles and sometimes like waves. It is also observed that the locations of the atoms do not always remain fixed, they are sometimes observed at one point, and at the next moment, they appear at some other point. In other words, it couldn't be said that subatomic particles have fixed locations; they appear to spread over different places. The solid objects, though appear to be motionless, still, and placid, but at the atomic level, these matter particles are full of a vigorous and violent flurry; electrons aren't only orbiting the atomic nuclei at an unbelievable high speed, they also jump from one atom to the other. Matter particles, because of the rotational motion of the electrons, might always appear to fluctuate like waves. Probably, for this reason, the atoms, from 1920 onwards, are considered "matter-waves." However, unlike matter particles, two waves, if made to collide, then after such a collision, neither would split into two or more pieces nor rebound in a manner in which the pieces of matter normally do. Similarly, waves cannot be picked up and shifted/transported like physical objects; moreover, it is not possible to either break the waves into innumerable pieces or melt them; however, in all of such cases, the individual atoms would continue to fluctuate as before. *Because of these facts, it appears to me that the waves by nature are very much different from the matter particles.*

Matter particles such as atoms, electrons, and protons, etc., though considered waves, must vastly differ from the waves of light or any other sorts of energy waves because the energy waves, after emanating out of their sources, continuously propagate outward spherically; on the other hand, various objects made of matter, normally remain stationary at their fixed places. As such, the particles of different objects would remain confined at definite limited places or regions. This is probably why matter particles are considered "three-dimensional standing waves" or simply "stationary waves."

By definition, the "standing waves" are the waves that are formed by the interference of two different waves that travel in opposite directions. These standing waves fluctuate between two points of zero displacements, which are known as nodes; the locations of these nodes remain fixed. The crests and the troughs or antinodes of such waves, due to their up-and-down undulation, are created after regular intervals between the fixed nodes.

Now, if the matter waves are created by the interference of two waves, then the proton situated at the nucleus of the hydrogen atom should comprise at least three stationary waves instead of 3 quarks. Similarly, another stationary wave shall orbit the nucleus in place of the electron. Further, to create 3 standing waves in the nucleus, there shall be at least 3 or a maximum of 6 sources of waves, which should emit waves propagating in the direction opposite of each other so that three standing waves could be formed in place of three quarks. Likewise, two separate revolving sources of waves would also be required to produce one standing wave representing a single electron. To fulfill this condition, two separate sources of waves must orbit the nucleus from different radii; however, this appears impossible, no such sources of waves have ever been seen within the atom. In case the proton is one source of waves and electron the other one, then the waves emitted by these two sources could never meet each other exactly at the electron; instead, they would cut each other somewhere at the midway. Several questions arise out of this possibility:

1. Wherefrom the waves that create the electron are generated, and how? Especially for the creation of the free electrons that break out from the atoms.

2. Do the sources of these waves also travel in an electric circuit along with the electrons flowing through that circuit?

3. Is it possible to produce an explosion like an atom bomb by simply breaking any wave?

4. How can the waves possess "mass" or absorb the same from the "Higgs field?"

5. How is it possible that when two waves, in the form of particles, are made to collide in a Particle Collider, they decay into two or more waves (particles) of different frequencies and masses?

6. What is that *thing or medium* that vibrates to form the matter waves?

The only possibility that emerges out of these questions is, "The subatomic particles are not formed by the interference of any sorts of waves; instead, such particles undergo periodic fluctuations." This possibility is supported by the *String Theory**, which envisages that everything in the world, including the subatomic particles, is made of vibrating strings. Further, in the year 2011, scientists did take a photograph of the shadow of the Yttrium (Yb) atom; besides, several photographs of the electrons were also taken at different laboratories. I feel, "Neither photograph of a wave can be taken, nor a wave can cast a steady shadow." This fact suggests that the particles are not, in fact, waves; they exhibit properties of waves because they keep fluctuating within themselves. This possibility further indicates that the so-called elementary particles also, because of their wavelike nature, might have some kind of inner construction. The said fluctuation might explain why the particles have dual properties of waves as well as particles.

---

* Please refer to chapter-7 for a broad overview of this theory.

Duality of the Particles and Waves

As against Sir Isaac Newton's theory of *"Light Particle,"* the illustrious scientist James Clerk Maxwell proved in 1861 that *"Light rays are actually electromagnetic waves, or the disturbances created in the electromagnetic field."* Probably, after this finding, the property of particles found in light-rays became secondary. Later, in 1900, Max Plank suggested that light and other radiations are not emitted in arbitrary quantities and at arbitrary rates; instead, they are emitted in packets or quanta (as they are called) containing certain amounts of energies; these packets at high frequencies have proportionately higher energies, which are not directly proportional to the frequencies. Instead, the energy is determined by multiplying the frequency by a fractional constant; that's why the energy of radiation doesn't increase very sharply at much higher frequencies.

After the decade of the 1920s, that is, after the formation of the theory of **Quantum Mechanics,** scientists started to believe that *radiation of energy has properties of both: waves as well as that of the particles, too, that is, there is a duality of waves and particles in the radiated energy.* If an incident light ray falls on any surface or collides with it, then it exerts a very feeble pressure on that surface (please search- *Crookes Radiometer*). This pressure is known as radiation pressure, and the same is solid proof that light rays have the properties of particles. Some scientists, by utilizing this property, are planning to make spaceships, which would be able to travel through the interstellar space at a very high speed, that too without using any fuel. Looking at the similarity to the sailboats, such future spaceships are called "solar sail," "photon sail," or "light sail."

Besides possessing the property of the particle, the light rays also have the properties of the waves too; anybody can observe that while looking at the shadow of any object, the shadow cast by the edges of the said object doesn't appear as clear and distinct as the edges of the object itself. Besides, a beam of light, after coming out of a small hole, goes on diverging. This property of light is known as "diffraction." Scientists have, based on this property, concluded that light propagates in waves.

The conclusion that matter-particles also behave as waves can also be deduced from another phenomenon. Analogous to the magnets, the molecules and atoms also attract other molecules or atoms, but this property somewhat differs from the magnetic attraction. The magnetic force draws magnetic materials till they come and adhere to the magnet that attracts them, whereas atoms, when reaching very close to each other, say at a distance of the atomic diameter, push away or repel each other. This phenomenon is known as "quantum deflection." Scientists, based on this phenomenon, have concluded that the subatomic particles have the property of waves; this can be proved by a very simple experiment, which is described below:

In case all the crests and the troughs of two sets of waves, which are moving together, lie at the same place and at the same instant of time, that is, they coincide with each other, then it can be seen that their energies get added up. Such waves are said to be moving in the same phase of time. In contradiction to this, if the crests of one set of waves are formed along with the troughs of the other set of the waves, that is, if both the sets of waves are in opposite phases, then their energies would cancel each other. This property of light rays can very easily be tested with the help of a very simple experiment, which is known as the *"two-slit experiment."* The equipment to carry out the above experiment comprises a partition-board having two narrow and parallel slits. To ensure that only a particular wavelength of light is emitted from its source, a source of light of only one color is placed on one of the sides of this board. When a screen is placed on the other side of this partition, then a series of bands of bright and dark stripes, which are also known as "fringes," is produced on the screen. An explanation of the formation of such bands is *"Different light beams while passing through different slits, have to move through different distances to reach the screen; therefore, both the beams of light having the same frequency, go out of phase."* *Accordingly, bright bands of light are formed at the places where the strength of two beams adds up; similarly, dark bands are formed at the places where the strength of such beams cancels each other.* This is known as "interference of waves," which is considered strong proof that light travels in waves.

If, in the above experiment, the source of light is replaced by a source of the beam of electrons or any other kind of particles, even then, we get exactly the same result. Based on the aforesaid experiment, it is concluded that sometimes particles also behave like waves. It is, however, amazing to see that if at a time, only one electron is fired, even then, similar fringes are formed on the screen. This could only be possible when the same electron could simultaneously pass through both the slits. This may look very odd and weird! When only one electron is fired, then it should pass through only one slot at a time, not through both the slits! Probably, based on the above phenomenon, scientists say that there is no place for common sense or general knowledge in the world of quantum science; this field is so strange, amazing, and mysterious that the things that appear to be utterly impossible may be found to happen practically.

In the analogy to the aforementioned phenomenon, it is also believed that even a single electron, when fired from an electron gun, may hit three or even more targets at the same instant of time. Scientists might have practically seen this happening; however, a common man fails to understand how it could be ascertained that only one electron was fired at a time? And in case only one electron was fired, then did it hit all the three targets simultaneously, or did it jump from one target to the next, and so on. In case the path of such an electron could be monitored by a "particle detector," "Attosecond technology," or by the camera capable of taking one trillion photo-frames per second, then could it not be possible to verify how a single electron can pass through two slits simultaneously? This should also be verified, how can an electron travel on three different paths simultaneously to hit three targets at a single instant of time? If this could be done, then it could also be seen whether a single electron, while marching onwards on its path, travels like a wave, i.e., it undulates **up and down,** or it travels in a straight path. *What I want to stress is-* **"We shall not accept any strange or weird phenomenon without thoroughly verifying the same from all the angles"**- *this is the most basic requirement of science, isn't it?*

Unification of the Natural Forces

Scientists believe that at the beginning of the universe that did begin with the so-called Big-bang, there was only a single force that is known as the "Super-Force." Immediately after the Big-bang, this super-force did divide into four natural forces, namely "the gravitational force," "electromagnetic force," "strong interaction," and "weak interaction."

All the aforementioned natural forces have different properties and strengths, and they are believed to interact with the matter particles through different kinds of force-carrying particles. Scientists have observed that the properties of the forces-carrying particles of the later three forces, excluding the gravitational force, undergo gradual changes under the increasing temperatures. Accordingly, the strengths of these forces are also affected by temperature. All the force-carrying particles are not mass-less; the carriers of the weak interaction, which are of "spin 1" type bosons, have a much greater mass than that of the carrier particles of the other forces; this is the reason that the range of effectiveness of this force, that is, its reach, is much shorter in comparison to that of the other forces. In 1967, Steven Weinberg and Abdus Salam, who is of Pakistani origin, suggested that the force carrier particles, which, at normal energies, behave differently, shall, on much higher energy, behave in an exactly similar way. During that era, it was not possible to achieve very high energies. However, this prediction was found very much correct in the next 10-12 years. Weak interaction and electromagnetic forces, at very high energies, became the same force. Based on this finding, it is also believed that these two forces shall unite with the strong interaction at much higher energy. Generation of the required level of energy is not possible in the present era. Therefore, it is not possible to verify this prediction in the present era or in the immediate future.

Normally, the electromagnetic force of the stars, even at their infinite heat, can be felt at the far-off distances, whereas the weak interaction can't reach out beyond the atomic radii. Therefore, the phenomenon of unification of the weak interaction and the electromagnetic force, at high energies (temperature), poses a question mark on the relation between the mass and energies of the subatomic particles because such a unification means that at much higher energy, the range of the effectiveness of the carrier particles of the weak interaction increases and becomes equal to that of the electromagnetic force; alternatively, the effectiveness of the E&M force decreases at higher energies. The fact that "The carrier particles of the electromagnetic force, at high energies, are capable of reaching out to the far-off distances" indicates that the force carrier particles of the weak interaction lose their masses at high temperature (at high energy) instead of gaining the mass. *This fact contradicts the prevailing convention that mass is the measure of energy; accordingly, it should increase with the increase of energy or temperature.*

I have no idea of the fact that how scientists view this phenomenon? However, I feel that this fact shall not be neglected or taken lightly.

The Antimatter and the Antiparticles

Almost 99% of the entire visible matter of the universe is concentrated in very hot stars like our Sun. Most extreme temperature and pressure conditions do exist within the cores of all the stars, where simpler atoms merge into each other to form heavier atoms. Atoms of all the elements, even at such high energies, retain their normal structures, due to which positively charged particles remain in the nucleus, and the negatively charged particles orbit them from the outside. This means that even at such high energies, the strong interaction continues to perform its work properly. It is clear from the above fact that such high energies are normal for the matter; however, the properties of all the elementary particles, including that of the force carrier particles, start to change at these extreme conditions. Scientists have observed that the strength of the *weak interaction* as well as that of the electromagnetic force, *gradually* increases at high temperatures, whereas the strength of the strong-interaction decreases. Scientists believe that all these three forces should become equally strong at a certain higher temperature; accordingly, at a certain very high energy, these three forces would get united or merge into each other, thereby they would become one and the same force; at the energy, which is

85

slightly higher than this limit, the speed of the elementary particles increases to such an extent that they break open the bonds of all of these three atomic forces and start to move in an uncontrolled manner, however, the quarks and gluons remain coupled together, though in the form of a soup of plasma. Such extremely high energy might have existed at the time of the creation of the universe; however, in the present era, such high energy is found either in gamma rays, and cosmic rays or the same is created in the particle colliders where antiparticles and antimatter, etc., are created in very small quantities.

The antiparticles that correspond to any kind of the normal particle have normal masses, but their electric charge and the color charge both are opposite of that of the normal particles. Similar to normal matter, the antiprotons and the antineutrons, etc., are formed by combining the antiquarks in a manner that is similar to the normal particles. The antineutrons also have no electric charge, which is similar to the normal neutrons, whereas the antiprotons possess a negative electric charge. The atoms of the antimatter are formed by a negatively charged nucleus surrounded by the positively charged antielectrons; the antielectrons are also known as *"positrons."* The antimatter is supposed to have the same properties as that of normal matter, but because of its opposite charges, it may annihilate the normal matter. We are very fortunate that at the time of the creation of the universe, the normal matter and the antimatter were not produced exactly in equal quantities; otherwise, the universe would have been annihilated or destroyed completely. The abundance of normal matter gives rise to a question- "Whether this is merely a matter of chance, a mistake of Nature, or is it resulted due to a very well planned Natural-selection?"

As we have already seen that the matter retains its normal structure, even at very high energies and temperatures found in the cores of the stars. This means that atomic structure remains unaltered till the natural forces that govern the structure of the atom perform their work in a normal way. If the energies of the particles increase further, beyond a critical limit, then properties of these forces change to such an extent that they cease to function properly; the structure of the matter breaks down beyond this limit of the energy. When the energy of particles reduces a little but remains just around this critical limit, then these forces try to reactivate and regain their normal properties. However, they might not succeed fully; at this limiting value of the energy, they might act erratically. Antimatter is probably created at such critically high energies. Here a question comes up *"Why only the normal matter is created in abundance, i.e., far more than the antimatter?"*

It seems to me that the answer to this question is probably hidden in a very old discovery. Scientists knew beforehand that the electrons are emitted during the decay of the radioactive substance Cobalt-60; this emission is called the Beta-Decay (b-decay). In 1956, American physicist Chien Shiung Wu of Chinese origin observed that when the nuclei of cobalt-60 are lined up in a magnetic field, they start to spin on their axes. She also observed that if the nuclei are made to spin in a direction opposite to their natural spin, then the emission of electrons due to the beta-decay increases manifold; that is, the rate of their decay increases. This experiment shattered the belief that the laws of physics always obey the *symmetry* *"P,"* that is, *"Laws are the same for any situation and its mirror image." **Besides this, the said experiment also indicated that Nature knows what it should do in adverse situations.*** Probably, because of this property, the antiparticles immediately after their creation would have decayed at a faster rate; this might be the reason that they are rarely found in Nature. Whenever the *mill of nature* spins in the reverse direction, for whatsoever reason, then the antiparticles are probably, produced not only at much slower rates but they also decay at a much higher rate. *The production of the antiparticles is probably, controlled by the well-set laws of Nature, not by any coincidence or by the mercy of Nature; this happened only as per the well-planned choice of Nature.*

Some Least-known Properties of the Light

For the past several thousands of years, mankind, ignoring the important property of light that a ray always travels in a straight line, has been pursuing an ambitious dream of becoming invisible by wearing an invisibility cloak. Although this seems to be impossible, even then, efforts are, perhaps, being made to divert the light rays in a way similar to the water current flowing around the obstacles that come in its way, so that the light rays, too, may go around a man or any other object, without blocking the vision of others. For the last few decades, scientists all around the world have been working in the field of invisibility. Probably, someone might achieve this goal within the next few months or a period of a few years.

In December 2007, a unique and unprecedented property of light waves was discovered by a group of Indian scientists led by Mr. Rasbindu Mehta. This team of scientists passed a laser beam through a soup of Nano-sized and Micro-sized magnetic particles; as soon as a magnetic field was applied to this soup, the laser beam passing through the said soup disappeared. Later on, when the said magnetic field was switched off, then a shine suddenly flashed out from the aforesaid soup. This experiment indicates that light waves can be trapped within an electromagnetic field for the desired period; these trapped waves can be released as and when desired. This experiment also proved that the energy of the trapped photons was neither destroyed nor their momentum to keep moving was lost or destroyed. This might be the reason why the said flash of light was produced after the magnetic field was switched off.

Einstein, as early as in 1927, envisaged that it is possible to trap the photons. Later on, in 1999, Lene Hau, a Danish physicist of Harvard University, observed that when light is made to pass through super-cool sodium atoms (cooled to almost 0°K), then its speed reduces to as slow as 17 meters per second; later, she also succeeded to totally stop a beam of light. In the year 2005, she succeeded in transforming light-waves into matter-waves and thereafter transforming them back into light-waves. All such experiments show that light, too, can be affected by either the electromagnetic force or by extremely low (super-cool) temperatures. Contrarily, *whenever any part of the sea freezes, then it freezes in a flat plane; in this case, any disturbance or ripple (waves) formed in water subsides by itself; when the ice melts down, waves don't form automatically. It is a matter to contemplate that in the aforesaid experiment, the medium of space-time didn't freeze; only the light-wave moving through this medium did freeze or convert into matter-wave. This seems impossible;* **how can a wave alone freeze without freezing the medium?** *This fact puts the wave nature of light under question mark; light, in the form of pure waves, cannot be affected by low temperature or a force like magnetism.* This fact indicates that light probably propagates as a shower of particles because only particles can be affected by such conditions, not the waves.

Positive and Negative Energies

Normally, atoms of various elements possess equal amounts of positive and negative electric charges, and as such, the resultant electric charge of an atom becomes zero. Thus, it could be said that at the level of the atoms, the zero-electric charge is, in fact, the summed-up value of equal amounts of positive and negative charges. In yet other words, "the summed-up *zero electric charges*" of any neutral atom can be divided into equal units of positive and negative electric charges. Probably, based on the above fact, Quantum Mechanics envisages that pairs of positive and negative energies are continuously created and destroyed in a vacuum or interstellar space. Scientists also believe that matter, too, is formed by the pairs of positive and negative energies; they also believe that matter is created by positive energy, whereas space is filled with an infinite amount of negative energy.

Scientist Paul Dirac, in the year 1929-30, suggested that the vacuum is an infinite sea of particles having negative energies. This sea, in honor of Paul Dirac, is known as the "Dirac Sea." According to the Quantum-Field-Theory, the vacuum is filled with the operators of creation and annihilation; the operators of creation have positive vibrations, and the operators of annihilation have negative vibrations. The operators having negative vibrations are supposed to reduce the energy of any particle, and the operators having positive vibrations increase the energy of such particles. These operators of the opposite nature are supposed to annihilate the vibrations of each other; however, it seems to me that the vibrations, which are in the opposite phases, shall only be able to do so. The concept of this theory is subjected to a question: "How any particle may have negative vibrations?" Whenever any real and physical object vibrates, then it can vibrate on both sides of a middle position, which is the position of its rest; only such vibrations, about the position of the rest of an object, can be said to be vibrations of positive nature. Normally, a negative sign is given to a number, or any other thing, when its value is supposed to be less than zero and also in the opposite of the positive direction. This concept poses a question mark on the ideology of negative vibration; the question is- "If any object has negative vibrations, then in what direction would its particles be displaced relative to their positions of rest; displacing any particle in a direction by an amount that is lesser than zero, doesn't seem feasible; they can vibrate in the opposite phase only." Another important question that comes up is: "Wherefrom these operators get the energy to vibrate continuously?"

Scientists, however, believe that negative vibration and negative energy both do exist in the world of very tiny particles. Quantum theory predicts that the vacuum is not simply an empty space; instead, short-lived pairs of particles and antiparticles, or the pairs of the virtual electrons and the antielectrons, are regularly created and destroyed in the vacuum. Scientists have practically observed that even the vacuum, too, can be excited by applying a magnetic field, i.e., when a magnetic field is applied across a vacuum, then these pairs reorganize themselves in such a way that they oppose the applied field. The strength of the applied field is thus found weaker than what it should have been. The reorganization of these pairs of virtual particles is called the "polarization of vacuum." I somehow feel that "It might be possible that the force carriers of the magnetic field, that is, virtual photons, might encounter resistance due to the permeability of the mediums like air and the walls of the glass container, etc., through which the virtual photons have to pass through. This, in turn, might be the reason that the field-strength was found weaker than what was expected." Therefore, the above aspect shall also be considered and evaluated while ascertaining the possibility of the formation of such pairs of the particles and the antiparticles.

Whatsoever be the reason for this phenomenon, scientists have reason to believe that such pairs are continuously created in the vacuum at the speed of light, and subsequently, they annihilate each other at the same speed. The pairs of the real particles and their antiparticles may also sometimes be produced in a vacuum. Probably, because of this reason, Swiss scientists in November 2011 achieved success in producing light from the vacuum. The success of this experiment is considered solid proof of this idea. However, in case such pairs are produced and annihilated regularly, then photons so produced shall always be seen, not on a single occasion.

×   ×   ×

We all know that some energy is required to break the solids into smaller pieces. This simple fact indicates that the molecules of the different kinds of matter are bound together with a very strong force; they have a great affinity towards each other. On the other hand, at the subatomic level, the matter particles *repel* each other from minuscule distances. It appears from both of these facts that two different kinds of forces do exist within the matter, which act in the opposite directions of each other; even then, they do not nullify each other; instead, they perform their duties independently. One of these forces attracts and holds all the particles together, whereas the 2nd force prevents them from merging into each other; individual particles thus maintain their independent existence. The state of balance between these two forces gives shape, size, and other properties to different kinds of materials. The forces acting within the atom perform different duties; the strong force and electromagnetic forces serve to bind different subatomic particles with each other, whereas the weak force breaks the unstable combinations of the particles. However, these forces don't oppose or nullify each other. Both of the above facts indicate that these forces though acting in the opposite directions, do coexist without nullifying each other.

Scientists believe that "One has to expend some energy to separate two particles or objects against their mutual attraction of gravitation;" therefore, the energy spent to separate them from each other is stored in those objects. Accordingly, it is also believed that the particles located far apart from each other have more energy than those located closer to each other. Based on the aforesaid ideology, it is interpreted that **gravity is negative energy**. At the same time, **Quantum Mechanics** envisage that **matter is made of positive energy; and interstellar space is filled with an infinite amount of Negative Energy.**

Somehow, I feel that all of the aforesaid ideologies, i.e., *the negativity of gravitational energy, the positivity of the energy of matter, the huge store of negative energy in the space, etc.,* contradict each other. In case all of the aforesaid concepts are correct, then the **positive energy of any congregation of matter** *(in an accumulation of matter, the positive energy of matter would mean the total quantity or mass of matter)* **shall**

90

annihilate the entire negative energy of gravity, or, at least, both the opposite kinds of energies shall nullify the effect of each other. As a result, such an accumulation of matter shall become a body having zeroed summed up energy, i.e., an accumulation of matter, regardless of its mass, shall not exhibit the property of gravitation; though this is similar to the zeroed electrical charge of an atom, this doesn't happen in reality. Contrarily, the gravitational force of an accumulation of matter increases proportionately with the increase of its mass. For example- the Sun is a collection of huge amounts of matter; due to that, it produces a very strong gravity field, and at the same time, it radiates an enormous amount of energy. Had gravity been negative energy, then the Sun shouldn't be able to radiate any energy; instead, energy from all-around shall rush toward the Sun. However, this never happens. *This fact signifies that something, somewhere, has gone wrong or misconceived. Similarly, if space is filled with negative energy, then no star or planet shall survive in space because the infinite negative energy of space shall annihilate the positive energy, i.e., the total quantity of matter or mass contained in all of such objects.*

The reality brought out above indicates that gravity is not at all negative energy; it never performs negative work; that is, it never causes anything to move in the direction opposite of the applied force. Had gravitation been negative energy, then *the more massive a celestial object would be, the easier it would be to take off a spaceship from that object;* contrarily, a spaceship needs far more energy to take off from the Earth as against the energy required by the same spaceship to lift off from a space station. Obviously, the Earth has much greater positive energy than the energy of the space station. In this context, it is a point to ponder that the far-off planets were never dragged away from the sun; instead, they, at the time of their creations, were formed at their respective locations, and as such, no energy had been expended to move them away from the sun. Of course, depending on their respective masses and distances from the sun, much work would be required to shift them to the sun; they, therefore, have huge potential energies, accordingly, when the sun tries to draw them closer, then some of their potential energy is converted into kinetic energy. However, their total energies remain unchanged. *Had gravity been really negative energy then, any star formed by the gravitational collapse of a large gas cloud should gradually lose almost all of its energy and eventually become a body having negative energy. In such a case, energy from all around shall rush toward such a star; this is, however, contrary to reality. In fact, the energy of the stars increases to such an extent that they radiate a huge amount of energy instead of soaking it. I feel that in the view of reality, the conventional concept of the negativity of gravitational energy shall be examined critically and with the utmost attention.* This is further explained in the chapter- 11.

In the above context, please refer to the chapter-5. "Relativity," wherein it is predicted that the ticking speed of time slows down in a high energy environment, i.e., near massive celestial objects. This prediction of relativity has been found correct; atomic clocks kept in the spaceships have been found running faster than the clocks that are kept on the earth's surface. This fact is irrefutable proof that massive objects do not possess negative energy.

One more fact, which signifies that Gravity is not negative energy, is: "*Some liquids such as Helium-3 and Helium-4, etc., when cooled nearly to 0°, Kelvin, exhibit a unique property of Superfluidity* **(the Fifth State of the Matter)**. *These liquids, at such a low temperature, behave as if they have zero viscosity. Such super-cool liquids exhibit the ability of self-propelling, i.e., flowing automatically even against gravity. This property is not resulted because of the negative nature of gravitational force; gravity doesn't push them away; instead, this phenomenon is resulted due to the capillary forces, i.e., due to the mutual attraction of the molecules." Although such super-cool liquids climb up or creep along the surface of their containers against gravity, they eventually drop down under the influence of gravity.* This fact reveals that gravity never repels anything under any condition, it always attracts, and thereby it always performs positive work. Or in other words, gravity is positive energy; **had gravity been negative energy, then it should have performed negative work, i.e., all kinds of materials shall move away against the gravitational attraction.**

The concept of the negative energy totally depends on the **Uncertainty-Principle,** which envisages that field strength of any energy-field can never have zero value because in that case, its value as well as the rate of change of its strength both, would have precise (zero) value, which is against this man-made law. According to the *uncertainty principle,* there must be a certain minimum amount of *uncertainty* or *quantum fluctuations in the value of the strength of any field.* Based on this concept, it is believed that the pairs of virtual particles and their antiparticles are always created in a vacuum. These particles, at any particular instant of time, appear together, and thereafter they move apart; immediately thereafter, such particles again come together and annihilate each other. Scientists believe that the chain of such an act of creation and destruction, of such pairs, always continues. **This concept raises several questions, "Wherefrom such particles are created, and wherefrom they gain energy to move apart against their mutual attraction?" Neither any particle can be created, and nor the energy to separate such particles can be created all by itself; both of these presumptions are against the laws. Moreover, the sum of the total energy of such pairs would certainly be zero, which is also against the spirit of the said Uncertainty Principle.** *When everything in the world is uncertain, then how could it be believed that the "uncertainty principle" does **certainly** applies to each and everything in the world? How can it be said firmly that pairs of "the particle and their* *antiparticles" are "**certainly**" created and destroyed in the vacuum?* ***A firm belief in this principle is totally against its spirit because everything is deemed uncertain.***

Further, during the period 1928 to 1930, scientists observed that a few particles sometimes move in a field of force in the direction opposite of that of the others, as if they possess negative energies. However, I feel that the vector direction of the energy of such particles shall not be decided by the direction of displacement alone; the direction in which the applied force intends to displace such particles shall also be considered. ***Since the capacity to displace any object, i.e., to do positive work, is called energy, therefore, the capacity to do negative work, i.e., to move the object in the direction opposite of the applied force, shall only be treated as negative energy.*** Any force, because of the different properties of different particles, would act on them in different directions; magnetic poles attract dissimilar poles but repel similar poles. Likewise, the same force would respectively act in different directions on the electrons and the antielectrons because their charges are opposite in nature. In any magnetic or electric field, the positrons would always be deflected in the direction opposite of that of the electrons. However, *both of them would certainly move in the direction of the applied force, not the opposite of it.* ***Based on the direction of the displacement of these particles, any one of them can be said to have negative energy in the reference frame of the other one; however, none of them moves in the direction opposite of that of the applied force.*** *This fact suggests that energies of such particles are though of opposite nature, none of them possesses negative energy.* Therefore, before deciding the vector direction of the energies of such particles, it should also be examined "Whether these particles are displaced in the direction of the applied force or opposite of it." ***If any particle moves in the direction opposite of the applied force, instead of moving in the direction of the force, only then could the same be said to possess negative energy; alternatively, the applied force can be said negative in nature. Moreover, since the mass of any particle is considered the measure of its energy, its energy cannot become negative until its mass also becomes negative, i.e., lesser than zero, which seems impossible. This point clearly tells that the energy of an object can never become negative.***

In case the above viewpoint of mine is found correct, then the possibility of the existence of negative energy would become very slender, or rather impossible. Anyhow, particles having energies equal in value but opposite in nature, such as the electrical and color charges, etc., do certainly exist. It is also possible that any other kinds of charges, which are not known to us to date, might also exist. *I personally feel that different sorts of virtual and real particles are probably made of extremely tiny charged particles, which possess charges of the opposite in nature. However, some kind of force might* *prevent them from coming very near to each other. If this is possible, then these particles having opposite kinds of energies might never meet each other directly. **Instead, the lighter particles of such pairs would orbit the heavier ones at very high speeds; the smaller would be the distance between them, the higher would be the speed of the particle that orbits the heavier one.** Probably, due to this reason, the energy fields of such a pair of rotating particles would seem to fluctuate very rapidly. In case this is true, then it may also be possible that the property of spin is produced in different subatomic particles because of this reason only.*

Photons, the carrier particles of light, are though considered the pairs of one unit each of positive and negative energies, their constituent energies of opposite nature, never annihilate each other. Instead, the photons travel through space for billions of years. *This fact suggests that the photons shall not be simple pairs of opposite kinds of energies; different constituents (opposite kinds of energies) of such pairs, despite their mutual attraction, might never come into physical contact with each other; probably, the quantum repulsion prevents them from doing so. As a result, the lighter constituent of the photon might orbit its heavier counterpart at an extremely high speed. (Please also refer to chapter – 4).*

×   ×   ×

4: Propagation of the Energy Waves

It is a well-known fact that the heat energy of any hot substance could be felt from a little distance, even without touching it. Such a transfer of energy over long distances is called radiation. We know that when the energy of any substance increases, then the speed of the electrons within its atoms increases proportionately. As a result, such substances start to emit energy waves. The frequency of the waves emitted from such a substance probably depends on the speed of the electrons. This means that the higher would be the energy of a substance, the more powerful energy waves would it emit. Energy waves are not necessarily emitted only from hot substances, even very cool substances also emit waves of low energies. However, the intensity of such waves is so weak that we fail to take any notice of them; only very sensitive instruments can detect such weak radiations.

Scientists, in the medieval age, believed that energy waves propagate through a medium called "Aether or Ether" This belief persisted for a couple of hundreds of years; it was ousted in the year 1905, by Einstein's Theory of *Relativity* (Kindly refer to Chapter-5). Since then, scientists believe that *energy waves, while on their voyage, propagate undulating through the fabric of space-time*. Based on the property of the Radiation Pressure that any kind of radiation exerts on any obstacle that comes in its way, the *quantum theory* introduced the concept of duality of particles and waves in light. Scientists generally believe that light waves are analogous to two-dimensional water ripples. *Contradicting this conception, light waves, after they are emitted from a source of light, propagate like a three-dimensional bubble that goes on continuously expanding steadily, but at a very high speed.* This is evident from the fact that even the dimmest light can be seen from all directions. Normally, light or any other energy wave doesn't travel in one or two rays, which is evident from the fact that a single ray can't be seen from all directions; this fact indicates that at least one spherical wave that spreads in all directions, must be emitted at a time. However, any such spherical wave, after passing through a hole, moves in a straight line like a *Ray*. This fact indicates that light might propagate in some other manner, not like a wave; this is discussed a little later:

Energy waves emitted from any source always spread radially, in straight lines; therefore, they seem to spread spherically, in all directions. These waves could be diverted by a prism or a mirror or get absorbed by blocking their paths. Once the energy rays are set on their way, they continue their onward journey to infinite distances until any object doesn't come in their way. The intensity of light goes on diminishing continuously; however, the same could never become zero. Energy waves, even after traveling for several billions of light-years, probably maintain their original frequencies, wavelengths, and speed, etc. (please also refer to chapter-15). If a wave on its way encounters an obstacle, then it exerts a very feeble pressure on such an object. This fact signifies that the light waves possess the property of particles too.

Propagation of the waves in three dimensions

Light, though propagates spherically in all three dimensions, scientists believe that light spreads in waves that are similar to the ripples propagating on the two-dimensional water surface; however, the water ripples can't propagate in three dimensions. The following properties of both kinds of waves reveal that light doesn't propagate like water ripples:

1. *Water ripples, while spreading on the two-dimensional surface, undulate up and down, i.e., in the direction perpendicular to the water surface or in a direction transverse to the direction of the propagation of the waves; this is not possible with the spherical waves. How can a spherical wave, while spreading in all three dimensions, vibrate in a direction perpendicular to the direction of its propagation? How and where will the crests and troughs of such a wave be formed, and how will its wavelength be measured?*

2. *The expanding water ripples never exert any pressure on the objects floating on the surface of the water; on the other hand, the so-called light waves exert a very feeble pressure on the objects that obstruct their way.*

Several real movies of the three-dimensional light waves expanding spherically or beams propagating in a straight line are available on the Internet, in which the *spherical light waves are seen expanding steadily, without vibrating in any way; such a movie poses a question* **"How a spherical wave can vibrate,"** *this is beyond imagination.* Quantum Science envisages that light has the dual property of waves, as well as that of the particles. This concept makes the matter more complicated; fluctuation of the waves can be understood; however, *it is beyond imagination how a particle while moving ahead can undulate up and down, i.e., perpendicular to its direction of motion, or in a "to and fro" manner; any particle, in the absence of any force acting on it can move only in a straight line with a steady speed, not undulating or swaying in any other direction.* The sound waves surely spread within the mediums like the air and water; however, such waves don't ripple through the two-dimensional surface of these mediums; they spread in the form of the fluctuating pressure-waves that spread spherically, but this type of waves cannot propagate through a vacuum or without a medium, i.e., through interstellar space. *This fact indicates very clearly that light waves are much different from either the water ripples or the sound waves.*

It is for consideration that no vibrating material or medium can persist in vibrating forever; such vibrations die off sooner or later, seemingly, due to the rigidity of the mediums and/or the internal forces acting between the particles of that medium. *Contrarily, the light waves propagating through interstellar space never die off, albeit continual expansion of the waves causes their intensity (number of photons per unit area) to attenuate persistently.* However, light traveling through space can travel through infinite distances of several billion light-years; whereas, the same rays lose all of their energy while covering a small distance of a few hundred feet within the seawater. Another point to note is *"The speed of light reduces within different mediums; on the other hand, its speed increases within a vacuum."* Based on these observations, it can be concluded *"Light, in the absence of any medium, propagates in a better way,"* in other words, *"The medium resists the free-movement of light waves or photons."*

Looking at the duality of *particle and wave* that is found in light, a question arises *"Though, in interstellar space, the particles can travel unrestrictedly in a straight line, how do such particles undulate while moving in an up-down or a to and fro manner?"* And, in case light propagates in waves, then how a spherical wave would undulate in the space, i.e., without any medium? *Any disturbance, fluctuation, or wave-like motion is not possible without a medium; no wave can propagate in the absence of a medium.* Einstein provided a solution to this question.

Einstein proposed* that the light waves ripple through the medium of **Space-time.** Although a vacuum comprises both space and time, neither of these two entities can vibrate or fluctuate to facilitate the propagation of light; therefore, an alternate possibility is discussed hereunder:

Any piece of the matter, when heated, emits light in the form of *a shower of photons* or energy waves. However, the number of photons so emitted might periodically increase and then decrease, and if the periodicity of this emission is regularly repeated, then this emission of photons would seem to vary about an average value like a wave. Such varying pulses of photons would spread from their source in the form of different co-axial expanding spheres that would not

---

* The details of Einstein's theory and its shortcomings are discussed in the next two chapters- 5 and 6

need any medium to propagate in the space. Moreover, such an emission would exhibit the dual properties of waves and particles, too. *However, such emission of photons in the two-slit experiment would not produce the characteristic pattern of the light and dark fringes; especially a single photon will surely fail to do so. This fact indicates that energy does not propagate in the above manner. This failed model of propagation of light* poses the question that *how does energy propagate* and *how it exhibits the property of the duality?*

Until the answer to the above questions is not obtained, it will not be possible to understand how does light propagate in interstellar space? An alternate solution to the above problem, as envisaged by me, is given hereunder:

Atoms are made up of different subatomic particles, which are even several thousand times smaller in size in comparison to the atom itself. The composite particles such as protons and neutrons are made up of three quarks, which are bound to each other by the *force carrier particles* named *"Gluons."* The bondages between these quarks are very flexible; this flexible bondage is known as *"flux tubes."* Probably, the constituents of all the composite particles, because of their flexible bonds and mutual repulsion/attraction, keep vibrating at their locations. However, this might not be the case with elementary particles like electrons. Although the smallest unit of negative charge is $1/3^{rd}$ of that of the charge of an electron, even then, electrons are not supposed to have any internal structure. However, during the collision between the electrons and positrons, the jets of gluons did also produce beside the jets of quarks **(See Chapter-3, "Pluto Experiment")**.

In the said experiment, the detection of gluons poses a question of *utmost importance: "Did gluons were simply bound to the electrons and/or antielectrons before they were made to collide?" Alternatively, whether gluons did produce due to the destruction of the internal structure of the so-called elementary particles?*

Whenever two composite/elementary particles are made to collide in a Particle-Collider, then different kinds of new particles having extremely short lifespans are produced; such particles produced due to the collision, immediately after their births, start to move on different curved paths. **This fact reveals that "These tiny particles must have a lot of angular momentum; otherwise, they should have adopted straight paths." The aforesaid probability suggests that "Either such particles have the inherent property to rotate about their axis, alternatively, the so-called elementary particles, too, might be made of still tinier particles, which revolve around their much tinier and comparatively much heavier (massive) companions."**

In the present era, scientists believe that electrons and other elementary particles don't have internal structures. On the other hand, *String Theory* (see chapter-7, for details) envisages *all sorts of* subatomic particles, energy carriers, and space-time, etc., are made of vibrating strings. If this theory is correct, then all the subatomic particles, including the force carriers and the *photons*, etc., must always keep vibrating; *perpetual vibration might be their intrinsic property.*

In the light of the string theory, it can be said **"If a shower or stream of such vibrating/rotating particles, is emitted from a source of any kind of energy, then these particles would travel in straight lines and spread radially in all the directions like an expanding sphere; at the same time, the inherent vibration/ rotation of these particles would bestow the property of the wave to such a shower of the rotating/vibrating particles."** *The waves, so formed, would behave exactly like the* **"de-Broglie"** *waves in which the vibrations remain confined within a small region, and this vibrating region propagates steadily in a straight line.* **Such a shower of particles would possess the dual property of particles as well as that of the wave; moreover, such a wave would not require any medium, not even "Space-Time," to propagate. Whenever a photon hits the edge of a pinhole or that of any other object, it would be deflected randomly due to its intrinsic vibration; accordingly, properties of "diffraction" and "interference," etc., would also be produced in light. This would also explain the reason why the rays coming out of a pinhole diverge within a limited angle. Various kinds of radiations are, probably, a succession of photons having different frequencies or energies.**

One more fact supports the aforesaid possibility: *In case the light waves propagate rippling through the fabric of Space-Time, then this fabric must have very high tension to support very high frequencies of light-waves. Accordingly, a very high amount of energy would be consumed by the light-waves to produce very high vibrations in such a vast and tense fabric. Therefore, the energy of any wave propagating through this fabric would continuously deplete at a very high rate, as a result, such a wave will die off after its energy is exhausted completely. Contrarily, the light ray have extremely long lives; this fact indicates that light is not a fluctuation or disturbance in any sort of medium. Moreover, had light been a wave, then it won't be possible to send the laser beams (which, too, are supposedly waves) on any straight path that too in only a single direction.*

I brought out in chapter-3, under the subheading "Positive and Negative Energy," that photons probably comprise a pair of positive and negative energies; a lighter particle in such a pair orbits the heavier one. In case this assumption of mine is correct, then, *while any photon moves along the axis of its spin, it would produce an illusion as if a helical wave is propagating. On the other hand, while any photon spinning in a vertical or horizontal plane would move on a straight path, then a pure sine wave would be ostensibly produced; however, each of such photons will create a ray of light. Such a rotating photon, after completion of one revolution, will move ahead by one wavelength. Accordingly, its speed of spin could be calculated.*

Any photon spinning at any other angle would appear like a wave that would apparently vibrate differently. If the prediction of scientists "When the energy of any particle increases, its speed also increases" is correct, then, whenever the energy of any photon will increase, the speed of its spin will also increase; accordingly, photons having higher energies will apparently create waves of higher frequency and shorter wavelengths; the inverse of this will also hold good. Thus, the rotational speed of very high-energy photons, such as those of Gamma Rays, might be found to exceed the speed of light.

While such a shower of photons would spread spherically, its intensity, i.e., the flux of photons per unit area, would though reduce persistently it would never die off. Whenever an obstacle would come in the path of such a shower, it would block the propagation of only a few photons; the rest would continue their onward journeys; this would result in casting a shadow of the said obstacle. Contrary to this possibility, whenever any spreading water ripple meets an obstacle, then its crest reappears almost immediately after that ripple moves past that obstacle; such ripples, because of this reason, can neither cast shadow nor propagate like a beam of light or light-rays. Moreover, contrary to light rays, which never spread at the rear-side of the reflector of a torch, sound produced by a loudspeaker, can be clearly heard at its rear side, too. The above facts indicate that all kinds of radiated energies might propagate in the form of a spherical shower of vibrating photons, not in the form of any two-dimensional ripple or spherical wave, which is produced in a medium.

The Doppler's Effect

How the speed of a moving light source can affect *the frequency of the light waves emitted from that source* was studied by Christian Doppler in the year 1842. Doppler discovered after studying the light coming from a binary star for a considerable time that the frequency of the light coming from any moving source changes *apparently* due to the relative velocities of such a source and the observer. This discovery is known as *"Doppler's Effect." In fact, the speed of any moving light source produces apparent changes in the frequency of light and sound, etc. However, such changes are not real; in fact, the frequency of these waves does not change at all.*

Whenever a moving source of the sound, for example, a car blowing the horn, moves toward an observer, then the said observer feels as if the sharpness of the horn is gradually increasing. Similarly, the horn of a car that is moving away from an observer sounds softer than actual. What actually happens in the former case, "The sound waves though move with their normal speed, every new sound wave coming from the car that was moving toward the observer, is emitted from a comparatively shorter distance." As a result, the next wave reaches the observer earlier than the previous one. The observer, therefore, feels as if the frequency or the sharpness of such a wave is continually, but gradually, increasing. In case the car moves away, then this process would be reversed; as a result, the sound would gradually seem to become softer.

Everybody probably knows that visible light is made up of seven colors. The Red color, which has the lowest frequency, always remains at the one end of the rainbow, and the violet color, having the highest frequency, appears at its other end. The light, split into its constituent colors, is known as "spectrum." Every element has its own and unique spectrum by which scientists can identify different elements. In 1924-29, Hubble observed that the light spectrum of different galaxies resembles that of our own galaxy, the "Milky Way," however, the colors in its spectrum are not seen at their normal places; they are shifted toward the red color. Shifting of colors in such a manner is called "Redshift," based on this, it was concluded that far-off galaxies are moving away from us at a very high speed. "Blueshift" was also observed in some of the galaxies, which indicates that such galaxies are moving toward us, or we are moving toward them. (*Please also refer to* **Chapter-15, Dark Energy,** *on my own viewpoint on the Redshift.*)

Emission of Energy from the Moving Sources

Several trillions of celestial bodies exist in this vast universe; all of them are moving at their own speeds. Since none of these bodies is at rest, the speed of any of these objects has to be measured relative to any other moving object. In other words, the velocities of the celestial objects have to be measured relative to that of the others; it is not possible to measure the absolute velocity of any one of them. Each of such moving light sources continuously emits a shower of photons or energy waves; every new wave follows the waves that were emitted earlier; all such waves propagate radially outward from their respective sources at equal speeds. In other words, a series of waves are continually emitted from all such moving sources; each wave is emanated from a new location of its source; each wave spreads from its point of origin. Photons are supposed to propagate with uniform, unaltered speeds. The apparent reason seems the photons are ejected at such a high speed that their speeds are not affected by the speeds of their sources; that is, no additional momentum is imparted to them due to the speeds of their sources, may these sources move with whatever speeds. Scientists believe that photons always move at their natural uniform speeds, may it be in the direction of the motion of the source or opposite of it.

Although the celestial bodies continually move away from that point from where any particular wave was originated, that point always remains fixed for that particular wave. *Any wave, after emanating out from a moving source, spreads out spherically from that very point from where it was originated. Once any wave is emitted out, all of its contacts with its source are cut off; it spreads outward spherically at its normal speed, in all the directions from that very point that acts as the center of such a spherical wave.* Therefore, whenever we see any wave, maybe after billions of years after its origin, it would appear to come from the point of its origin; accordingly, its source would appear at that very point and at that very instant of time when that wave was originated; however, the source of such waves would continue to move on its own path with its own speed. Each (individual) wave emitted from such a moving source, at different points in time, would appear like an expanding sphere; each of such spheres would seem to expand from its respective point of origin. Any such wave would always carry the information about the time when it was emitted out from its source; it would not give any further information about its source. However, the subsequent waves, emanating from their respective new point of origin, would be seen one after another, and as a result, such a light source would be seen moving with the same speed at which it was actually moving when the said waves were originated. Accordingly, the source would be seen at a far-remote point in the past, wherefrom those waves were emitted. In case the source is destroyed or ceases to shine in our present time, even then, this couldn't be noticed at present; our future generations would see this event when the rays emitted from such a source, at the time of its destruction would reach the earth.

As per the conventions prevailing in the present, the speed of the source doesn't add up to the speed of the photons that were emitted out from the moving source; light waves, therefore, always move with a fixed uniform speed. On the other hand, the source also continues to move at its own speed. As such, light, in the reference frame of the speed of different moving objects, should appear to move at different relative speeds. However, the speed of light is so high that this fact could only be verified when a moving object either moves through a considerably longer distance or the speed of the moving object is considerably high. For example, the distance between the Sun and the Earth is about 150 million kilometers. Light waves, moving at a speed of approximately 300,000 kilometers/second, take a time period of about 8.33 minutes to cover this distance. In the same time interval, the Earth that moves at a speed of 30 kilometers/second, that is, about 10 thousand times slower as compared to the speed of light, also moves through a distance of about 15 thousand kilometers, that is, about 0.05 light-seconds.

By the time light travels through a distance of 5 Km on the Earth's surface, the Earth, moving at a speed of 30 Km. per second, also moves ahead by about ½ meters. The light, to cover this extra distance, takes an extra time period of about 1.66 nanoseconds, which is though a very small time-interval, the same is still measurable, provided the speed of light can be measured in one direction only; however, this is considered near-impossible.

Danish astronomer *Olaus (Ole) Roemer* observed in 1676 that when the earth was moving away from the planet Jupiter, then the time interval between the successive eclipses of any particular moon of the planet Jupiter was about seven minutes larger as compared to a similar event, but when the earth was moving toward the planet Jupiter. He reasoned that when the earth was moving away from planet Jupiter, then the light, emitted from each successive eclipse, had to travel a greater distance to reach the earth, whereas, when the earth was moving toward Jupiter, the light had to travel a comparatively smaller distance. Olaus Roemer was the first man who, based on the above observation, measured the speed of light fairly accurately. Apart from measuring the speed of light, it can also be inferred from the aforesaid observation that the apparent change observed in the time gap between the successive eclipse of the moons of planet Jupiter clearly indicates that the speed of light, too, undergoes an apparent change in the reference frame of the speed of a moving celestial object such as the Earth or Jupiter; however, no real change would occur in the speed of light. This can be understood from the following examples:

In case any source of light is moving at an absolute speed of ½ the speed of light, then any wave emitted from its source would, after 100 years, spread as a sphere having a radius of 100 light-years measured from its point of origin. In the same time interval, the source would also move ahead in the direction of its motion through a distance of 50 light-years. This means that in this direction, the relative velocity of the light waves, in the reference frame of its source, would be ½ of its normal speed, whereas, in the opposite direction, the relative speed of the light would appear 1½ times its normal speed.

Now, if two different light sources are moving independently in the opposite directions, both at ½ the speed of the light when measured from an imaginary point that is at rest in space, then both of these sources, relative to each other, would move at the speed of light. In such a scenario, a common man would think that light emitted from any of the said light sources would never reach out to the other one. However, this is, in fact, not correct; any wave, after emerging out from any one of these two sources, would spread spherically from the point of its origin at its normal speed. On the other hand, the second source, in the reference frame of the same point, would move at only half the speed of the light. Now, since the light wave emitted from the first source would move at double the speed of the second source, it would certainly reach out to the said moving light source after some time. However, when observed from the second light source, a very large redshift would certainly be observed in the said wave. Now, if one out of these sources is at rest, and the other one is moving at the speed of light, then the light rays emitted from the stationary source would never be able to reach the other source because both the wave and the second object are moving at the same speed. However, light emitted from the moving source would certainly reach the stationary object because the point of its origin of the wave and the second object both will remain stationary.

×   ×   ×

People generally believe that any source of light, which is moving with the absolute speed of light, couldn't be seen because light rays won't be able to emanate from any such object. However, there could be two possibilities: 1) Speed of light would not at all be affected by the speed of the source. & 2) Contrary to the aforementioned conventions, the speed of the source would be added up to the speed of the light emanating from it. We will first discuss the earlier possibility; the second possibility would be discussed later though the same is against the accepted conventions:

Let's analyze the former case first. In this case, all the light waves emanating from such a source would surely radiate out from their respective points of origin and spread spherically in all directions at their normal speeds; this is depicted below in Fig 1.

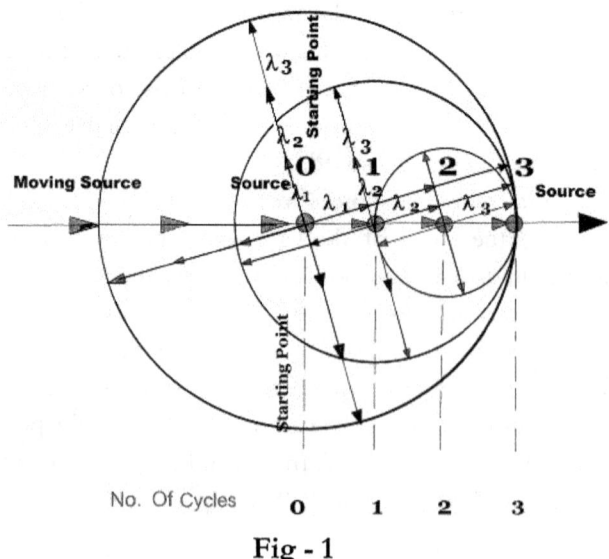

Fig - 1

Different points of origin of three consecutive waves that did emanate from a moving source, which is moving at an absolute speed of light, are shown in the said Fig-1, wherein the source starts moving from the point "0." This source, after each frequency of the wave, i.e., after the wave completes one cycle, would move ahead by one wavelength. At the same time, each wave emanated after each cycle would also spread spherically by the same distance, i.e., by one wavelength. Consecutive three cycles of the waves emitted from the said moving source and corresponding three wavelengths λ_1, λ_2 and λ_3, etc., are also shown in Fig-1. After completion of each cycle of the waves, both the source of light as well as the waves emanated from it would move ahead by one wavelength each.

Since the source and the first wave both began their journeys together, both of them, after completion of the first cycle, would travel through the distance of one wavelength each. Accordingly, after the completion of one cycle, the source and the wavefront of the first wave would arrive at point 1 simultaneously; immediately thereafter, the source would emit the second wave. During this cycle, the source, as well as the wavefronts of the first and second waves, both would again move through one wavelength each. Accordingly, after completion of the second cycle, the source, as well as both of the wavefronts of the first and second waves, would simultaneously arrive at point 2. Similarly, at the end of the third cycle, the source, and also the wavefronts of the first, second, and third waves would move to the end of the third wavelength; that is, they would arrive at point 3.

Thus, it is clear that all the wavefronts moving in the direction of the source would never be able to move ahead of their source; the source and all of the wavefronts would always move together. In other words, the relative speed of the light in the reference frame of the said source would be zero. However, in the opposite direction, though the waves would move with their normal speeds, their wavefronts in the reference frame of the moving source would appear to move at double the speed of light. Subject to the condition that the speed of light is not affected by the speed of its source, these waves wood surely be able to travel in the direction opposite of that of the said source; however, the time interval between two successive waves would be two times the normal time gap. Therefore, it might not be possible to detect the said waves due to a very large redshift.

Based on all of the above examples, it could be concluded that the light travels with different relative speeds when viewed in the reference frame of different moving objects that are moving at different speeds. Such objects, whether moving at very high or very low speeds, the law for them shall always remain the same. However, the speed of light is almost infinite times greater than that of the objects moving with the speeds that appear normal to us. Accordingly, light moves past such moving objects almost instantaneously; therefore, we fail to take notice of the important fact that "Do the light waves while moving past any moving object really move past that object with its normal speed, or do they move at the definite relative speeds in the reference frame of the speeds of different moving objects?"

To have a better idea of the fact "What is the relative speed and how it changes when viewed in different reference frames," let us take up a case of a passenger traveling in a train who is playing with a ball by tossing it up and catching it back. Since all the passengers are apparently stationary relative to the train, it would appear to all the passengers, including the passenger who is playing with the ball, that the ball when tossed up, moves vertically upward and after reaching a certain height, it falls back vertically downward into the hands of the said passenger. Now, this is a matter to ponder that when the said passenger was moving ahead with the train, then had the ball traveled in a true vertical up and down motion, then it should have fallen at the initial location from where the said passenger tossed it upward, not in the hands of the said passenger. Since all the passengers were stationary in the reference frame of the train, it would appear to them that the ball was moving up and down in a true vertical line. However, a stationary observer looking at the ball from outside the train would see that the ball was traveling in a lob, that is, going upward from the initial location on a parabolic trajectory and then falling back in the hands of the same passenger who also moved ahead by a few meters along-with the moving train. Similarly, in the reference frame of the sun, the ball would move ahead by a few kilometers with the Earth. It could thus be seen that different reference frames give different information; all such information(s) are only partially correct for any particular reference frame, not comprehensively correct for all the reference frames.

In the aforesaid example, the ball, in the reference frame of the train, was moving in straight upward/downward motion; however, the said ball was simultaneously moving ahead with the train with the same speed due to its inertia. Therefore, in the reference frame of the Earth, its resultant or final speed would be the vector sum of both of these two speeds. However, this picture changes when we talk about light, whose speed is almost, repeat almost, not actually, infinite times higher than all the known things that move on the Earth. Therefore, light moves past all the other things within almost negligible time interval so that we fail to realize that during the same time interval, such moving things also did move through a very small distance, which can't be taken notice of being almost negligible; therefore, we feel that light moves past the moving objects, at its normal, unaffected speed.

×   ×   ×

It is a general belief that the speed of the photons is not affected by the speed of their sources. However, nobody knows what would happen when the speed of the source approaches the limit of the speed of light because no object moving at this speed has ever been seen. Therefore, let us examine what might happen when the speed of the source would get added to the speeds of the photons, though this is totally against the prevailing conventions. In this special case, the photons moving in the direction of the source will travel at twice the speed of their normal speed, whereas, in the opposite direction, they would not be able to emanate from the source; in this direction, their speed relative to the source, would be zero.

Since no direct means are available to trace or directly observe any particular ray of light right from the point of its origin to its destination point, it is not possible to find out exactly whether or not the speed of the light is affected by the speed of its source. I, somehow, feel that this fact may probably be determined by the help of a geostatic satellite.

The geostationary satellites orbit the Earth at the height of 35,786 kilometers or 22,236 miles above the Equator; these satellites orbit the Earth in the direction of its spin, at a speed of 3.07 kilometers (1.91 miles) per second. At this speed, they complete their one orbit around the earth, in exactly 23.93446 hours, i.e., exactly in one day. As a result, they always appear stationary in the reference frame of their footprint formed on the ground. Light waves take approximately 0.12 seconds to travel from the earth to these satellites. Whereas, the Earth, during this time-interval, moving at a speed of 30 km/sec, moves by 3.6 kilometers in its orbit; in addition to this distance, the satellite, too, moves by approximately 0.36 kilometers in its own orbit around the earth. Therefore, during this time interval, it moves by a total distance of 3.96 kilometers. The speed of the Earth is far below the speed of light; therefore, its speed might not have any effect on the direction of the light beam sent toward the satellite. However, this fact may be verified roughly by sending a pulse of the laser, or a radio wave, vertically upward, toward the satellite. Now, in case the speed of light is not affected by the earth's speed, then by the time the said beam would reach the required altitude, its target would move ahead by about 3.96 kilometers; therefore, it would miss its aim. In case it is found that the beam missed its target by a distance lesser than 3.96 kilometers, then it may be concluded that the speed of light is also affected by the speed of its source. This can be verified by sending a beam towards such a satellite at such an angle that after reflecting back from the satellite, it is received back on the earth. In case the distance between the footprint of the satellite, and the point where the beam was received back, is found greater than the distance between the footprint of the satellite and the point of origin of the beam, then, and only then, it can be inferred that both the speed and direction of the light beam emanating from a moving source, are affected by the speed of its source. On the other hand, if both the distances are found exactly equal, then it would be proved that the speed of light is not affected by the speed of moving sources that are moving at much slower speeds as compared to the light.

We can verify the aforementioned property of light by carrying out a very simple experiment on the Earth's surface; the procedure of this experiment is described below:

In a vast open land, draw a line in the direction of the Earth's motion and mark a suitable point "A" on this line. Next, draw a perpendicular line passing from the said point "A" and mark another point "B," which is located on the said perpendicular line, 3 km away from the said point "A." Alternatively, since it is difficult to draw such a long perpendicular line, send a powerful laser beam from the point "A" at an angle of 90° to the direction of the Earth's motion, i.e., in the direction where the point "B" is supposedly located. In this case, the said beam, propagating at the speed of 300,000 km per second, would take a time interval of 10,000 nanoseconds to travel through the distance of 3 km, whereas, in the same time interval, the Earth, moving at the speed of 30 km per second, would move ahead on its orbit by approximately 30 cm or one foot. As a result, the said laser beam would miss the said point "B," instead, it would pass through another point "C" that would trail the point "B" by about one foot. Now, place a mirror at point "C," facing exactly toward the said point "A." Next, project a laser beam from point "A" at 90° to the direction of the Earth's motion. This time, the said beam, moving on a straight line, would hit the said mirror at 90° and reflect back in a line perpendicular to the said mirror. By the time the reflected beam will cover the distance of 3 km, the Earth will again move on its orbit by another distance of 1 foot. Accordingly, the reflected beam will pass through a point "D" that will trail the said point "A" by 2 feet.

In case the result envisaged in the aforementioned experiment is obtained, then it would be concluded, **"The speed and direction of light/laser beam is not affected by Earth's speed; light persists in propagating in an unaltered direction at its normal speed, and the Earth also persists in moving at an unaltered speed."**

5: The Theory of Relativity - An Introduction

Up till the Middle Ages (Medieval Period), it was believed that our Earth is the center of the universe, roots of numerous other misconceptions too, had reached very deep into our minds. During the fifteenth/sixteenth centuries, *Copernicus* and *Galileo* tried to break some of these misconceptions. The invention of the *telescope* brought a similar revolution in the world of *cosmological science,* as would have been brought by the stone weapons in the lifestyle of the primitive man. *Galileo* was the first man to tell that *"All the celestial bodies are in **relative motion** concerning each other; none of such an object is stationary."* He further told that *"A passenger traveling with steady speed within a closed cabin of a ship, would not feel the speed of the ship;"* this is the **core idea of Relativity.** Seventy to eighty years thereafter, **Sir Isaac Newton** suggested that the **gravitational force** holds all the celestial bodies together. *In that era, people believed that the universe was eternal, endless, and it was all the same all over; it always maintained a steady state; all the stars were held in their respective locations due to uniform gravitational force acting from all the directions.* It was also believed that the universe was in the steady state since the eternal times, and the same will remain so in the endless future too.

In the sixteenth/seventeenth century, it was discovered that the Earth is not the center of the universe; it orbits the sun, which is an ordinary star situated at the inner edges of one of the spiral arms of our galaxy the Milky Way. This galaxy is a congregation of nearly two to four hundred billion stars; the Sun is just an average-sized star out of them. The whole of this congregation of stars, including the sun and its family of planets, is rotating about the center of the Milky Way, which is known as the *"galactic center."* Great chaos was created by this discovery; people earlier believed that the earth was at rest, but this belief was shattered after the aforementioned discovery. Up till that time, the distances of other celestial bodies were measured relative to the earth, but now no place in the whole of the universe could be said to be at rest. As a result, the concepts of the *"position of absolute rest"* and the *"absolute God Time"* were shattered. In that era, the wave-nature of light was also discovered; it was believed that light propagates through a medium named *"Aether,"* *which* is present everywhere in interstellar space. During that era, different scientists made some new discoveries and removed various misconceptions; thereby, they gave a new direction to the ever-developing world of science.

Background of the Theory of Relativity

One of the greatest scientists, James Clerk Maxwell, proposed in the year 1864 that light, electricity, magnetism, etc., all are the disturbance produced in the same substance. He also suggested that light, too, obeys the laws of electromagnetism. After the introduction of Maxwell's above discovery, everybody believed that light propagates rippling through a medium named "Aether" or "Ether," which was supposed to permeate throughout the universe; it was also believed that the universe is stationary with respect to Aether. Accordingly, Aether, in the reference frame of the moving Earth, should appear to move the opposite of Earth's motion. As far back as 1887, the duo of American scientists Michelson and Morley, to verify the truth of **Aether** compared the speed of light in the direction of the earth's motion in its orbit, as well as in the direction perpendicular to it.

As illustrated in Fig-2, Michelson, and Morley, using a half-silvered glass plate "P," divided a beam of light into two independent beams running at 90° to each other. These beams were then made to reflect back, by two mirrors M-1 and M-2 that were placed at an angle of 90° to each other, at an equal distance "l" from the said half-silvered glass plate "P." Both of these beams were seen together through a telescope. *Michelson and Morley proposed that the speed of the light waves traveling in the direction perpendicular to Earth's motion will remain unchanged, whereas the second beam of light, while traveling in the direction of the moving Aether, would travel through the **fixed distance** existing between the glass-plate "P" and mirror "M-1," at a speed that would be different from that at which it would travel in the direction opposite of the moving Aether. The justification for this proposition was* **"Time gained by light waves from traveling downwind, would be less than that lost traveling upwind."** Accordingly, these two beams, because of different speeds, would go out of phase. As a result, when both the beams would be seen together, then interference fringes would be produced; however, no fringes could be seen. ***Thus, this experiment utterly failed to prove the existence of Æther; or rather, this experiment proved that no such medium does exist in reality.***

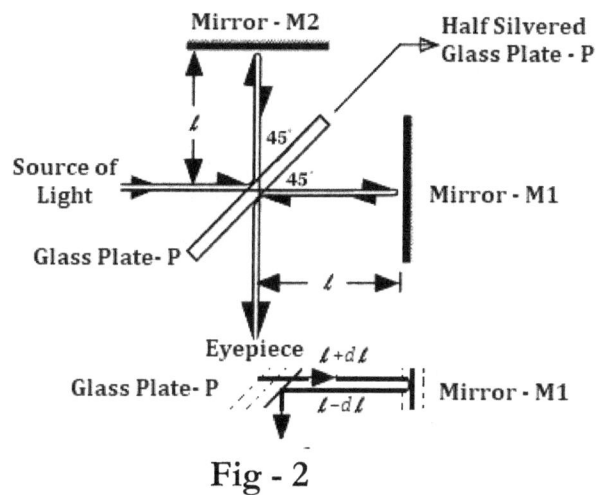

Fig - 2

Although this experiment failed to prove the existence of the Aether, even then, scientists continued to believe that light waves propagate rippling through Aether. **This belief of scientists led to the conclusion that light, irrespective of the speed of the moving Aether, travels at the same speed in all directions.** This meant that light, even in the reference frame of any moving object, persists in moving with its unaffected speed of approximately 300,000 km/second. In other words, light doesn't have any relative speed in the reference frame of the speed of different moving objects. Later on, in 1895, Dutch physicist, **Hendrik Antoon Lorentz,** proposed to justify the conclusion deduced from the aforesaid experiment that the "null" result obtained by Michelson and Morley was resulted due to **the contraction** of their apparatus due to the drag of the moving Aether.

Michelson and Morley did believe that light, between the plate "P" and mirror "M-1," would always travel through a **fixed distance***, but its speed would vary due to the relative motion of the Earth and Aether. Although the distance between the plate "P" and the mirror "M-1" was fixed, the light didn't move through the equal distance in both directions; as shown in Fig. -2,* **while the light beam, after emerging out from the plate "P" did move in the direction of the Earth's motion, the mirror "M-1," and the plate "P" both also did continue to move in the same direction;** *consequently*, this beam had to travel some extra distance *(dℓ)* to reach the mirror M-1. On the other hand, during the return journey of the same beam, the plate "P" continued to move further ahead; therefore, light, which was now running in the opposite direction, had to travel through a lesser distance, lesser by almost the same amount *(dℓ)*, or rather, a little lesser than that. Thus, **instead of moving through a fixed distance,** this beam, while running in both the directions, had to travel through **unequal** distances. However, the summed-up distance was almost 2ℓ, that is, equal to the distance through which the other beam did travel in the direction perpendicular to the Earth's motion. In fact, light did travel at the same speed in both directions but through different distances. Accordingly, the effect of the Earth's speed on this beam was almost nullified. This is the reason why *the null result was obtained.* Clearly, the result obtained had no relation *to the contraction of the equipment. Unfortunately, nobody ever realized this simple and ordinary fact; instead, Lorentz created a misconception to justify an impossible, or rather, a wrong conclusion. Thenceforth, a fallacious belief was formed that the distances contract at the speed of light and time dilates (please also refer to chapter-6). This misconception can only be removed when the speed of light is first measured in the direction of the Earth's motion, and then in the opposite direction, not both ways simultaneously, i.e., measurement of the speed of light shall be taken in the direction of Earth's motion independent of the measurement, taken in the opposite direction, not both ways at the same time.*

Based purely on the constancy of light's speed that was established by the aforesaid experiment, Einstein, in the year 1905, proposed the famous theory of *"Special Relativity."* This theory rendered both *"Absolute Time"* and *"Aether"* unnecessary. A comprehensive introduction to the said theory, excluding its mathematical angle, is given below…

The Special Theory of Relativity

This theory is based on the following postulates (unproved speculations):

1. "Speed of light is not altered by the speed of its source,"

2. "Nothing may travel faster than the speed of light," and

3. "Laws of science should be the same for all the freely moving observers who are moving at whatever speeds relative to the reference frame of the speed of the source of light."

Probably, based on Michelson-Morley's experiment, Lorentz's assumption, as well as based on the above-mentioned postulates, Einstein suggested that *"Speed of light, irrespective of the speed and direction of its moving source always remains the same for all the observers." In other words, different freely moving observers may take different readings of the distance traveled by a light beam and the time taken by that beam to cover that distance; however, when the speed of light would be calculated on the basis of different measurements taken by different observers, every one of them would always get the same result.*

116

As per this theory, nothing except the energy waves, which have zero mass, can travel at the speed of light; no massive body can attain this speed. Einstein also suggested that matter and energy are interchangeable into each other; he established the famous equation $E = mc^2$. The atom bomb was also based on the above equation. When an Atom bomb is exploded, it releases immense energy by converting very little amounts of matter into very large amounts of energy.

In 1905, *Einstein* suggested, "Space and time are not two different and independent things; they combine to form a continuum or an altogether different object known as *Space-Time,* in which any of these entities can't be identified separately." *This idea revolutionized the concept of distances. Since the speed of light was widely regarded as finite and fixed, it became possible to define long distances in the term of time, for example, light-year or light second, etc. Einstein suggested that interstellar space is not an ordinary vacant space; instead, it is like a fabric woven from the **warp and weft** of space and time. The roots of the idea of space-time, are very, very old; 'Incas, who are the aboriginal inhabitants of the Andes, did also believe in the continuum of space-time, which, in their language, is known as "Pacha" or "Quechua."* Arthur Schopenhauer in the year 1813, and Edgar Allen Poe, in 1848, respectively, brought out similar ideas in their books. The famous book, "The Time Machine," written by H. G. Wells, which was published in 1895, was also based on a similar idea. *However, Einstein introduced an entirely new idea "Light waves propagate through this fabric of space-time in a similar manner in which the water-ripples spread on the surface of the water."*

Earlier, in the year 1865, one of the greatest scientists, *James Clerk Maxwell* demonstrated that all electricity, magnetism, and light are disturbances in the same medium; accordingly, it was envisaged that light propagates or ripples through an interstellar medium named Aether. After the introduction of *Einstein's theory of Special Relativity,* the imaginary medium of "Aether" was replaced by the *fabric of space-time.*

The aforementioned mode of propagation of light has been accepted by all scientists all over the world. However, as brought out in chapter- 4 of this book, this mode of propagation of light is not capable of clarifying the point "In case, the light really propagates in ripples or waves, then how is the *duality of waves and particles* produced in the light? Moreover, how light, after emerging out from a small opening, moves in a beam of light?" Apart from these two points, one more point that needs consideration is "A disturbance in any medium subsides or calms down after a short time period; it can't continue for indefinitely long periods. Even the biggest tsunami, or disturbance created in the seawater due to an asteroid impact, dies off within a few hours; contrarily, the light wave (a disturbance in space-time) coming from stars located as far away as 13.2 billion light-years can still be seen." Similarly, the light produced from a dying star can be seen even billions of years after its death. It doesn't seem possible that the relics of a disturbance that has died off billions of years earlier, at a far-off location, may still be spreading out. A question, therefore, arises that "Is light really a disturbance in the fabric of space-time, and does the relics of such a disturbance, which has already died off at the source, can continue to spread out even after its death?" This seems impossible to me. These questions negate the possibility of light being a disturbance; further, in case light propagates in the two-dimensional fabric, then how can it spread in three dimensions? These facts put the very existence of space-time under question. There seems only one solution to all these problems that light probably propagates as a shower of spinning photons. (kindly refer to Chapter- 4).

In the above scenario, it is up for consideration that though the theory of relativity is entirely based on the concept of the continuum of space-time, the existence of space-time is not supported by any direct proof; even different scientists have different ideas of space-time. Einstein believed that time is **Smooth and Continuous**. However, this idea has now come under dispute. Scientists of the present generation, based on quantum physics, believe that time is not continuous; in case time is divided into very tiny fractions, then it will become granular in construction, i.e., made up of very tiny intervals of time. Moreover, Italian physicist **Carlo Rovelli** has envisaged that **Time** is not a separate entity; it starts as a result of the interaction between **Heat-Energy** and **Matter**.

Anyhow, regardless of the above contradiction between **Relativity** *and* **Quantum Mechanics,** *the* **Special Theory of Relativity** *envisages that the "time" for those objects that move at higher speeds lapses at slower rates.* Accordingly, a moving clock shall run at a slower pace when compared to a stationary clock. And if by any means, a moving object could be made to move at the speed of light, then the time would come to a total halt, i.e., it will totally cease to lapse for such an object; in other words, the speed of ticking of time for such an object would become zero. This theory also envisages that such variations in the ticking speed of time would depend only on its speed; such a change in the ticking speed of time would not depend either on its acceleration or direction of its speed. Since the rate of *lapsing of time* slows down for the moving objects, *scientists probably believe that such objects might move a little ahead in time, i.e., such objects would enter into the future;* this effect, in the case of the slow-moving objects, would be so meager that it could not even be noticed; any change in the ticking speed of time could be noticed only at the speeds that are very close to the speed of light.

Looking at the interconvertibility of matter and energy, **_Einstein_** further envisaged "*Mass of the moving objects would increase in proportion to the square of their* *respective speeds."* Accordingly, the mass of an object should increase at a much more rapid rate as compared to the increase in its speed. As a result, the amount of energy required to increase the speed of such an object would increase exponentially. Since the energy requirement at the near-light speed increases exponentially, the mass of the objects moving at this speed would tend to become infinite; accordingly, infinite energy would be needed to raise their speeds any further. Now, since it is not possible to produce energy at an infinite rate, the speeds of all the moving objects except for the mass-less energy waves must be confined to the limit *"lesser than the speed of light."*

The need for a New Theory

Sir Isaac Newton, in the theory of gravity, established that all the celestial bodies are held up with each other by the attractive force of gravity. It meant that as soon as any one of these bodies moves from its place, the gravitational force acting between all other celestial bodies would change instantaneously without any loss of time. It directly meant that the speed of gravitational force is infinite (Chapter-10 of this book gives a better idea about the speed of gravity). Since *the supposedly infinite speed of gravity violates* **the Special Theory of Relativity,** *wherein it is envisaged that nothing can move faster than light,* **Einstein,** in 1915, came out with a new theory to eliminate this anomaly. This new theory is known as *"The General Theory of Relativity,"* a broad overview of which is given below:

The General Theory of Relativity

In **Special Relativity, Einstein** envisaged that the universe is constituted of **space and time,** both of which are interwoven to form a continuum, manifold, or fabric of **space-time;** all the celestial bodies are supposedly moving over this fabric of space-time.

As per this new theory, space-time is not truly flat as was believed in his earlier theory, the *"Special theory of relativity,"* instead, the space-time is warped below different celestial objects due to their respective masses. The name **Special Relativity** was given to the earlier theory because it deals only with the special cases in which the objects and/or the observers move only in straight lines and at constant speeds. When the speed changes or diverts from its straight path by even a slight margin, then the **Special Relativity** loses relevance; here, the *"General Relativity"* comes into the picture **because it can explain all the general cases, that is, any sort of motion.**

In the new theory, the **General Relativity, Einstein** *proposed that at different places, the distribution of matter and energy causes the fabric of* **space-time** *to warp; the mass of each and every celestial body, whether big or small, warps the fabric of space-time or causes it to sag down under different objects; as a result, a curved depression or a deep-well, which resembles the interior surface of a bowl, is supposedly created in this fabric. The size and depth of such a dimple depend on the mass and size of the body that causes the space to warp.* **Einstein further proposed that "Gravitation is not a force like other forces. The warping of space-time draws the smaller objects toward the centers of the massive objects that caused space-time to warp; different celestial objects, in the absence of any force, move as if they are falling freely. Accordingly, the smaller bodies move toward the nearest massive body in straight lines; while doing so, they push against space-time; in turn, space-time also pushes back these moving objects with equal force and compels them to follow the curvature of the warped space. Thus, the smaller objects orbit the massive ones due to the warping of space, not due to the force of gravity."**

Einstein introduced the aforesaid concept of gravity to simply negate the higher than light-speed (infinite-looking speed) of gravity; however, this idea was not based on any direct evidence or dependable observation; it was merely speculation without any proof. Moreover, not even a single object under the micro-gravity environment persisting within a spaceship has ever been observed to move like a free-falling object. *Although this theory enjoys worldwide acceptance, a question arises on this speculation,* **"In case space-time is really extremely flexible, then how could the wall of the said deep-well push back the moving planets, and thereby compel them to orbit the Sun. Instead, the moving celestial objects shall be able to deform this wall infinitely and continue to move on their straight paths."** *Had space-time been extremely flexible, the same won't be able to push back the objects moving through it. Further, if the gravitational force has been there due to the consequence of the warping of space, then the intensity of the force of gravitation, within such a deep-well, shall always remain the same everywhere; it shall not vary with the varying distances from the center of the massive objects.*

×   ×   ×

In 1905, when Einstein proposed the **Theory of Special Relativity,** it was believed that our galaxy, the Milky-Way, constitutes the entire universe, which is spread over an almost two-dimensional plane; accordingly, it was believed in that era that the fabric of space-time is spread in two dimensions only. However, almost 20 years thereafter, astronomer **Hubble** discovered that the universe is spread in a huge three-dimensional region that expands far beyond the Milky Way. This discovery indicates that space-time is not like a two-dimensional fabric. Instead, it is spread in a three-dimensional region. In such a case, *space-time would not be able to warp beneath any massive object, like a two-dimensional fabric**. *Anyhow, Einstein* (who was ignorant of the fact that *the universe is extended over almost an endless three-dimensional region)* proposed *to negate the infinite speed of Gravitational Force,* that gravitation is not at all a force; instead, the planets orbit the Sun due to the warping of space-time. *However, in case the celestial objects really move along the inner surface of the deep-well formed around different massive objects, then the smaller objects like the meteors, in the absence of a real force, shall never be able to fall on the surface of massive objects like the earth; instead, such objects shall always keep moving along the curved-space. However, this never happens.* Further, *different planets (from Mercury to Neptune) orbit the sun from different distances but in almost the same plane (Ecliptic) as if all the planets are ostensibly rolling on a flat plane, not on the curved surface of the warped space. The aforesaid fact i*ndicates: "In reality, the indentation formed in space (around the Sun) is not deep enough to compel the planets to go round the Sun." Moreover, the fact "These objects don't move at different altitudes of the indentation formed in space" means "Neither the planets distort the wall of the said indentation nor the said wall pushes back the planets;" this, in turn, means that curved space doesn't compel the planets to move on orbits around the Sun. Anyhow, if *Relativity* is correct, then each planet is moving on its definite orbit along a definite curved region of space, which pushes back the planets and thereby compels them to follow the curvature of space. On the other hand, various other celestial objects like comets, asteroids, meteors, and even the man-made spaceships/probs, etc., freely pierce through these curved regions of space, from time to time, as if the same curved-space, which compels the planets to move on their respective orbits, doesn't resist the movement of other objects or rather desist celestial objects from piercing through the warped space; how space-time can be so selective?

Although this prediction of **Einstein** *is purely based on the concept of the continuum of space-time,* **the physical existence of such a continuum has never been proved; to me, this prediction seems purely hypothetical.** *In case the prediction of relativity is correct, then the Earth shall cast a dimple in this fabric whose size shall be, at least, equal to the orbit of the moon. In case this is correct, then how are the man-made satellites orbiting the earth*

---

* This possibility is further discussed in Chapter-11, under "The alternate possibility of the formation of the Blackholes."

*from a distance of only a few hundred kilometers, i.e., from a distance that is about 100 times smaller in comparison to the distance between the Earth and the Moon? How are the satellites orbiting the earth without being pushed back by the walls of the said dimple? How the inanimate fabric of space-time can be so choosy, so discriminate that on the one hand, it compels all the moving celestial objects like planets, asteroids, and even small meteors, etc., to follow its curvature, and on the other hand, the man-made objects like satellites, or the space-crafts, etc., orbit the earth from the distances from where space-time can't compel them to follow its curvature, in such a case, it shall not be possible to establish a satellite in the pole to pole orbits **because Relativity envisages that the fabric of space-time warps below the massive objects, i.e., in the horizontal plane, not in a vertical plane.***

A doubt comes up on the aforesaid concept "If gravitation is not a real force, then it will mean that all the stars and other celestial bodies are randomly spread in the universe without any compulsion to obey any law;" however, the local dimples formed in space-time beneath each and every celestial object would decide their orbits. In **Relativity,** the curvature of space-time at any particular place can be calculated according to the distribution of matter and energy existing at that place, but according to the best of my knowledge, there is no formula to calculate the size and the depth of the dimple so formed. This is a well-known fact that different galaxies do attract nearby galaxies to form a local group of galaxies; similarly, local groups of galaxies, too, attract nearby groups of galaxies to form still bigger groups of galaxies. This fact reveals that a large number of such dimples, measuring several hundred-thousand lightyears, must exist in the universe; however, no means are probably available to find out the depth of such dimples; accordingly, it is not possible to find out the degree of curvature or the angle of the inclination of such dimples. Based on the fact that the outer limit of the solar system is extended up to the Oort Cloud, which surrounds the Sun from a distance of about one and a half lightyear (please refer to the Addendum-1). This fact indicates that the diameter of the indentation cast by the Sun in space-time shall be more than one lightyear; on the other hand, the maximum distance from the Sun from where the light is observed to deviate indicates that the depth of the said dimple shall be almost negligible as compared to its diameter; thus, the dimple cast by the Sun shall be almost flat, accordingly, its curvature and inclination, both shall be too small to compel different planets to orbit the Sun at a very high linear speed. In reality, all the planets orbit the Sun in almost a flat plane, which is known as "Ecliptic."

Ignoring the above fact, **Einstein** envisaged that the celestial objects are not attracted toward the massive objects due to the **Gravitational force.** *Instead*, they are drawn toward the massive objects due to the warping of the fabric of space-time. However, such a phenomenon seems impossible to me because this fabric is subjected to a very high tension; it is supposed to continually expand at the speed of light. Accordingly, the outer edges of such a dimple formed beneath a star shall always tend to stretch outward, i.e., the edges of the said dimple are not free to warp toward the center of the massive body. In such a scenario, how the objects located in the warped region of space may, in the absence of a force, slide down the slope of the said dimple, or all by themselves, drawn toward the center of that object due to the warping of space. A series of questions come up on the aforesaid presumption:

1. *Wherefrom the planets, in the absence of any force, get the energy to move toward the massive objects,* **that too, in a particular direction?**

2. *How the extremely flexible walls of the deep well can compel any smaller celestial object to follow its curvature?*

In reality, the process of the creation of gravitation due to the warping of space is analogous to the process of the creation of air pressure within an inflated balloon in which the air is squeezed from all around; accordingly, the air pressure everywhere within the balloon always remains the same. *In case gravitation is produced by the warping of space, then the atmospheric air shall be compressed by warped space from all-around with equal force. Accordingly, the air pressure all over the air blanket of the Earth shall always remain constant;* **its pressure shall not vary with the varying altitudes.** *Moreover, the mass of an object can though cause the sheet of space-time to sag beneath that object, but it seems impossible for this sheet to cover that object from all-around* **like a balloon**. *Therefore, the effect of gravitation that varies with variations in the distance from the Earth's surface cannot be produced due to the formation of such a deep well in space.* **Had the effect of gravitation been produced due to the warping of space, then, while traveling upward in a lift, we shall feel pressure on our heads, not on our feet.** These two facts clearly indicate that a definite force is pulling all the objects toward the center of the Earth; nothing is pushing them down from the overhead. Clearly, the planets are orbiting the Sun due to some other reason.

In case the wall of the deep well, which is formed below the sun, really pushes back the planets and thereby compels them to orbit the sun, then only those planets, which move along the said wall, and thereby tend to deform it shall be able to orbit the Sun. However, in reality, all the planets of the solar system orbit the sun in almost the same plane; therefore, it is totally impossible for each one of them to come in contact with the wall of the said deep well and thereby follow its curvature. This fact clearly indicates that the planets don't orbit the Sun because of the warping of space-time.

This theory gives rise to yet another controversy, of which nobody has ever taken any notice, this controversy is: "Since the fabric of space-time warps beneath massive objects, *both the warp and weft of this fabric, i.e., the threads of space and time would not pass through the region that lies above the warped portion of space, i.e., inside the said deep-well formed in the space; accordingly, the entire three-dimensional region enclosed within such a deep-well, including the object that has caused space-time to warp, shall not have any time or space,"* however, *such a consequence of the said warping of space-time seems absolutely impossible.*

All the questions raised above indicate that **Einstein's concept of warping of space-time and thereby creating the effect of gravitation doesn't seem sustainable because nothing can move unless it is acted upon by a force.** *This point is further discussed at length in this chapter as well as in the forthcoming Chapters 6, 10 & 11.*

Anyhow, this theory, without considering the aforementioned possibilities, envisages that *"Different celestial bodies like our Earth and other planets, etc., are neither held together by the Gravitational force nor they orbit the* **Sun** *due to gravitational force; instead, they orbit the Sun due to the warping of space-time."* Einstein proposed that **gravity is not, at all, a force like other forces, it is merely a consequence of the warping of "space;"** accordingly, *more the curvature of space the more would be the effect of gravity.* Einstein further suggested that all the things that move through space-time have to follow its curvature; even the rays of light, while passing through such curved regions shall fallow the curvature of space; accordingly, light-waves shall deviate from their straight paths.

Further, similar to the proposition made in the *"special theory of relativity"* that time slows down at high speeds, the theory of *"general relativity"* predicts that the speed of the *"ticking of time"* shall slow down in the higher energy areas such as the strong gravitational fields; the reverse of this is also correct; time should run faster in the weaker gravity fields. The fact that *"the clocks carried by a satellite or a spacecraft are found to run faster than the clocks placed on the Earth's surface"* is considered irrefutable proof of this prediction.

The prediction that **"The ticking-speed of time shall slowdown in the areas of higher energy concentration"** *contradicts the idea of* **quantum mechanics** *that* **gravitation is negative energy. Had gravity has been negative energy, then the gravity-fields in the close vicinities of the massive objects shall become weaker energy-area; accordingly, the clocks located near these objects shall run faster; however, this is against the reality.** In fact, the clocks located on the surface of the earth or any other massive celestial object are found to run at slower rates.

The conception of the "Theory of Relativity"

This theory is broadly based on speed or rather *the speed of light;* however, the concept of this theory differs a little from our general conception of speed, which is based on our day-to-day observations. For example, the passengers traveling in a train or a car might have noticed that the speeds of the vehicles approaching them from the front or rear appear much different from their real speeds. For example, if one of such vehicles is running at a speed of 40 kilometers/hour and the other one is running at a speed of 60 kmph, then while they approach each other, their relative speed (speed in the reference frame of each other) would appear to be 100 (60+40) kilometers/hour, whereas, while the vehicles follow each other, their relative speed would be 20 (60-40) kilometers/hour. However, if their speed is measured relative to the earth's surface or any other stationary object, then no change would be observed in their speeds. It is clear from the above example that *the change in speed* observed by moving passengers was relative to each other, which was merely an *illusion,* not a reality. Even then, based on the apparent change noticed in their velocities, their respective future positions can easily be calculated.

Above mathematics holds good within the limit of the speeds that seem normal to us. However, scientists believe that these rules start to change at speeds very close to the speed of light. They believe that light travels past the moving objects or observers, regardless of their speeds, at its normal speed of 300,000kilometers/hour. This belief that light-waves always travels at a steady and constant speed is though 100% correct; scientists *believe that the speed of light doesn't change even in the reference frames of the moving objects too; in other words, no apparent change occurs in the speed of light due to the relative motion of different moving objects.* Here comes the twist, *if the speed of light is not at all affected by the speeds of other moving objects, then something else must change elsewhere; what is the thing that changes at this speed?* To solve this puzzle, scientists have concluded that the very structure of the universe, i.e., the warp and weft of the *fabric of space-time,* start to change at the speed of light or speeds very close to it. *Scientists believe that both the lengths (space) as well as time, measured by different observers,* do change *at this speed, time dilates, and distances or objects contract.* These changes enable different observers to measure distances and time differently. However, this notion attracts a question- *"Does the structure of the Universe change only for the light waves? This question has come up because, at the same time-instant, its structure remains unchanged for all other slow-moving objects?" In case so, then it would simply mean that at the same time-instant the universe has two different structures for different entities that move at different speeds.* However, this is totally impossible.

Contrary to the aforesaid concept of Relativity, different examples given in the previous chapter-4 reveal that the speed of light is also apparently affected by the speeds of other moving objects. In case the readers feel so, they may refer back to these examples; there cannot be two different sets of rules for similar situations.

Practical Applications of Relativity

The Relativity Theory has very wide applications in cosmological science. Hundreds of satellites were sent into space during the last 5 to 6 decades. We have sent manned rockets to the Moon, landed various probes on Mars and other planets. Our spaceships and probes have reached Jupiter, Saturn, Pluto, and even beyond. The spaceships on their way had not only avoided collision with any planet, but they also utilized the gravitational force of the Moon, planet Mars or other planets, etc., to obtain the power to move ahead. All this could only be achieved by very accurate and precise calculations of time. The slightest mistake in such calculations, amounting to even the smallest fraction of a second, could have resulted in any disaster. Such precise calculation of time could only be achieved by the equations of *Relativity.*

Besides space, the *Theory of Relativity* is practically serving humankind on the Earth also. GPS or the Global Positioning System, which is based on *relativity,* has been found very useful worldwide in finding correct routes and correct locations on the Earth's surface, as well as in air and sea navigation, etc. This equipment works with the help of different satellites that orbit the Earth. Presently, almost 2,500 to 3,000 satellites are orbiting the Earth, out of which 30 satellites are exclusively used for Global Positioning System. The satellites used for this purpose are orbiting the Earth at different angles with a speed of 14,000 kilometers/hour and at the height of 20,200 kilometers or 12,550 miles above sea level. These satellites, moving at this speed, complete one orbit every 12 hours. Angles of the orbits of these satellites are so selected that at least 4 to 6 satellites could always be viewed from any place on the Earth. The atomic clocks carried by these satellites send radio signals to the Earth continuously. These signals convey very accurate information about (1) the exact time of the origin of that signal & (2) the exact location of the satellite in its orbit from where the signal was originated. When these signals reach the receiver carried by a moving vehicle, then based on the time taken by the signal to reach the moving vehicle, the receiver mounted on such vehicles calculates the distance between the satellite and the receiver. Likewise, the receiver, by measuring the angle cast by at least three satellites on the receiver, calculates the exact location of the vehicle, its speed, and the direction in which the vehicle is moving.

As envisaged in *General Relativity,* the atomic clocks carried by these satellites are supposed to run faster than the similar clocks placed on the Earth's surface because the strength of the gravitational field diminishes at such a high altitude, whereas, as per the provisions of *Special relativity,* the same clocks should click at slower rates because of their very high speeds. Now, in case the above clocks were showing exactly the same time that existed before launching the said satellites, then, after establishing these satellites in their respective orbits, they surely would show different times; as a result, the GPS would not function at all. Therefore, for deciding the exact locations and speeds of all the receivers, it is necessary that the clock carried by the satellites and the other ones placed on Earth's surface must always show exactly the same time. In order to deal with this problem, all the clocks carried by the satellites are adjusted before launching the satellite, in such a way that after the satellite is established in its orbit, the clock carried by it would show exactly the same time as is shown by its counterpart placed on the earth's surface. The correction to be made in these clocks is decided on the basis of the complicated equations formulated by Einstein. In the present era, GPS has become an essential part of the life of almost everyone; hundreds of million people are finding their destination places with its help. The proper functioning of this equipment is considered solid proof that the theory of relativity is absolutely correct; it is not a fake in any way.

Relativity predicts *"The ticking rate of time* for any moving object depends on both its *speed* as well as the *strength of the gravitational field around such an object;"* the success of *GPS* is considered indisputable proof of this prediction. However, though both the strength of the gravitational field and the liner speed (not the rotational speed) of the spinning Earth have different values at the bottom of the sea and at the highest peak of the Himalayas, the rotational speed of the Earth or its orbital speed around the Sun doesn't undergo any change in any of the above places. Based on the above fact, it seems **"Time doesn't lapse at different speeds at these places,** *because the concept of time on the earth is linked with the Earth's speed of completing one spin around its axis and one orbit around the Sun."*

Under the above scenario, the variation in the ticking speed of time, in the reference frame of the *speed* of an object as well as in the *strength of the gravitational field around that object* is discussed in the next chapter under the sub-heading **"Analyzing different** *Predictions* **of Relativity"** (please refer to the prediction nos. 1 to 4).

127

Contradicting *Einstein's* prediction, the above-referred section of this book concludes *any variation "in the speed of the moving Cesium-clocks, and/or the strength of the gravitational field around these clocks" results in an increase/decrease in the speed of the electrons revolving around the atomic nucleus of the Cesium-133 atom, as well as in the orbital-radius of such electrons; these changes recast the measuring-scale of the moving atomic clocks. As a result, these clocks take different readings of the same time-span when viewed from another reference frame that differs from the standard reference frame set on the Earth.*

I strongly feel that the aforesaid *speculation of mine* shall be thoroughly verified at the highest level to establish the truth.

×   ×   ×

At this juncture, it seems necessary that we shall sort out what the *"time"* is so that we may understand *Relativity* in a better way: -

All the units to measure the time, i.e., minute or hour, etc., are based on the spinning speed of the earth. However, the spinning of the earth is not the only event in this endless universe; any movement anywhere in the whole of the universe is a separate and independent event; the time interval between any two events can be measured in any of the units of time. Based on this fact, it can be said that the *time,* and the rate of its lapsing, is the concept of the *interval* between the occurrence of two or more events. Although time is not a physical entity, the same seems to run or expand continually because the earth and numerous other celestial bodies persistently rotate about their axes and revolve in their orbits.

Relativity predicts "the *warp* and *weft* of *space* and *time* constitute the universe," both of these constitutes have the inherent property to expand. Accordingly, scientists believe that the universe is continually expanding* with time, and all the stars are moving away from each other. In case, this belief is correct, then *it may mean* "the locations of all the celestial objects are decided, or rather predestined by the time." In other words, the paths, speeds, and the locations of all the stars, at any time-instant, are decided by the lapsing of the time, i.e., every movement of the stars is time-bound, every phenomenon in the universe is decided by the flow of time. If this is so, then whatever has happened in the past, whatever

---

* The alternate perspective of the expanding universe, as described in chapter-15, "Dark Energy," may also be referred to.

is happening in the present, or whatever is destined to happen in the future, shall be fixed very rigidly that can't be altered. Although it seems impossible to change the past, the present or future events may be altered. For example, the Near-Earth Asteroids normally pass by the earth from very safe distances, however, in case, an asteroid, at a certain point in time is found to be on a collision course with the earth, and if, by appropriate efforts, we successfully divert or destroy such an asteroid, then the imminent future would be altered. This example indicates that time does not govern the paths and speeds of the celestial objects, instead, the movements of these objects produce the concept of time, as well as the rate of its lapsing.

In case, the conclusion arrived at in the preceding para is correct, then the rate of lapsing of the time may be defined in the reference frame of any particular activity or the rate of the occurrence of any such event. However, if all the activities going on in the whole of the universe, come to a total halt, i.e., all the activities in the whole of the universe are suspended forever, then the concept of lapsing of the time will lose its relevance; time will stop, freeze or cease to lapse. But if any activity restarts, anywhere, maybe after an infinitely long time interval, then the time will continue to run even during the said time period when no activity did happen. I feel that time is not a physical entity, it is merely a conception of the interval between the happening of any two or more events. The lapsing rate of the time is directly linked to the rate of the happening of the different events, not with the speed of the light, as was envisaged by the all-time great scientist *Einstein.*

6: Analysis of The Relativity Theory

Relativity enjoys utmost importance in the field of cosmological science; however, this theory is too complicated for a common man to understand. In case distances at the speed of light "contract," and the time ceases to "lapse," then, at least, the light should not require even the smallest fraction of time to travel from one place to another, contrary to this possibility, *everybody knows that light, too, takes billions of years to reach us from the furthest stars, even then this distance doesn't shrink a bit.* To a common man, the distance of one kilometer always remains fixed; likewise, a time interval of one hour also remains fixed. *How the distances measured on the surface of the earth could contract? And, if at all, the distances do really contract, then how could such distances retain their original shapes and sizes for human beings? How could any substance have two different sizes at the same instant of time? Since the Sunrays always continue to propagate from all the sides of the earth, why does its size not shrink continually, and if it doesn't shrink continually, which it never does, then why should this happen only while taking the measurements?* All such questions seem beyond the perception of a common man. Such matters, being extremely complicated, the common man accepts this idea as it is without even cudgeling his brain and without even any concern. Unfortunately, I, myself, have never studied this theory systematically; however, the more I tried to understand this theory, the more I became skeptical about its correctness.

The question that troubles me the most is - *"How the light could zoom-past a moving object at the same speed at that, it moves past the stationary objects, i.e., why doesn't it have any relative speed in the reference frame of the speed of such moving objects?"* Does the light not differentiate stationary objects from those which are moving with a definite speed? Does the speed of light increase while zooming past a moving object, or do such objects freeze at their places for a very short while? All of the aforesaid possibilities under question do not seem feasible. However, scientists all over the world, based on Michelson-Morley's experiment, believe that "time dilates at the speed of light, and the distances do contract." I, somehow, feel that probably a misinterpretation of the aforesaid famous experiment has created some confusion somewhere at some level. This is absolutely correct that light always moves at its own speed, but contracting of the distances (space) or objects, seems unbelievable and impossible to me, because *in case the drag of the moving Aether, or light, can shrink the trailing edge of an object, then such drag shall elongate the leading edge of the same object, by the same amount; does the drag of the Aether act on*

one edge only? My perception on this point is- *"Light, because of its very high speed of about 300,000 kilometers/second zooms past any object within a negligible fraction of the time, which couldn't even be noticed; therefore, it appears to us that light has moved past the moving objects with the same speed at which it normally moves; however, these objects too, without any change in their lengths or speeds, persist in moving along with the light. Therefore, such objects, too, travel through extremely small distances, which are beyond measurements. We, accordingly, feel that* **light always moves at a constant speed- even in the reference frame of the moving objects too."** *A possible solution to the above enigma is given below:*

Speed of Light and the Basic Mathematics

All those speeds at which the known objects generally move around us seem normal to us; however, light moves almost infinite times faster than all such objects. This could be understood by the following example: *If we imagine that an observer is moving at 99% speed of light, then light rays would zoom past this observer with 1% of its normal speed, not with an unaffected speed of 300,000 kilometers/second. In other words, the light would move past this observer at a relative speed of 3,000 kilometers/second or 10,800,000 kilometers/hour. Normally, we, on the surface of the earth, can't see beyond 5 to 10 kilometers due to the curvature of the Earth, as well as due to different obstacles that come in between. Therefore, this 1% speed of the light would be beyond the visibility limit of any observer. Another way round, the light would take a minimal time of 1.65 micro-seconds to cover our range of visibility, say about 5 kilometers (on Earth's surface), which is too small for human beings to take any notice of. Therefore, the observer might feel that light has moved past him at an infinite speed. Now, if another observer moves at a speed of 99.999% speed of the light, then he will find that light is moving at a speed of 0.001% speed of light, that is, 3km/Sec or 10,800 km/hour, which is ten times slower than the earth's speed. Even this speed, too, might seem infinite to him. However, if the observer moves at 99.9999999% speed of that of the light, then he would find that light is moving at a speed of 3 meters/Sec or 10.8 km/HR.* The observer would now easily identify this speed. All this is the *magic* of the relative speeds.

Relativity envisages that different observers, who are moving at different speeds, might take different measurements of the distances traveled by different light beams because such distances might contract or shrink to different amounts at different speeds. Similarly, the clocks carried by them might also show different readings because the time may lapse at different rates for the observers moving at different speeds. Despite these differences, they would still find the same result. However, I feel that this is not at all weird or magical. Similar results would also be obtained by the most fundamental laws of basic mathematics. **This is very simple, provided it may be kept in mind that both the beam of light and** *the observers would be moving independently at their respective speeds, i.e., they would move at different relative speeds concerning each other.* No matter how slow the observer was moving, his speed can never be neglected because he was also moving while taking the measurements. If this simple fact is kept in mind, then we will always find the correct results. This is explained below, in brief:

The Computation of the Speed of Light

In case an observer, while at rest, observes that a beam of light, while traveling through a distance "d," takes a certain time-duration "t," then the speed of light would be denoted by the expression C=d/t, where the speed of light is denoted by the letter "C," which is a universal constant.

Now, if the observer is moving at a speed of 10% speed of light, then the speed of light relative to him would be 90% of its normal speed, that is, 0.9C. Therefore, what he would calculate would be the 90% speed of light instead of its 100% speed. Accordingly, in the time period "t," the beam of light, moving at this relative speed in the reference frame of the moving observer, would travel a comparatively smaller distance. This might be considered as a contraction of the distance; however, the said distance "d" would not be altered at all; instead, the observer would also move through a distance of 0.1d. Accordingly, the observer would find that the beam has traveled a distance of only 0.9d (d − 0.1d). Accordingly, he would find that 0.9d/ t = 0.9C; calculating back from this figure, he will find C=d/t, which is the normal speed of light.

In case the observer, while moving at the above speed, allows the beam to travel a distance "d" ahead of him, then the observer would find that beam has taken comparatively longer time to travel a distance "d;" this may be regarded as dilation of the time; however, light moving at a relative speed of 0.9C would take a longer time of t/0.9 to travel a distance "d" ahead of him, which is in fact, greater than "d." When the speed of light would be computed using this set of readings, he would find 0.9C=d/ (t÷0.9) or C=d/t, as before. The other way round, since the observer is moving at 10% speed of the light, he would also move through a distance of 0.1d, in the time period "t." And as such, the beam, in order to move through a distance "d" ahead of that observer, will have to cover an actual distance of 1.1d. In order to move through a comparatively longer distance, the beam will also take a longer time of 1.1t. Thus, the observer, while computing the speed of light, will find C= (1.1d)/ (1.1t) or C=d/t, as before. The same would be true for all different speeds of any other moving observer who is moving at whatsoever speed.

Based on the above example, it could be concluded that whatsoever might be the speed of any entity; its speed neither causes the time to dilate nor the distances to contract, both of them retain their normal values; instead, the light, in the reference frame of different observers or objects, apparently moves at different relative speeds. This relative speed of light is not real; it is only an illusion. Light, in the reference frame of space, or the point of its origin, would always move at its normal speed in any direction.

Interaction of the Speed of Light with that of the speed of the other Moving Objects

It is absolutely correct that light moves at its own speed while surging ahead of other moving objects; however, other moving objects also persist in moving with their respective speeds that would remain unaffected by the speed of the light. *As a result, during some of the natural phenomena, the locations of some very distant natural objects appear to have changed ostensibly because by the time light reaches our eyes, the Earth also travels through a substantial distance; a similar effect is also observed during a few experiments performed at different laboratories.* The results of such observations suggest that, like other moving objects, the light, too, does move with different relative velocities in the reference frame of the other moving objects. A broad description of some of such observations is given below.

Aberration of Light

After Copernicus proposed in 1543 that the Sun is the center of the universe, a debate broke out among astronomers that if the earth really moves around the sun, then the stars shall appear to deviate from their places. To verify this idea, James Bradley and Samuel Molyneux, during the year 1725, began to observe the movements of a star named Gamma (g) Draconis. They observed that this star occupies the true or mid-position during June and December, respectively. In June, it starts to shift toward the north; it touches the peak of its displacement in September. Thereafter, it turns back and again passes through its mid-position in December. Thereafter, it reaches the maximum southward displacement in March. These observations established that this star oscillates between its peak-to-peak displacements every six months. In 1729, Bradley suggested that the earth moving in its orbit occupies the position of "Quadrature" twice a year. Quadrature is a position of any celestial object wherefrom it casts an angle of 90° or 270° on the line joining the sun and the earth. The earth, from each of these points, moves either directly toward the star or in the direction opposite of it. The said star appears at its true position "S" from both of such positions; this is depicted in Fig-3. When the earth, from any one of these points, moves ahead on its curved path, then a transverse component is generated in the speed of the earth, which acts in the direction perpendicular to the light coming from the star. This transverse speed gradually increases till the earth doesn't reach the line joining the Sun to the star. Thereafter, this transverse speed starts decreasing and becomes zero at the second quadrature.

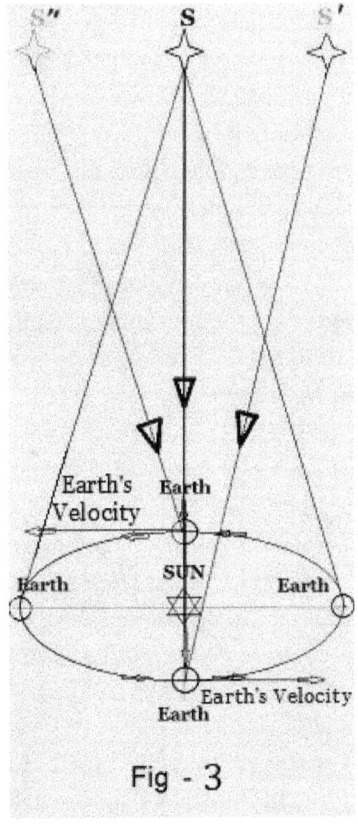

Fig - 3

As shown in Fig-3, the star, from both the Quadrature, is seen at its true position that is denoted by S. As the earth moves past the left Quadrature, a component of speed perpendicular to the incoming rays is also produced in its speed. The light beam coming straight toward the observer would, therefore, not be able to reach the observer directly; instead, another beam would be seen by him, which was emitted previously in the direction where the earth and the coming ray would reach together, at a future point of time. The star would, therefore, appear to shift in the direction of the resultant speed of the earth's transverse speed and that of the light. When the earth would arrive at the line joining the star to the Sun, the star would appear displaced to its maximum position S'. Thereafter, this apparent displacement would gradually reduce and become zero at the second (right) Quadrature. As the earth moves further ahead, the star would be seen to displace in the opposite direction, and at halfway to the left Quadrature, its apparent location would again shift to its maximum displacement denoted by S", but in the opposite direction.

A similar effect is produced when the raindrops are seen from a moving car. If no wind is blowing, then the raindrops shall fall in a perfectly vertical direction;

however, the faster the car moves, the more inclination is observed in their direction. Aberration is also a similar phenomenon; the shower of photons comes directly toward us, but due to Earth's speed, these photons appear to come in the direction of the resultant speed of the light and that of the transverse component of the Earth's speed. *Accordingly, the varying displacement of the star depends on the value of Earth's transverse speed; however, this change in the direction of the light is merely an illusion.*

The SAGNAC EFFECT

In 1913, French physicist Georges Sagnac, with the help of a half-polished mirror, did divide a beam of light into two separate beams. Next, with the help of 3 or more mirrors, he made both of these beams propagate on two opposite branches of a closed path, such as a square or a circular loop. Both of these beams, after traveling through equal distances but in opposite directions, were made to meet on a screen at the same point. This equipment was mounted on a platform that could be made to rotate either in the direction of propagation of any one of these two beams or in the direction opposite of it. At the beginning of the experiment, while the equipment was at rest, no fringes did produce on the screen. This fact did indicate *"While the equipment was stationary, both the light beams did travel through equal distances due to which the light waves in both of these beams did move in the same phase."* As soon as the equipment was made to rotate in the direction of any one of these split beams, interference fringes did appear on the screen. *It is conspicuous from the above fact* *"The light waves in the rotating apparatus were traveling through unequal distances; as a result, these beams did shift out of phase."* *In turn, this fact clearly indicates* *"The speed of rotation of the equipment did add up to the speed of light; accordingly, the beam traveling in the direction of rotation of the equipment did cover a longer distance because its path was also moving forward, whereas the second beam did travel through a shorter distance; its path was moving opposite of light."*

It is explicit from this experiment that the speed of light too ostensibly changes in the reference frame of the moving objects. Had the conventional belief that light moves past moving objects with the unaffected speed been true, then irrespective of the fact that the equipment was stationary or rotating, the light should have taken equal time to travel across both of the paths because both the paths were exactly equal in lengths. In case the *lengths really contract, or the time dilates* at the speed of light, then, regardless of the fact "Whether the equipment was stationary or rotating, such a contraction/dilation should have happened equally in both the paths;"

136

accordingly, both the beams should have remained in the same phase. *The phase-shift noticed in this experiment shows very clearly* **"The speed of the path moving in the direction opposite of the direction of the light beam was added up to the speed of light, and the speed of the second path was subtracted from the speed of light."**

It is clear from both "Aberration of light" and "Sagnac Effect" that though light always moves with a constant speed of 300,000 kilometers/second, its speed while moving past the moving objects, is also subjected to a relative change in the reference frame of the speed of the moving objects. Depending upon the directions of their relative motions, this apparent relative speed can be calculated in a normal way.

ROSSI & HALL Experiment

Before discussing this experiment, let us acquire the basic knowledge of the elementary subatomic particles, the **"Muons."** The muon has *a unitary negative electric charge, which* is equal in value to that of the *charge* of the electrons, but they are about 200 times more massive when compared to the electrons.

The duo of scientists **Rossi** and **Hall,** in the year 1941, based on the aforesaid property of muons, proved that "The **muons** when moving at very high speeds live longer in the comparison of the stationary muons; the stationary muons decay at much higher rates." In other words, Rossi and Hall proved that "At high speeds, **time** ticks at a slower rate; alternatively, the lifespan of the muons extends (dilates) at high speeds." This experiment is considered irrefutable proof of **Relativity.**

Muons are very unstable particles that do not exist in normal conditions; in a particle collider, they are generated when subatomic particles are made to collide at very high energies. In nature, muons are produced when **cosmic rays** directly collide with an atomic nucleus. **Cosmic rays** are a shower of high-energy particles such as protons, neutrons, atomic nuclei, etc.; this shower comes from numerous unidentified sources that are located very deep in space. The **muons**, immediately after their creation, start moving almost at the speed of light, in the same direction in which the cosmic rays were moving. Muons are so energetic that they can penetrate through the rocks buried very deeply; they are found even at a depth of 700 meters below the earth's surface. Almost 50% of the muons decay within 2.2 microseconds (microsecond means one-millionth part of a second) of their creation, each muon decays into **one electron and a pair of neutrino and antineutrino**. On average, muons, during their lifespan, can travel up to a distance of about 660 meters, or about 2,200 feet.

Rossi and hall, at the height of 6,300 feet, recorded the total number of muons in a time period of one hour, that is, their "flux," arriving at a detector, which was placed on a mountain. A scintillator counter, comprising a thin plastic sheet that was clad with an iron sheet, was used for this purpose. It was assumed that the muons, having lesser energy, wouldn't be able to penetrate through this cladding; they, therefore, won't reach the counter, whereas more energetic muons would instantaneously pass through the scintillator counter without any restriction. Only those muons could be counted, which immediately after passing through the iron cladding, would lose all of their energies and come to rest. Such muons, while losing their energies would give a flash of light, and while decaying, they would produce another flash of light. Both of such flashes would be recorded in the said scintillator counter. In this experiment, the scientist duo recorded the decay of a total of 568 muons in an hour, out of which only **27 muons exhibited a rest-life of 6.3 microseconds or more.**

Rossi and Hall argued that in case the **moving muons** and the **muons at rest** both have the same lifespans, then only **the aforesaid 27 muons**, *having a lifespan of 6.3 microseconds*, should, after traveling the distance of 6,300 feet, reach the second counter kept at sea level. However, against this assumption, as many as 412 muons were recorded at sea level. Based on the *result of this experiment, it was concluded:* **"The muons moving at higher speeds have higher life as compared to the life of the muons at rest."** *In other words,* **"The lifespan of the muons dilates at near-light speeds."** Based on the higher count of 412 muons, instead of the expected count of 27 numbers of muons found at sea level, the speed of muons was back-calculated, *which was found 99.5% of the speed of light.*

This experiment is considered solid proof that **"The life-span, of the muons, dilates at the near-light speeds,"** *and also that* **"The ticking rate of time slows down at the near-light speeds."**

I, however, feel that the aforesaid conclusion was deduced without considering the following points:

1. It was assumed that when the muons would come to a rest, they would decay instantaneously; however, *photons don't decay even when they are made to stop within the super-cooled sodium atoms or in a magnetic field (chapter-3). This fact indicates that speed has no link with the lives of the photons.* On the other hand, the rest-life of the muons was found as low as 2.2 microseconds. *Compared to this rest-life of the muons, the real photons (light), whether at rest or while moving at slow speed, have infinitely long lives.* This fact indicates, **"The lifespan of muons while at rest or moving, shall not be linked to their**

speed or the state of rest; there might be some other reason for their short lives."

2. While calculating the lifespan of muons at rest, it was assumed that whenever any muon would reach the counter, it would emit a flash of light. Thereafter it would emit another flash of light while decaying. *However, it was impossible to differentiate whether the second flash of light was emitted by any decaying muon, or it was emitted by a new muon entering the counter; no fool-proof method for the same seemingly exists.* **Thus, the total number of muons recorded at the hilltop, having almost an equal amount of energy, does not seem 100% accurate.**

3. The assumption that only 27 muons will reach the sea level was based on the count of those muons, whose lifespan at the hilltop was found exceeding 6.3 microseconds. However, the number of high-energy muons, which passed through the first recorder unrestrictedly, was not considered. **Those muons that passed through the first counter unrestrictedly would have lost some of their energy before reaching sea level; they would lose some more energy while penetrating the iron cladding of the second counter. Probably, such muons, too, might have decayed within this recorder and would have been recorded along with others. Therefore, the assumption of Rossi and Hall that only 27 muons would reach sea level doesn't seem realistic and logical.**

4. Cosmic rays penetrate deep into the atmosphere; accordingly, on their way, they persistently generate muons at different altitudes. Each of the muons so generated has different levels of energy; each one of them decays after different time intervals and after traveling through different distances. **No method exists to determine whether the Muon that decayed after penetrating the iron cladding of the recorder decayed because it did come to rest, or else, it did decay after completing its normal life; this could be purely a matter of chance.** *This fact indicates:* **"The count of muons, which was recorded at the hilltop or sea level, doesn't truly represent the true count of the muons of equal energies." Therefore, this should not be concluded: "The speed of muons has any relation with their lifespans because it cannot be decided which muon did decay because it did come to rest or which one of the muons decayed after completing the normal life."**

5. Cosmic rays persistently produce muons at different altitudes, which is evident from the fact that muons are found even 700 meters below the Earth's surface. ***Accordingly, new muons might also produce below the altitude of 6,300 feet where the first recorder was placed or even much lower altitudes.*** Such newly produced muons, depending upon the fact whether the collision between the cosmic ray and the atomic nucleus was a direct head-on collision or the same was a partial, side-on collision, have different energies and life spans, such muons would surely travel up to different distances. As such, *this cannot be ascertained at what time, or what altitude, all the 412 muons, recorded at the sea level, were produced? Probably, all of them did produce at an altitude much below the altitude of 6,300 feet.* ***Because of this possibility, it is not necessary that all the 412 muons, which were recorded at the sea level, reached the second recorder after covering a distance of 6,300 feet, or they lived a longer life due to their near-light speeds;*** *instead, most of them might have produced at a later point in time, at much lower altitudes.* ***Because of the above possibilities, it doesn't seem proper to link the lifespan of the muons recorded at the second recorder, to their "speeds" or to the "dilation of time.***

In view of the above discourse, it appears that the data collected in the above experiment doesn't truly represent the true lifespan of the muons while they are either in motion or at rest; the result obtained in this experiment, therefore, can't be considered to be realistic or reliable.

This phenomenon shall also be viewed from an alternate perspective: -

Since muons have high masses as well as speeds, they must possess extra energy in comparison to electrons because the mass of a particle is considered the measure of its energy. As per the law of "conservation of energy," neither any particle can be created nor destroyed; t*his law attracts a question:* ***"How and wherefrom muons are produced?"*** *However, the* ***"String Theory"*** *(chapter-7) envisages any particle might be converted into another particle by changing its energy, i.e.,* ***by altering the frequency of the vibrating strings that constitute such particles.*** *Accordingly, the energy of cosmic protons, during the course of the collision with the atomic nuclei, would have been transferred to the electrons. Thereby, such electrons would have converted into muons. The energy of impact probably causes these muons, after they are knocked off from the parent atom, to move almost at the speed of light. Such conversion seems possible because electrons and muons differ only* *in their masses.* **Though this assumption of mine contradicts the belief that muons are elementary particles, the aforesaid assumption might explain the "creation and decay" of muons as well as a host of other unstable particles (see Chapter-7 under the sub-heading: "Some Alternate Possibilities").**

Now, if the mass of the muons, i.e., their extra energy, was resulted due to the energy transfer caused by the collision between cosmic proton and atomic nucleus, then this extra energy might liberate at any moment. When such muons would collide with other particles, they might de-accelerate, change their direction of motion or come to a total halt, which might result in the release of their extra energies. Moreover, since the moving muons continuously ionize other particles, which come in their way, they gradually and regularly lose their energies. Accordingly, the muons, after liberating their extra energies, convert back into electrons. Likewise, if their kinetic energy is lost due to the resistance of the path or due to the head-on collision with some other particle, they may lose their extra energy and convert back into electrons; **the very existence of muons depends on their extra energy.**

In case the above speculation of mine is correct, *then the decay of the muons is resulted due to the liberation of their extra energy. If this speculation is correct,* **"The lifespan of the muons shall not bear any relation, whatsoever, with either their speeds or with the dilation of time, etc."**

Analyzing different Predictions of relativity

(1) The Effect of Gravitational Field on the Ticking Speed of Time

General Relativity predicted in 1915, "The time, in the high-energy regions such as in the vicinity of massive celestial bodies like Earth, would tick at a slower rate because *when any wave would move away from a region of higher gravity, it would lose some of its energy, due to which, its frequency would go on reducing."* As a result, the time gap between two subsequent wave-fronts would increase. Any observer looking from space would feel that every event down on the earth is taking more time than usual.* This assumption was found correct in 1962, when two clocks, one mounted at the top of a water table and the other one placed at its bottom, were viewed simultaneously from a satellite; in this experiment, the clock placed nearer to the earth was found to run at a slower rate. *I believe that the clock placed on the earth's surface was not viewed through the water; otherwise, the speed of light while passing through the water was sure to have slowed down. Moreover, since light takes about one nanosecond to move through a distance of one foot, the extra time taken by the light to travel an extra distance equal to the height of the water-column was also taken care* *of,* if not, *then the conclusion deduced from this experiment would surely, have gone wrong.*

This prediction seems to contradict Quantum Mechanics, which predicts that gravity is negative energy (chapter-3); if this prediction of Quantum mechanics is correct, then the clocks shall run faster near the massive objects because the ray going away from such an object will gain energy.

Anyhow, if this prediction of Relativity is correct, then *the energy of light waves, while going away from the stars that are a million times more massive than the earth, must reduce proportionately.* The reduction of energy (frequency) of a light wave, due to this reason, is known as *"gravitational redshift."* Black holes are the best example of this idea; they even capture the light-rays completely. Scientists of the later generations believe that ticking speed of time shall totally stop in the very close vicinity of a **Black-hole,** this would mean that *"Any object just before falling into a black-hole, shall always appear to hover over the black-hole, as if, it has become still at that particular place."* **However, this idea contradicts reality; none of the objects that were swallowed by the black hole, which exists in the galactic center during the past millions of years, could still be seen; it seems impossible that light from such objects (swallowed by a black hole) can continue to come to our eyes; such rays too shall be captured by the infinite gravitational force of the black hole.**

In case the gravitational redshift is a reality, then we might not be able to make an exact estimate of the energies of the massive stars; their actual energies might be far greater than what we can estimate from the earth. The information collected through the light waves, which are coming from different stars, would go wrong; the temperatures and masses, etc., of distant stars as estimated based on the frequencies of the waves received at the earth, might be far below their actual values. This possibility gives rise to a few questions, *"Does the frequency of the light of distant stars, when observed directly from the Earth's surface, is higher than the frequency of the same light when viewed through the "Hubble" or any other telescope installed in the earth's orbit, where micro-gravity persists?"* The *measurements as suggested above shall be taken to establish the fact.*

(2) Relation between the Speed of the Moving objects and the Time

Scientists, based on the discovery made by Michelson-Morley, *believe "Time, at the near-light speeds, dilates and the distances contract;" they* also believe that when the speed of any moving object increases, then the time, for that object, starts to lapse at a slower rate, due to which, such objects move ahead in time, i.e., into the future. When the speed of any moving object touches the speed of light, then time for that object comes to a total halt, i.e., the ticking speed of time for such an object becomes zero. *However, if this prediction is correct, the time for an object moving at the speed of light really stops, then the said object shall cease to move because no event can happen in the still time.*

Anyhow, if the above belief of **Einstein** is correct, then *time shall run at different speeds relative to different fast-moving objects.* The time of even the slow-moving objects shall also be affected; however, the time-shift for such slow-moving objects is said to be too small to take any notice of. Anyhow, *such a shift of time shall be cognizable in the case of those objects, which are moving for very long periods at comparatively much higher speeds. This would mean that every moving object in this vast universe shall reach out to different points in the future; as a result, all the objects **moving at different speeds can't remain in the same instant of time,** or every moving object should move into a different point in the future. Depending upon the speeds of such moving objects, the time gap between them would continue to increase with time.* This might appear very thrilling; however, if this can really happen, then the entire world would turn into a *"Time-Machine!"*

In fact, we can see only the past of different stars; we can't see either their present or the future. *Obviously, the light coming from either the past or present events only can be seen; it is not possible to see any happening or the future objects beforehand, no matter how close in the future, they might exist; the reason being **"How-come the rays, which have not yet emerged out of an object, can reach our eyes;"** this is impossible. Moreover, we can't touch that object even if it lies at a fraction of one nanosecond ahead in the future because it won't exist at that particular instant of time when anyone would try to touch it.* In fact, when a light beam, which is emitted out from any object, reaches us at a later point in time, only then we can see the past of such objects, in this case too, we can't touch the objects that exist in the remote past and are beyond our reach. In order to physically touch anything, it is necessary that such a thing and the person trying to touch it both shall coexist in the same place and at the same point in time. *This fact indicates that if we* can *touch any moving object, then such an object should necessarily exist in that particular time instant when it was touched. Objects lying in different points in time, i.e., the objects which do not exist in the same place and the same point in time, can't be touched by any person who exists in any other point in time.*

We all know that in reality, moving vehicles that move at different speeds do sometimes collide with stationary objects or with any other moving vehicle; *this fact, if viewed in the perspective of the previous paragraph, would reveal that all such moving objects* were precisely in the same point in the time, that is, all such moving vehicles didn't exist in the future, not by even the smallest fraction of a nanosecond; **otherwise, they could not have collided**. This should also be true for any object moving at any speed, not only for the objects moving at speeds very close to the light speeds.

Light is the fastest moving thing known to mankind; probably this is the reason why time is considered to run at the speed of light. However, light doesn't move at a constant speed within different mediums such as water, glass, diamond, etc., even it can be totally stopped within a magnetic field or by supercooling (kindly refer to chapter-3); but neither the time runs at different speeds in the aforesaid mediums, nor it can be stopped. In fact, time is not a physical entity; it is merely a concept of the rate of happenings of different events. No matter if an event is happening at higher than light speed, its rate of happening can still be measured. Scientists believe "In the beginning, the universe was expanding at a speed much higher than the speed of light, but since new events did continually happen during that very point in time, the time was also running at the same speed, i.e., higher than light-speed instead of becoming standstill or even negative." It is clear from this fact "The speed of time is independent of the speed of light." In case time slows down at higher speeds, then any fast-moving person would feel that everything behind him is taking a longer time period to happen, but the events ahead of him would appear to happen higher than the normal speed. If he could move at the speed of light, then only a single wave coming from behind would always accompany him; accordingly, he may feel that time has apparently stopped, but this would be only an illusion. Since his movement through space would be a continual series of new events, time would continue to run even for him, too; it won't stop to lapse.

Although, as brought out in chapter- 3, the light rays could be trapped by super-cooling the medium or within a magnetic field, the same is not possible with the ticking of time; this fact suggests that time moves independently of light. For example, while observing light rays coming from distant stars, we see the glimpses of time of the origin of those rays, but neither the time of origin of those rays could reach the earth nor the time of those stars would freeze; the time for those stars would continue to tick. It is a fact that when any light ray moves with normal speed or it is deflected by a mirror or totally stopped, the time continues to lapse, as usual; obviously, time, in such an event, can't be stopped or diverted. Moreover, time doesn't stop for even the light; light does take some time to move through a certain distance.

Further, as brought out in chapter-3, under section "Visualizing the Electron Moving in its Orbit," mankind has succeeded in making a camera that is capable of making a video of the moving light-waves; naturally, this camera is capable of taking snapshots at speeds higher-than-light speed, but not faster-than-time. This fact proves that the postulate "Nothing can move faster than light" doesn't hold good for, at least, the aforementioned camera. When a man-made camera can take snapshots at a speed faster-than-light, then speeds faster than light shall not be impossible in Nature too. In other words, "The speed of time is not limited to the speed of light; it may tick even at the rate faster than the rate of the light."

(3) The Ticking Speed of the Time; the Combined effect of the Speed and Gravitation

Our concept of time is based on the Earth's rotation on its axis and revolving around the Sun. However, when a spaceship, moving at the speed of light, moves past any planet, then neither the speed of spin/revolution of that planet changes a bit, nor the speeds of our Earth or Sun, would, at all, change. *This fact indicates that the speed of any moving object won't change the time of the entire universe.* Anyhow, Relativity predicts that the speed of the lapsing of time for any spaceship that moves at the speed of light would freeze; accordingly, the said spaceship would reach its destination at the same point in time when it was launched. Whereas during the journey period of the said spaceship, the time of the destination planet would have continued to lapse; as a result, the spaceship will arrive-at at its destination in the *past* of that planet. On the other hand, said spaceship, too, because of its speed, would continue to move ahead in time; as a result, the spaceship would reach the planet in the *future* relative to the said planet. This possibility poses the question that *"Which one of these two predictions is correct?"* Anyhow, in both of these cases, the ship shouldn't be able to land on that planet because of a long-time gap between the individual time of the planet and that of the spaceship; instead, it should disappear or be lost in time. Anyhow, in reality, several man-made probes have landed on different planets; this fact indicates that such a spaceship and that planet both were in the same instant of time; no change did occur in their respective times.

Contrary to the possibilities discussed above, it was seen in the example of GPS (Chapter-5) that the ticking speed of the atomic clocks, carried by the satellites, does really shift by a few nanoseconds. However, the ordinary clocks aren't capable of measuring time so accurately; as such, *CESIUM-133* clocks are used for this purpose. On the other hand, as elaborated in the previous paragraph, the speeds of spaceships do not affect the ticking speed of time, at least, of the other objects. *This controversy raises a few questions: 1) which one of these two ideas is correct? & 2)* what is the mystery of the time-shift? In order to solve this mystery, it is first necessary to understand how the atomic clocks work; only then could it be understood that *how and what changes do occur due to the change in speed and/or strength of the gravity-field that exists around such clocks?*

Atomic clocks are based on a special property of the Cesium-133 atom. Cesium-133 is an isotope (variety) of the element "CESIUM." In these clocks, the *"Hyperfine Transition"* of the Cesium-133 atom is counted very accurately. *Please don't get panicky by the term "Hyperfine Transition;" a very simple overview of this scientific term is given here.* We all know that electrons orbit the nuclei at different fixed energy levels, which are known as orbitals. These energy levels, because of the interaction of energies of the electron and the nucleus, may, sometimes, split into two or more energy levels; the energies of these split energy levels differ from each other by very minute margins. Such split orbitals are called the *"Hyperfine Structures,"* and the jumping of an electron from one hyperfine structure to the other one is called the *"Hyperfine Transition."*

The nucleus of Cesium-133 atoms comprises 55 protons and 78 neutrons; accordingly, 55 electrons orbit its nucleus at different energy levels, out of which 54 electrons orbit in the five inner orbitals, leaving a single electron in the outermost 6th orbital. All the electrons occupying the five inner orbitals form the stable pairs by aligning their magnetic poles in opposite directions of each other. Therefore, the *Hyperfine Structure* is formed in the last, i.e., the 6th orbital, where a single unpaired electron is left alone. In the atomic clocks, Cesium-133 atoms are cooled to almost the absolute zero-degree K, thereafter, they are excited by continual exposure to the radiation of microwaves having a frequency equal to the natural frequency of the Cesium-133 atom. As a result, the lone electron, orbiting in the last orbital, absorbs some energy from this radiation and jumps to a hyperfine structure having higher energy. Thereafter, it liberates this extra energy and jumps back to the original energy level. This electron, due to continual exposure to microwave radiation, repeatedly transits from one hyperfine structure to the other; this phenomenon is known as *Hyperfine Transition.*

In Cesium clocks, the transition of an electron, as described above, is repeated exactly 9,192,631,770 times or roughly 9.2 billion times in one second; accordingly, this Cesium-133 atom absorbs or radiates microwaves of the same frequency, i.e., say 9.2 billion times every second. *It is believed that this transition count (of Cesium-133 atoms) always remains constant in the reference frame of the earth's speed and its gravity field, not anywhere else.* Accordingly, the ticking speed of time at the Earth's surface is determined by this count; the duration of 1 second is decided by very precise counting of the frequency of this radiation. In order to adjust the time of these clocks, a signal of this radiation is sent to an oscillator, where the ticking speed of the clock could be adjusted by adjusting this signal.

In the perspective of the aforesaid phenomenon, let us find out: "How the ticking speeds of the clocks carried by the satellites are affected by the change in speed/gravity field." The electrons lose some energy in the weaker gravitational field prevailing in space. As a result, they jump from a higher energy level to a lower energy level, that is, a little closer to the nucleus. On the other hand, they gain some energy from the very high speed of the satellite; as such, they move away from the nucleus to a higher energy level. A combination of both of the above effects might compel the electrons to occupy a slightly higher energy level than what they occupy at the earth's surface **(in order to verify this idea, any change in the diameter of the electron's orbit can be detected by attosecond technology).** *Consequently, the electrons might take a slightly longer time to complete their orbits, which, in turn, might reduce the transition count of the electrons. As a result, counting of the fixed number of, say, 9.2 billion transitions* **(this count is fixed only at the earth's surface, not in space***) might take a slightly longer time period as compared to* **"one second." As a result, the rate of ticking of time seems to slow down, though time continues to tick at a constant unchanged rate.** *Similarly, when the lone electron orbiting in the last orbital loses some energy due to any change in the* **speed of the satellite and in the intensity of gravity,** *then it would go a little closer to the nucleus; consequently, the number of transitions per second might increase. As a result, counting of the fixed number of 9.2 billion transitions will be completed in less than 1 second, due to which, time would appear to tick at a faster rate;* **this means that the very scale of the atomic clocks for measuring time, would be altered, though no change in the ticking speed of time would take place. These clocks, therefore, under any other reference frame, other than the Earth's speed and the strength of its Gravity field, would measure the same time period differently. This is analogous to the old-fashioned wall-clocks, in which the scale for measuring the ticking rate of time was affected due to variations in temperature and subsequent variation in the length of the pendulum; "Had this speculation of mine been wrong, then all the satellites would have gradually moved in the future and become invisible."** *Anyhow, the correctness of this speculation can be verified by the following fact:*

For the last 14 billion years, that is, since the inception of the universe, various stars located in our galaxy have been orbiting the galactic center at an average speed of about 250 km/Sec, or so. *If the time really slows down at higher speeds, then during the said time period of 14 billion years,* **all these stars, in comparison to the Galactic Centre, should have, by now, moved ahead in the future by about 500 years or more.** Similarly, electrons in the Hydrogen atoms orbit the nucleus at a speed of about 2300 km/Sec. Therefore, by now, the electrons, *moving at this speed, should have, relative to the atomic nuclei, moved ahead in time by about four-hundred-thousand years. Accordingly,* in the present, both the **Galactic-Center** *and the* **atomic nucleus,** *respectively,* **should not exist at that very point in time in which the stars and the electrons are moving in their respective orbits (around their respective non-existing centers);** *accordingly, the rotating constituents of both of these structures, and their cores both would be located at different points in time, (both of them would not exist simultaneously at the same point in time). Therefore, by now, these components should become independent units, which are located at different time instants. As such, they shouldn't exert any force on the other one or influence it in any way.* **Since both the electrons and the atomic** *nuclei won't exist simultaneously at the same time instant, therefore, electrons within an atom shall not orbit a nucleus that does not exist at the same point in time. This possibility poses a question "What force will cause the electron to move in the orbit, and around what object?"* Similar would be the case with the Galactic Centre and the stars orbiting it. *Had the time-shift been possible, then the structure of the entire universe and that of the matter as well would have shattered long back.* Contrary to the above possibility, the matter and the universe both have successfully maintained their existence right from the inception of the universe. *This fact indicates that the ticking speed of time is not at all altered by the speeds of the electrons or that of the stars;* in other words: *"This prediction of Relativity seems incorrect."*

It seems from the foregoing discussions that the "time" is not at all altered due to any variation in the speeds of the moving objects or due to any variation in the intensity of the gravity field. Instead, it is the scale of the atomic clocks for measuring the ticking speed of time, which undergoes the alteration., Even in such a possibility, the importance and utility of Relativity or the Atomic Clocks won't, at all, be diluted because the precise calculation of time is required to dock any spaceship with a space station or land a probe on any planet. Such calculations could only be done by the formulae of the relativity theory, also, such a precise measurement of time is possible by the atomic clocks only. However, we might have to modify our existing concept of "time" and the capacity of time to tick at variable speeds.

(4) Slowing-down of the Process of the Aging

In Relativity, the concept of aging is linked with the belief that the rate of ticking of time depends on the speed as well as on the strengths of the gravity fields. However, as discussed herein before, any change in the ticking speed of time seems impossible. It is also believed that at the speed of light, the process of aging of the living beings would come to a total halt, even if they travel at this speed for infinitely longer time periods. However, in such a case, the heartbeats of such living beings would also come to a total halt due to the freezing of time. I feel that since the aging process of living beings depends on a biological process, the aging process can't come to a halt until this biological process is not totally suspended, i.e., the beating of the heart, flow of the blood through the veins, and breathing, etc., are not stopped. We, the ordinary people, believe that the earth is at rest. However, the earth, the Sun, the Milky Way, or any other galaxy, etc., or any of the celestial bodies in the whole of the universe, are not stationary; all of them are moving at considerably higher speeds. The entire universe is supposed to expand at the near-light speed; in case this is true, then we are also moving at the near-light speed; at least, our speed can't be considered negligible as compared to that of the light. *However, our aging process does not seem to be affected, in any way, by these different speeds at which the earth is moving.* This fact gives rise to a question: *"Do the higher speeds really affect the process of aging?"* No proof for the same is available.

We have, for a very long time period, conceived a preconception, rather a misconception, that light and time travel at the same speeds. This is the reason why it is believed that time totally stops at the speed of light. However, if a light beam is deflected or its direction is reversed by a mirror, then the time prevailing at the source of that beam, on the earth, or anywhere else in the universe, couldn't be changed or reserved. *This fact signifies that only the glimpses of time, not the time itself, travels with the light; as brought out earlier, the speeds of both are independent of each other.*

The effect of speed on the process of aging could be understood by an example of a manned spaceship, which is sent, at the speed of light, to an exoplanet located about ten light-years away from the earth. During the entire journey period of that spaceship *(if at all, the spaceship could move in the frozen time)*, any of the passengers of the said spaceship would see only one wave of light coming from behind, and therefore, he may feel that the time has totally stopped. Now, in case the said spaceship, after a period of 10 years, reaches the destined planet, *then this would mean that during the entire duration of the last ten years, the ship, on its path, had been moving ahead by 300,000 kilometers each second; accordingly, new events would have persisted in happening continually all the way. Since time is simply a concept of duration between the happening of two or more events, or more accurately, the rate of their occurrence, it would mean that the time also kept running all the way. Had the time halted, then the ship couldn't have moved from its place because no event can happen in the frozen time. If time didn't stop for the said spaceship, then how could it have stopped for the astronauts sitting inside? If the time had stopped for the astronauts sitting within the spaceship, how could it have kept ticking with normal speed outside the said ship? Different times can't exist outside and inside of such a spaceship simultaneously.* It could, of course, so happen that due to the extreme pressure produced due to a very high rate of acceleration, the astronauts might be crushed to death. In that case, time will surely stop forever for such passengers; however, *if they reach their destination, alive or dead, then it could not be said that the ticking of time did stop for them at the speed of light.*

Probably the said spaceship, on attaining the speed of light, might disappear from the sight of the earthly people; the radio contact might also be totally snapped. Consequently, the fact that "W*hat did happen to the passengers of the* *said spaceship"* could only be explored after that ship returns from its journey. If the time did really stop during the journey period of the said ship, then, though a time period of 20 years would have elapsed for the people living on the earth, the passengers of the said ship would still be at the same time and age when they started their journey. In other words, the astronauts, after returning to the earth, would find themselves 20 years lagging behind the current time. However, w*hen it is not possible to see or touch any object located at different points in time, then how would the astronauts be able to see the earth or land on it?* If, at all, they succeed in landing on the earth, then how the people living at two different points in time could see, meet, or talk to each other? Could two different times prevail on the earth simultaneously? In the meantime, the earth would have gone 20 times around the sun and 7300 times on its own axis; how this time gap could be bridged? No clarification on these points could be given at this moment because no journey at the speed of light has yet been performed; nobody knows the truth. However, if such journeys are performed in the future, then the only possibility that I can visualize is: "A*ll **the astronauts, in that case, would grow older by 20 years; no real effect of the speed on the aging process of living beings seems feasible to me.***"

(5) The relation between the Mass of an object and its Speed

Relativity envisages that matter and energy are interchangeable; accordingly, it is believed that the mass of any moving object should increase due to the extra energy acquired by it at higher speeds; as a result, more energy would be required to increase its speed any further. Thus, the requirement of energy to increase its speed would increase exponentially; when the speed of an object approaches the limit of the *speed-of-light,* then the mass of such an object as well as the amount of energy required to increase its speed both would become infinite. In other words, an increase in the energy of any object shall result in an increase in its mass because the mass of an object is the measure of its energy. This shall also be true when extra energy is added to an object in any other form, i.e., by heating such a piece of matter or exposing it to strong radiation, etc., because the laws of science shall always remain the same, isn't so? Probably, this aspect has never been verified or reported by any of the sources.

In case the mass of an object really increases due to an increase in its speed, *then the reverse of this shall also be true; the mass of any object shall decrease by decreasing its speed; if this presumption of mine is correct, then the energy of an object, that is, its mass might become zero at the absolute zero speed. The aforesaid possibility indicates that the matter might decay either in the position of the absolute rest or when its intrinsic energy becomes zero.* However, in case the particles don't decay at the position of the absolute rest, then their intrinsic energies, that is, their masses, might remain intact; in that case, the mass might not be directly affected by the speed. Further, scientists believe that the entire universe, including all the galaxies, all the stars, etc., is expanding almost at the speed of light; *this belief of scientists defies the prediction of Relativity that "Nothing except mass-less energy waves can move with the speed of light." Further, no effect of such a high speed is seen on the masses of the stars or the living beings living on the earth; this fact contradicts the prediction that the mass of the moving objects increases at very high speeds. Both of these anomalies in the aforementioned predictions of Relativity indicate that "Either the universe is not expanding at near-light-speed or such a high speed does not affect the mass of the objects."*

As we know, the light, while hitting a surface, exerts a slight pressure on that surface; this means *"Light particles possess some mass; a particle without mass can't exist."* However, in reality, a photon moving at the speed of light doesn't acquire infinite mass. This fact challenges the aforesaid prediction.

(6) The Fabric of the Space-Time

Relativity envisages that interstellar space is not an ordinary vacant space; it is filled with infinite amounts of energy and matter, both of which are distributed all through the interstellar space in a random manner. In fact, the universe is supposedly a manifold or a combination of three different entities, namely: *energy, matter, and space (distances)*. *Einstein*, in his theory of relativity, suggested: "Time is not separate or independent of space, both of these entities are interwoven to form the continuum of space-time, which is very strong as well as flexible." This continuum, like other physical things, can be stretched, folded, contracted, or twisted like a fabric. *Einstein* proposed that light waves propagate through this fabric like the ripples propagating on the surface of the water. He further envisaged "The rotating masses cause this fabric to drag-along, or twist around them; accordingly, *light shall travel faster in the direction of the Earth's spin and vice versa.*" however, this proposition doesn't match the result obtained from the Michelson & Morley's Experiment. He also envisaged that the fabric of space-time has an inherent property to expand; accordingly, *the far-parts of this fabric are moving away from us, at higher-than-light speed*. However, had all of the aforesaid predictions been correct, *then light shall travel at twice its speed in the direction of expansion of the universe, and on the other hand, it should not, at all, be able to move in the direction opposite of the expansion;* however, this does never happen. Anyhow, scientists of the modern era believe that all the stars and galaxies are moving away from each other like the *spots made on an expanding balloon,* such spots also grow bigger with the expanding balloon; *however, galaxies and planetary systems don't seem to grow bigger at the speed at which the universe is supposed to expand. This fact poses a question mark on the very existence of the so-called fabric of "Space-Time."*

Based on the flexibility of the fabric of space-time, it is believed *"In the near future, the long-distance space-travel could be completed within a very short time period, by pulling or contracting this fabric or by folding this fabric and matching any particular time-zone with another time-zone."* This myth is equally popular amongst scientists as well as science-fiction writers both earnestly believe that this belief is true. *However, in case this fabric really exists, even then, this presumption doesn't seem practicable because: "How is it possible to stretch or fold this fabric along with billions of stars, to a distance of hundreds of light-years?" Moreover, "How the ends of this fabric could be held because nobody knows where do the edges of this fabric exist, and how to hold them; where would the person holding it stand?"*

It is also believed that the sheet of space-time is spread only in two dimensions. *If this belief is true, then all the galaxies shall lie in the same plane in which the Milky Way is located. In case even a single galaxy is seen outside of this plane, then it could be inferred that space-time is not spread only in two dimensions; instead, it is spread in a three-dimensional region.* Two other facts: 1) *"Light spreads even in the perpendicular direction to the galactic plane."* & 2) *"Satellites can orbit the Earth either over the equator, over the poles, or at any angle in between,"* indicate that *"Space-time is not spread merely in two dimensions only."*

The sheet of space-time is considered so flexible that it wraps below every celestial body due to its individual mass; it is supposed to warp below even the tiniest objects like meteors. On the other hand, this fabric is thought to be so strong and stiff that it carries the mass of the entire universe. *However, the fact that "All the visible stars are located in a flat plane (galactic plane)" indicates that even the collective mass of all the stars, planets, etc., doesn't cause this fabric to warp or sag. Further,* the fact: "Various small objects, such as meteors and asteroids, etc., keep falling on the earth from time to time" puts a question on *its strength, as well as on the concept that "The warped space compels the smaller celestial objects to orbit the bigger ones;" in case the celestial objects move along the curved surface of the deep-well, which is formed in the fabric of space-time, then meteors/asteroids shall persist in moving in their orbits, they shall never fall on the Earth? This fact puts a question mark on the very existence of this fabric and the truthfulness of this idea.*

(7) Space-Time And The Gravitational Force

Relativity envisages that, unlike other forces, gravitation is *not a real force; it* is merely a consequence of the warping of the *fabric of space-time*. It is also believed that though the planets move toward their nearest star in a straight line, they appear to move in orbits because they follow the curvature of the space-time that warps around such a star, not because of the force of gravity.

The aforesaid concept can be practically explained by the help of simple equipment that looks like a "Trampoline" (toy equipment used by kids to jump). This equipment comprises a flexible sheet mounted on a strong frame; when any heavy object is placed at the center of this sheet, then that sheet warps below that object, as shown in fig-3A. Another piece of equipment made of plastic or metal sheet, like the one shown in Fig 3-B, is also exhibited in most of the Science-Museums; such equipment comprises a funnel-shaped curved cone having a central opening. To demonstrate how planets orbit their stars, a coin is rolled down along the slope of any of the aforesaid equipment in the transverse (tangential) direction to the curved surface, not directly toward the central opening. As a result, the said coin starts to move along the curved surface of the equipment in a circular path. Although the sheet of space-time doesn't exactly warp as depicted below, it is believed that different planets move in their orbits around different stars because they follow the curvature of the warped space-time, in a similar manner in which the coins move on the curved surface of the said equipment.

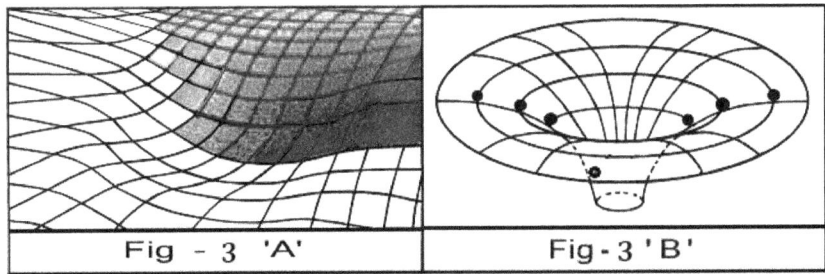

| Fig – 3 'A' | Fig - 3 'B' |

At a glance, the coins moving in different circular paths in the warped portion of the aforesaid equipment may convince anybody that the planets do orbit their stars due to the warping of space-time beneath different stars. However, on contemplating a little, a host of doubts would come-up on this concept; some of such doubts are discussed below: -

A. In reality, all the planets orbit the Sun *in almost the same plane; such* a condition can't be simulated in the aforesaid experiment. Moreover, such an experiment can't be carried out with a piece of equipment made of an ***extremely*** flexible fabric.

B. Though any single star may warp space-time as shown in Fig- 3A/3B depicted above, it is beyond the imagination *"In what shape space will sag beneath a pair of Binary-Star or a Triplet/Quadruplicate of Stars? How would a planet move in the indentation made by a triplet/quadruplicate of stars? Each of the stars in a multiple star-system shouldn't be able to make independent indentation having uniform curvature; in this scenario, it is not clear that how different stars of a multiple star system would orbit each other around a common center of mass, that too, from different distances; or, how the planets of any such star, would orbit its star?"*

C. Both *space and time* permeate in all three dimensions; accordingly, massive bodies shall displace the continuum of space-time all around themselves as per their shapes, as is done by any object that is submerged in water or air. Further, since no force acts on any star from underneath, the sagging-down of the fabric of space-time under any star, like a two-dimensional fabric, neither seems logical nor feasible.

D. In fact, in both of the equipment depicted in Fig- 3A/3B, the pull of gravity doesn't allow the coins to climb up the inclination of the said equipment. As a result, the initial momentum of such coins compels them to change their directions continually; accordingly, such coins move in circular paths along the surface of the pit. *Ironically, with the direct help of gravitational force itself, it was ostensibly proved "Gravity is not a force."* In case such an experiment is carried-out on-board a spaceship, then, in the absence of any force acting from the beneath, the coin would fail to move along the curved surface of the equipment. This fact clearly tells: "Different planets, instead of following the curvature of space, orbit the stars due to the *Force of Gravity.*"

E. *Since no force pulls down the massive stars from the underneath, the fabric of space-time shall not, at all, sag beneath them; however, in case the fabric of space-time really warps beneath different stars and planets, then, in such an eventuality, this fabric won't pass through, or touch, any celestial object, under which the sheet of space has warped. Furthermore, the light would not be able to spread out from even the luminous objects, which caused this fabric to warp; because the light is supposed to propagate through space-time. However, in reality, nothing like this ever happens. This fact indicates very clearly, "The fabric of space-time doesn't warp beneath any of the celestial bodies."*

F. In case celestial bodies follow the nearest thing in a straight line, then a question arises that "What force causes different planets to move; or where from they get the energy to move continuously in their orbits, that too in a definite direction; not in a random manner?" *This seems impossible being against the laws of science; nothing can move without applying force.*

G. In case the curvature of space-time compels the planets to orbit the sun, then the same space-time while expanding at near-light speed shall also cause the size of the solar system, or the Milky Way, etc., to expand with the same speed. In that case, the distances between different planets of the solar system, and that among the nearby stars, shall physically increase, and the same shall be seen very clearly. However, nothing like this has ever been observed. This fact raises a question- "Whether warping of space really compels the planets to orbit the Sun?"

H. In case the gravitational force is really a consequence of the warping of space, then no effect of gravitation shall be felt in the flat portions of the space-time that extends beyond the outer edges of the indentations made by different massive objects. In case the aforesaid assertion of mine is correct, then different galaxies located far-away from each other shall not attract other galaxies because a flat portion of space might exist between distant galaxies. Contrarily, the galaxies located within any group of galaxies attract each other, even from far-off distances.

I. Relativity predicts, "The warping of the fabric of space-time produces a tendency to draw every entity toward the center of the massive objects; this tendency produces the effect of Gravitational Force (chapter-5 under General Relativity)." In that case, the strength of this force shall remain uniform within the whole of the interior of the deep well produced in space; the condition of micro-gravity shall not exist anywhere within such a deep well; however, this never happens. Moreover, the warping of space shall not be able to produce any such force within the solid crust of any celestial object. Further, as illustrated below, the entire mass of such an object will be counterbalanced by the upward thrust produced in the bottom-most point of such an indentation. Therefore, the maximum effect of gravitation shall be produced in the lower hemisphere of different celestial objects. In such a case, this force will never have uniform strength because when we move up along the slopes of such an indentation, the strength of the gravitational force would gradually reduce; both the value of the force so produced and its direction will go on changing from place to place. Moreover, no gravity could be produced in the upper hemisphere; this defies reality.

155

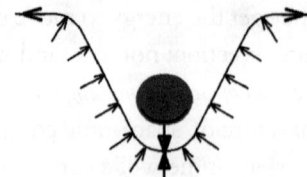

J. Any celestial body of a definite mass shall warp the fabric of space-time with a definite and uniform curvature. Accordingly, any small object that follows such a curvature made in space-time shall move in a definite and perfectly circular path around that massive object, which caused this fabric to warp. However, defying this possibility, most of the far-off celestial objects like the comets, etc., adopt elliptical to highly eccentric and highly elongated elliptical orbits. This fact indicates, "In case such objects follow the curvature of space, then neither the indentation formed in space have uniform shapes, nor the objects moving through it, follow its curvature." Apart from this fact, different probes, sent to the Moon or Mars, gain more energy for the onward journey by jumping from a smaller orbit into gradually increasing elliptical (not circular) orbits around the Earth or Moon; this feat is achieved by increasing the speed of such a probe. This fact raises a few questions: "Do the celestial objects carve multiple curvatures in space, which are like different layers of an onion? Do the outer layers push any object moving through them, with greater strengths because it holds the probe moving at higher speed?" These questions have come up because when speed is increased, the same probe seems to penetrate through an inner layer and move on to the next outer layer that pushes it back with more strength to hold it on that particular layer. This possibility contradicts the truth; the strength of these layers shall reduce gradually due to gradually increasing distances from the Earth or the Moon.

K. Several small to medium-sized black holes also orbit the galactic center of the Milky Way. Some of such black holes, such as Cygnus X-1, Cygnus X-3, V404 Cyg, GROJ 1655-40, etc., continually snatch away gas from the stars orbiting them from the close quarters. Such infalling stellar wind (gas) forms a two-dimensional ring before it is swallowed by the respective black hole. In case, as believed in Relativity, the space-time really warps beneath any massive stellar object, then any object moving through such warped space shall follow the curvature of space. *In such a case, the gas falling into any black hole should invariably fall in the shape of a curved cone, not in the shape of a two-dimensional accretion ring. This fact clearly tells that no conical indentation is formed beneath any stellar object, not even beneath a Blackhole.*

L.   In case the warped space really compels celestial objects moving through it to follow its curvature, then only the outermost planet/star, orbiting in a two-dimensional plane, should be able to move along the curved space. Whereas the remaining planets shall either keep hanging on the same plane or settle down at different distances/altitudes on the wall of the deep well, these distances would correspond to the sizes of the orbits of the respective planets; in that case, such planets shall not be seen spreading over a flat plane (the Galactic Plane), instead, should appear to move at different altitudes, on the curved space, in a manner similar to that in which different coins, in the equipment shown in Fig 3B, move at different altitudes. Accordingly, the shapes and sizes of the indentations made in space-time shall be seen very clearly. Contrarily, different planets of the Solar System and various stars in different spiral galaxies are seen to spread over almost flat surfaces. This fact raises a question: "Whether different celestial bodies follow the curvature of space in a single plane without touching it, alternatively, does space-time is capable of compelling celestial bodies to follow its curvature even without pushing them back?" However, both of these alternates seem totally impossible.

I feel that in case the stars and planets are not moving directly on the curved surface of space-time, then neither would they push the space-time, nor shall the curved space be able to push them back, and thereby compel them to follow its curvature and move in their respective orbits. Therefore, the distribution of stars on a single flat plane within the Milky Way suggests that space-time isn't warped, at least not in the shape as depicted in the said Figures 3A, or 3B. The aforesaid fact raises a question: *"Do the massive bodies really warp the fabric of space-time, or is the gravitational force really a consequence of the warping of the fabric of space-time?"* The fact that nothing could move without any force gives rise to the same question again, *"What force causes the celestial bodies to move in definite orbits, that too in a definite direction only?"* *This shouldn't be possible in the absence of any force.*

Einstein probably envisaged that the straight paths on which the planets move along the warped space are like closed rings. According to his concept, the celestial bodies that are trapped in these curved paths follow the nearest massive thing in a straight line; however, the curvature of these paths pushes them back persistently and thereby compels them to follow these closed ring-shaped paths.

As a result, such a celestial body, though moving ahead in a straight line, appears to move in circular orbits. This is analogous to the toy trains moving on fixed circular tracks, where the train, though it moves in a straight line, appears to move in a circular path because of the curvature of the track.

In the aforementioned case, these curved paths should necessarily be very stiff and strong so that they may exert adequate pressure on the celestial bodies that move through their respective curved paths and thereby compel them to follow the curvature of these paths. How strong these paths should be made in the warped space can be understood by yet another example: If an ordinary ball is rolled down the slope of the kid's slide, then it will easily roll down along the curvature of the slide. However, what would happen if that ball was replaced by a steel ball weighing 5-10 tons or by a very heavy ball fired from a cannon? In such a case, the slide would probably, be totally destroyed. Masses of different celestial bodies moving through space are millions of trillion times greater, and they are far-heavier as compared to any cannonball. Moreover, they move with a much higher speed. This simply means that the paths made in space-time should be extremely strong and rigid; contrary to this requirement, small objects like meteors very easily penetrate through this so-called sheet of space-time. *This fact attracts a question "Can two nonphysical things like an empty place, or merely a distance (space), and the time interval between two or more events, when combining together, become so strong and stiff that it may withstand the pressure of massive celestial bodies, moving at very high speeds?" It is clear from the example of the toy train that the train, while moving on the steel tracks, easily follows the curvature of the track, whereas the same train can't run on the paths made on the non-rigid surface of the water. Similarly, the direction of a moving ship can't be changed by any path made on the surface of the water.*

Relativity envisages "The moving celestial objects regularly produce gravity waves, which carry away the energy from the moving celestial objects. As a result, the planets are supposed to gradually spiral inward and finally fall into the stars they do orbit." On the other hand, Relativity also envisages that "The celestial objects orbit their respective stars because they follow the curvature of the deep-well created below the stars." In case this prediction is correct, then the size of the respective orbits of different planets would keep shrinking gradually and persistently. Such a shrinking, in turn, will result in a gradual reduction in the sizes of different planetary systems, as well as in that of different galaxies too. This means the whole of the universe shall continually keep shrinking. On the other hand, scientists believe that the universe is continually expanding at the speed of light. If this prediction is correct, then the sizes of different galaxies as well as that of our solar system shall continually increase; however, no such increase/ decrease in the size of the solar system has ever been observed.

In the absence of any solid proof of the existence of the fabric of space-time, it is hard to believe that gravitation is not an independent force. If gravity is not an independent force; instead, it is the consequence of the warping of space, then what is the explanation for each of the following questions/possibilities?

1. Why do different objects accelerate while falling toward the Earth, whereas the same object keeps floating under microgravity conditions within a spaceship.

2. The region of microgravity starts at the height of about 400 Kms. and beyond; had gravity has been resulted due to the warping of space, then neither the intensity of gravity should vary due to any variation in the altitude, nor the air/water pressure should vary due to the variation in the height of the air/water columns. (Please refer to Chapter-5)

3. Why is the atmospheric air, while the earth moves in its orbit, not left behind due to its inertia; with what force is it bound to the earth? Why is it not blown away by the Solar Wind?

4. Why all the massive celestial bodies are spherical in shapes? Why don't they acquire the shape of the deep-well carved in space, as is done by the water stored within a bowl?

5. What force restricts us from lifting heavy loads? This is certainly resulted due to a force that acts in a downward direction.

6. Had the atmospheric pressure been created due to the warping of space-time, then this pressure shall always remain constant. However, in reality, the atmospheric pressure decreases with increasing altitude and vice versa. This fact indicates that a force is certainly pulling the blanket of the atmospheric air toward the center of the Earth.

7. In case the celestial bodies, while moving in their orbits, push the fabric of space-time, and in turn, the space-time also pushes back these bodies with an equal force, then *"Why the air blanket of the earth, and also, the creatures living on Earth's surface, are not crushed by the immense pressure created between these two?" Further, in case the atmospheric pressure is created because of this reason, then why does it decrease at higher altitudes; the atmospheric pressure so produced (due to the reaction of the warping of space) shall always remain constant. (Kindly refer to chapter- 5)*

8. Different nebulae (gas clouds) are normally hundreds of thousands of times more massive than a star. Such a huge mass must make a huge and deep indentation in space. Accordingly, a nebula, like all-other celestial bodies, shall spin about its axes, like a *single body;* several stars shall also orbit any of the nebulae. I am not aware whether any such activity has ever been observed. However, numerous rotating regions are regularly produced within different nebulas. This fact implies that curved space doesn't force celestial objects to spin or go around in orbits.

9. Whether the man-made satellites are established in different pre-existing geodesics existing around the earth, i.e., calculation of their paths does not depend on Earth's gravitational force? At present, thousands of satellites are orbiting the Earth at different altitudes and different angles. It is not feasible to establish such satellites at different altitudes in a single bowl-shaped indentation created in the fabric of space-time, that too at different angles to each other.

10. How do the high tides occur in the sea? Why do the solid surfaces of "Io" and "Europa" (the moons of planet Jupiter, which directly face this planet) bulge out by about 100 meters? How is such a tidal effect produced without any force, that too, on the solid surfaces of the said moons?

11. Sometimes, asteroids or Comets, whenever they happen to come very close to a massive planet like Jupiter or even Earth, etc., do break apart into several pieces. This is not possible without any force; no object can break apart into pieces until and unless it is acted upon by a force like Gravitation, the curvature of space can't compel such objects to break into pieces.

12. The Gravity-field of the sun extends up to the outer edges of the Oort cloud (please see annexure-1); accordingly, the radius of the deep-well, which is formed below the Sun, shall be around 1.8 light-years. This distance is about 100,000 times greater than the distance between the Sun and the Earth. Moreover, the size of the Earth is almost 1.3 millionth of that of the Sun; accordingly, the Earth or any other planet shall not be able to touch or push the inner walls of the indentation cast by the Sun; as such, the said indentation won't be able to compel different planets to orbit the sun in a single plane. However, if strings of different lengths

are tied to different pieces of stone of different masses; and if someone, by holding the other ends of these strings together, swings these stones in a circular motion, then depending upon the rotational speeds, all these stones will revolve in almost a same plane. This fact indicates "The attractive force of the Sun causes all the planets to move in a single plane; the curvature of space doesn't play any role."

13. Relativity envisages that different planets move toward the nearest stars in straight lines; however, it is not clear what force causes them to move? In case celestial objects, in the absence of any force, move like freely falling objects, then why don't the objects that float freely within any spaceship continuously accelerate all by themselves? Nothing can move all by itself; a force is essentially required for this purpose. Moreover, when light emitted from any star, which is located in the center of the deep-well, travels radially in all directions in straight lines, without following the curvature of the space, then why can't the planets also move in straight lines toward the Sun, without following the curvature of space. This fact tells that "Neither any curvature is formed in space nor any moving thing follows it."

14. In case the gravitational force is merely a consequence of the warping of space-time, then the Sun's gravity field should have almost equal strength all along the orbital path of the earth. Contrarily, the strength of the gravitational field at any space station, which is located a few hundred kilometers away from the earth, is almost zero. This means the depth of the dimple carved by the Sun is almost zero at such a distance from the Earth; such a shallow curvature cannot compel the Earth to orbit the Sun. On the other hand, the moon orbits the Earth from a distance of 380,000 kilometers. This means that the strength of the gravity field around the Earth, at an almost equal distance from the Sun, is not uniform; thus, at this distance from the Sun, there must be at least three different values of the curvature created in the space, but this seems impossible. A question, therefore, arises: "Does the earth really orbits the Sun due to the warping of space?"

15. The far-parts of space-time are supposed to expand at the speed of light; the further away these parts are, the higher is the speed of their expansion. However, this uneven, but very high speed of the expansion of different parts of space-time, doesn't distort or rather elongate

the galaxies or the planetary systems. This fact indicates that "space-time, while expanding, doesn't push celestial bodies;" this fact further suggests that "space-time doesn't compel celestial bodies to move in their respective orbits."

16. In case the moon orbits the earth due to the warping of space, then the Earth will also cast a deep well in the inclination of the bigger indentation made by the Sun; accordingly, as shown in the sketch depicted below, the moon shall orbit the earth in an inclined plane. In that case, no eclipse of the sun or the moon would ever occur. Moreover, whenever the moon would cross the orbital paths of the earth, then it may be distracted by the bigger indentation (made by the sun), within which the earth moves in its orbit. Both of these possibilities indicate that neither the sheet of space-time does exist nor its curvature compels celestial objects to move in orbits.

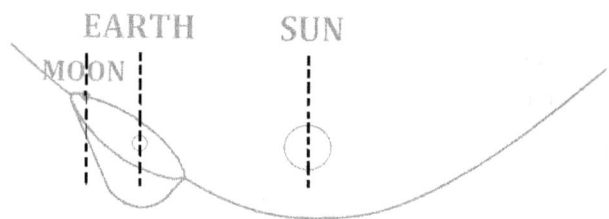

17. Relativity envisages that planets, while moving through space, do lose their energies by way of the emission of the gravity waves; therefore, they gradually spiral toward their stars. This presumption contradicts another presumption that "Planets follow the curvature of the space warped by their respective stars;" the reason for the same being the mass of the star doesn't change due to the depletion of the energy of any of its planets; accordingly, the planets, following the curved space of a fixed size, shall always move in a fixed orbit. Therefore, in case the energy of any planet really depletes, then such a planet, instead of spiraling inwards, shall follow the curved-space on a fixed orbit, but with gradually reducing speeds; eventually, when its energy is completely exhausted, it shall stop moving, i.e., it shall come to a total halt at one location. This is, however, totally against reality.

18. In case gravity is not a force, then how the planet Neptune causes irregularities in the path of the planet Uranus? Astronomers have also

observed that any massive planet while orbiting its star may cause the star to wobble. This fact casts a question "What force compels such a star to wobble? This phenomenon is impossible in the absence of any independent force.

19. The fact that sometimes two galaxies do collide with each other attracts a question- "What force causes them to move toward each other? Don't they move on the curved *space of fixed shape?" Moreover*, after such a collision, stars of both the galaxies are first scattered in an uneven and irregular manner; thereafter, they start to orbit the combined galactic center of both the galaxies, with different angles, which continually go on changing during a long transient period that is required to form a bigger galaxy; such a transient period spans over a few hundred million years. Such changes in the orbits of different stars are known as "Tidal-Waves." *Now, in case stars really orbit galactic centers due to the warping of space, then while two different galaxies approach each other, the curvature cast by them shall gradually adopt a new shape, due to which the tidal waves shall not continue for such a long time-periods, different stars shall gradually, and smoothly, adopt new orbits, not in such a violent manner. The aforesaid possibility indicates that stars do not orbit the galactic center due to the curvature of space; there shall be some other reason for the same.*

20. How the universe comprising different galaxies was formed? Did the gas clouds collapse without any force like gravitation? How the stars, in the absence of gravitational force, are created, or how do they end up in supernovae or even in the black holes? (For further details of the above-mentioned phenomena, please see Chapters-8 to 11).

Quantum mechanics envisages that *the gravitational force, too, acts through its carrier-particles known as "Gravitons" (gravitons could not yet be detected)*. However, in view of the questions listed above, scientists should make it clear *"whether gravity is a force, or it is a consequence of the warping of the fabric of space-time;"* only one of these two concepts can be correct not both.

(8) Deflection of Light near the Massive Bodies

As brought out earlier, the fact "Different planets orbit the sun in an almost flat plane" indicates that the warping of space doesn't compel the planets to orbit the sun. However, as predicted in **relativity***, the light does really deflect while passing from a close distance from the sun. This fact indicates that "In case the light was deflected by the warping of space, then the* *maximum curvature of space is created near the surface of the sun, i.e., at a radial distance of about 700,000 km from the center of the sun, not lesser than that." On the other hand, about 300 binary asteroids in the asteroid belt orbit each other from a distance of around 100 Km or so; this fact indicates that the curvature of the warped space near such asteroids shall be physically greater than that created near the Sun because the smaller is the radius, the greater is the curvature. Contrary to this fact, it is believed that the more massive the object would be, the greater would be the curvature of the warped space. This controversy indicates that the curvature of space is probably determined in some other way.*

*In the flat space, while moving gradually toward the sun by one kilometer at a time, the circumference of the circle formed around the sun would reduce by $2pi\ r - 2pi\ (r-1)$, i.e., by $2pi$ kilometer, but **Einstein** envisaged that in the warped space, the diameter of such a circle would diminish by 0.99999999 times of $2pi$ kilometer. Accordingly, a funnel-shaped curvature will be created in space, similar to that illustrated in Fig. 3B. In that case, the maximum curvature in the warped space shall be created far below the sun. Accordingly, the light should deviate from quite a distance from the sun, not from its close vicinity; however, this possibility does not match the physical observations. Moreover, such warping of space has no explanation why do the planets orbit the sun in an almost flat plane?*

*In **Relativity,** the curvature of the four-dimensional space, at any particular point and at any particular time, resulting due to the presence of the mass-energy and its linear momentum (stress-energy), is determined by Einstein's Field equations or tensors. However, countless meteors, asteroids, comets, and various planets, etc., are orbiting the sun from different distances, but in a single plane. Therefore, all such objects, whether massive or tiny, shall follow the different curved paths made on that flat plane corresponding to their respective orbits; they can't move on the interior wall of a single bowl-shaped dimple at different altitudes. If this theory is correct, then the orbits of different planets and other objects so determined shall always conform with reality; moreover, different presumptions of this theory shall always maintain coherence with each other, but in this case, they seem to contradict each other.*

In whatsoever manner the curvature of space-time is determined, the effect of the same shall always conform to the reality, which we observe physically. Accordingly, if the light rays are really deflected due to the curvature of space, then they shall follow this rule everywhere without fail; the light rays shall invariably deflect near each and every celestial body, and also, at every warped region of space-time. However, light neither bends due to warped space-time near planets (it doesn't bend even near the biggest planet Jupiter) nor anywhere else along their orbital paths (geodesics). This fact indicates that *either light is not affected by each and every curvature of space or space-time doesn't warp near smaller objects.* Since light deviates only from very close distances from very massive bodies, not everywhere, ***the bending of light, due to the warping of space-time, is not a hard and fast rule that holds good in*** *the entire universe; there seems no direct relation between the mass of an object and the angle of deflection of light caused by the warping of space. This fact also indicates, "Light deflects only near those bodies, which have a mass greater than a certain limit, not lower than that limit."*

The fact that "Light bends only at very close distances around extremely massive bodies like the Sun" indicates that "Only the extremely potent gravity fields can deflect the light, the weaker ones are not capable of doing so." This means that "The massive objects, instead of carving very big and deep indentations, warp very small portions of the fabric of space-time, that too, only in their very close vicinities." This fact also indicates, "The region of maximum curvature of such indentations shall be located very close to the outer surface of such objects." This fact suggests that "The indentations, made by the massive bodies like the stars, are neither deep nor big enough to compel the far-off planets to orbit them. Moreover, since a large distance exists between the furthest planet and the said star and it (the star) deflects light from very near to its surface, the said star would cast a very shallow dimple in space. Accordingly, the planets located at a far-off distance from a star might not be orbiting the said star due to the warping of space because the curvature of such an indentation at the far-off distances might become too small and far-inadequate to compel the planets to orbit their respective stars."

*In case the light is a periodic disturbance (wave) that moves through the fabric of space-time, then such a disturbance coming from the backside of a non-luminous celestial object shall never be able to jump over the indentation made by such an object, **and thereby the light shall not be able to move in a straight line** because as soon as that wave would leave the medium of space-time by jumping-over the indentation, its very existence would be terminated; no ripple can exist without a medium. Accordingly, as shown below, such a disturbance must continue to ripple along the curvature of the warped space. Therefore, the light shall reach our eyes from underneath that object; accordingly, we shall be able to see the objects located at the other side of the non-luminous objects; this is similar to the propagation of light through a bent or curved optic fiber. Similarly, if space really warps below different stars, then such stars would though emit the light, but light so emitted would not find any medium to propagate because the warp and waft of space and time won't pass through such stars. Therefore, we shall not be able to see them. **Contrarily, massive objects always obstruct our vision, and we always see the light coming from different stars/planets. This fact indicates that neither the light follows the curvature of space nor ripples through the medium of space-time; there must be another reason for the bending of light or its propagation in the straight line.***

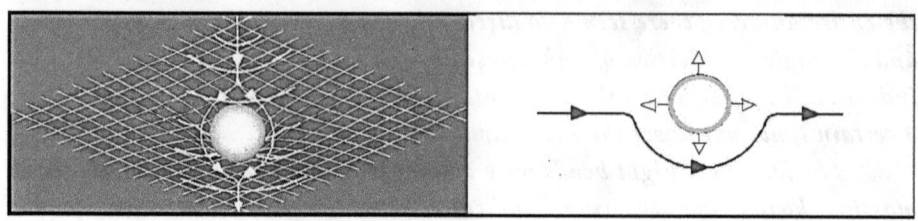

Anyhow, in case the light-waves really follow the curvature of space-time, then as illustrated in the sketch given below, a light-wave running towards a deep-well, made by a point-sized massive body, such as a Blackhole, would, following the curvature of such a deep-well, bend in the direction away from the center of gravity of the said massive body, instead of bending toward its gravity center; contrarily, the light passing nearby a black-hole is never seen to deflect in this manner. This fact further indicates that light doesn't follow the curvature of the warped space; there must be some other reason for the same.

Ray of Light

Next, since the depth of any of the so-called deep-well, made below any point-sized massive object, depends on the mass of such an object; the heavier (massive) would be the body, the greater would be the depth of the deep-well it would cast. Accordingly, the maximum curvature of space-time would be created just below the deep well; the point of maximum curvature will be created as far away from that body as the depth of the deep well. *Thus, as shown in the sketch depicted below, black holes would cast infinite curvature at a far-off point that would be created at a distance equal to the depth of the deep well, **not at its close vicinity.*** Accordingly, ***any black hole shall trap light from a distance equal to the depth of the deep-well that it would carve in space, not from its close vicinity;*** *however, this is against the reality. This fact indicates that massive bodies do not warp space, or light waves do not propagate through space-time.*

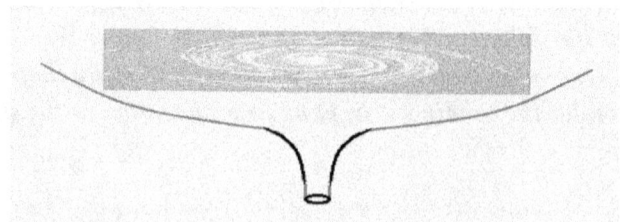

In 1860-65, James Clerk Maxwell established the wave nature of light; since then, it has been believed that light waves can't be affected by any force like gravitation. Scientists still have faith in the aforesaid belief. *Contrary to this belief, some Indian scientists have succeeded in trapping photons in the magnetic field (chapter-3, "Some Least-known Properties of the Light"). In case the magnetic force is capable of capturing the so-called light waves, then why can't any other force have an influence on the light?* **This possibility suggests that gravity, in the capacity of an independent force, might also cause the light waves (a shower of photons) to bend or deflect, albeit from a very close distance only, because this force loses strength very sharply with even a very slight increase in the distance.**

(9) The Limit of the Speed

The *Theory of Relativity* envisages that 1) "Nothing can move faster than light," and 2) "Only the mass-less waves can move at this speed." Contrary to this belief, the cosmic rays, which are, in fact, a shower of sub-atomic particles like protons, neutrons, nuclei of Hydrogen and Helium, etc., are found to travel at speeds that are very close to that of the light. Likewise, Muons, which are 200 times heavier than electrons, also move at near-light speeds. The jets of particles that are emitted from the quasars also travel at the speed of light; these jets have a composition similar to that of cosmic rays. Even the entire universe, along with all the galaxies, stars, planets, etc., is also believed to expand at the near-light speed. **This fact indicates that massive particles and even massive objects too can move at the speed of light.**

We all know that even light, whose speed is highest among all the known moving things, can't escape the black holes, whereas the effects of the gravitational force of the black hole, as well as that of its magnetic force, both are surely felt from very large distances. This fact puts up the question: **"How does the magnetic force escape the black holes? Does the magnetic force or the virtual photons of magnetism move at higher-than-light speed?"** *This possibility doesn't seem feasible.*

Normally powerful pulses of radio waves are emitted from the poles of the celestial objects known as **Pulsars**; these objects are very compact stars, which are many times heavier than the Sun; these stars rotate on their axes at very high speeds. These stars emit the said radio pulses after regular intervals. Scientists believe that *"The spinning magnetic poles of these stars produce rotating polarized currents, which, in turn, produce the said radio pulses."* Some scientists believe that *"Though these pulses travel at the speed of light, the source, from which these pulses are emitted, moves at a speed 6 times higher than the speed of the light."* Present-day scientists now believe that *"The supermassive* *black holes also rotate at speeds faster than the speed of light; moreover, the outer edges of the universe are believed to expand at a speed faster than light-speed." All such possibilities indicate that "***Faster-than-light speeds may be possible in Nature."***

As per the information available on the Internet, "Some scientists have developed a piece of equipment, which is capable of transmitting radio signals at speeds faster than light-speeds." ***Further, during November 2011, neutrinos were also found to travel at speeds faster than light.*** During the same period, that is, December 2011, MIT, USA, exhibited a camera *(Chapter-3, "Visualizing the Atom and the Electron Moving in its Orbit")* which can take photographs at a speed faster than light speed. The said camera points to a possibility, *"When a camera developed by mankind can work at the speeds faster-than-light, then* **Nature** *may also be capable of doing the same." **In case the aforementioned speculation of mine is correct, then the speed of light is not the limit; speeds higher than that may also be possible, at least, under certain conditions.***

In this context, one more fact needs consideration: As soon as a ripple is formed on the surface of the water, then the molecules of water start to rise in the form of a sine-wave, until they reach the crust of the ripple, thereafter, they start to move in a downward direction and reach the trough of the ripple, the first cycle of such a wave is completed when the water-molecules rise from the trough and reach the normal level of the water, thereafter, this cycle is repeated again and again; in this way, the wave so formed, continues to propagate on the water surface. This means that by the time the ripple propagates through the distance equal to one wavelength, the molecules of the medium are displaced by a comparatively much larger distance. In other words, the speed of displacement of the medium is greater than the speed of the wave propagating through it. The same is also true for the light waves, too, and the displacement of the medium of space-time. In case this proposition of mine is correct, then this prediction of ***Relativity would go wrong*** because, during the process of the formation of the light-ripple, space-time would be displaced at a speed much higher than that of the speed of the light. Thus, the very idea of propagation of light, in the form of waves, would defy the prediction that nothing can move at a speed that is faster than light speed.

(10) Relativity and Propagation of the Energy Waves

General Relativity envisages that "All kinds of energy waves propagate rippling through the fabric of space-time;" such waves are considered similar to water ripples that move on the two-dimensional surface of the water. This idea is accepted worldwide *because it is said to successfully explain "How the waves propagate in space and how they bend near massive bodies."* Despite the worldwide acceptability, this concept doesn't give a clear picture of **"How light propagates in all three dimensions or how the spherical waves would undulate while propagating in three dimensions?"** Further, the fact that *"The light waves propagate in all three dimensions"* very clearly tells that the prediction *that space-time is spread only in two dimensions* is unrealistic because **propagation of light in three dimensions is not possible in a *two-dimensional fabric.*** This fact clearly suggests that **"Space-time (if at all it does exist) is not like a two-dimensional fabric, it should permeate in all three dimensions."** However, in that case, four-dimensional *space-time shall not be able to warp beneath the stars or the planets like a two-dimensional fabric*. And, in case *the fabric of space-time won't warp, then it won't be able to compel different planets to orbit their respective stars*.

In the above context, it is worthwhile to point out that some of the possible anomalies in the concept of propagation of light were highlighted in chapter-4 of this book; in addition to the said anomalies, a few more anomalies in the conventional concept of propagation of light, are discussed below: *All such anomalies put a question mark on the very existence of the fabric of "space-time."*

Before we proceed further, a brief introduction to the existing concept on the formation of the black holes seems necessary. (Kindly see chapter-11 for further details).

Scientists believe that "When a massive star collapses due to its own gravity, then the curvature of space-time, in the immediate vicinity of such an object, begins to increase gradually; finally, such an object suddenly shrinks to zero-size; as a result, the curvature of the space-time surrounding that shrinking star becomes infinite." This concept seems to contradict the prediction, "Gravitation is not a force;" accordingly, "If the mass of the collapsing star remains unchanged, then the curvature of warped space shall also remain unchanged." However, scientists believe that such a star will continue to shrink as if a sort of chain-reaction is set up in which the mass of the star causes space to warp, and in turn, the warped space squeezes the star so that it persists to become more dense; and as a result, the curvature of warped space increases gradually. This increasing curvature causes the star to shrink continually until the black hole is formed; however, setting up such a chain reaction seems impossible. Moreover, the infinite curvature of space necessitates that "Both the circumference and the radius of the indentation formed in space around such an object shall necessarily become zero in size." The "zero sizes" of such an indentation would mean that **no such indentation is formed, in other words, no such indentation would exist in space-time**. *Anyhow, ignoring this possibility, scientists believe that such a collapsing star (the black holes under making) sucks in the fabric of space-time from its immediate surroundings so that a hole having infinite curvature is ripped open in this fabric, all around the newly formed black hole. Scientists further believe that the light rays that happen to pass through*

169

*such a hole are trapped within the infinite curvature of space, where such rays keep moving in a closed loop and keep hovering in the said hole forever, i.e., within the region of space of the infinite curvature. The outer edge of this indentation made beneath the black hole is known as the "**event horizon**;" even the medium of space-time doesn't exist inside the said* **event horizon. In contradiction to this belief, the supermassive black holes practically catch light from a distance of several million kilometers. This fact indicates that neither the diameter of such a big circular zone of space can have zero value, nor this region of the warped space, having such a large radius, can have infinite curvature.**

Although scientists believe that "The medium of space-time doesn't exist around the black hole." even then, its magnetic force finds the way out to spread beyond the event horizon; this fact **contradicts** the concept that a medium is necessarily required for the spreading of magnetism. This fact further suggests that "The medium of space-time is not necessary for the propagation of the virtual photons of magnetism." In case virtual photons of the magnetic field emanating out from the poles of any black hole don't need the medium of space-time to spread all around, then this should be possible for the light-carrier photons or the real photons too. And if this is possible for light, then the so-called waves of any other kind of energy, too, shall not require the medium of space-time for their propagation. Because of this possibility, an alternate mode of propagation of light, without any medium (as was discussed earlier in Chapter-4), shall be thoroughly examined at the highest level. The secret behind the *duality of particle and wave nature of energy-radiation,* as was proposed in the said chapter-4, shall also be scrutinized thoroughly.

In case predictions of *Relativity* that 1) *light waves propagate through the fabric of the space-time,* & 2) *the said fabric of space-time is itself expanding at almost the speed of light are both correct,* then the speed of the light shall be doubled in the direction of expansion of the universe because whatever distance the light will travel in the direction of expansion, space-time will also expand by the same amount in that direction. On the other hand, whatever distance the light would travel in the opposite direction, the expanding space-time would drag it back to the point of its origin; it won't be able to propagate in the direction opposite of the expansion of the universe; it will become stagnant in this direction. Contrary to this possibility, the light spreads in all directions at the same speed. *The fact that light moves spherically in all the directions at equal speed indicates that either the light waves do not ripple through the space-time, or else the space-time is not expanding, as is believed.* Both of these assumptions of the Relativity cannot be simultaneously correct, and therefore, this point needs thorough scrutiny.

Scientists believe, "Three-dimensional space-time adopts the curvature of the earth, at least, in its near vicinity." Accordingly, if light waves really ripple through space-time, then light, traveling closer to the earth's surface, should follow the curvature of the fabric of space-time and propagate on a curved path that would be parallel to the earth's curved surface. Accordingly, light rays emitted from any event happening behind us shall, following the curvature of space, come back to our eyes after a very short time period; however, this never happens. The straight line of the vision of the living beings is solid proof that "Light **does not** follow the curvature of space-time;" we can't see the far-off objects due to the earth's curvature. Contradicting ***Relativity,*** the aforementioned fact tells very clearly that ***"Light, i.e., the shower of real photons, always propagates in a straight line; neither does it travel along the curvature of space-time, nor does it follow its curvature."*** *All the facts, which prove that the Earth is spherical in shape, also prove that* ***"Light travels in a straight line."*** *Moreover, this fact raises a question:* ***"Does the light really need the medium of space-time to travel, or does any such medium really exist?"***

Further to the discussion made in the previous para, since the fabric of space-time bears the mass of all the galaxies and other celestial bodies, it becomes necessary that this fabric should be under very high tension, which should be the same or fixed all over. In this context, a question arises, "In case the tension of the sheet of space-time always remains fixed all over, then how infinite numbers of different kinds of the energy-waves, having different frequencies, propagate through this sheet simultaneously and at the same time-instant, that too without distorting or interfering with each other?" This seems impossible to me because everybody must have experienced that in a complete pandemonium of a crowded hall, it is very difficult to hear and clearly distinguish, or follow, any particular sound. This fact points toward another possibility that: had different radiations coming from all the possible directions, been *waves, undulating in the fabric of space-time*, then it would have been impossible to differentiate all these radiations from each other and identify any individual radiation distinctly. ***This is possible only in the condition that was predicted in chapter-4, i.e., different kinds of energy-radiations do propagate in the form of fast-moving photons; each kind of radiation is a shower of photons that vibrate/spin at a particular frequency***. This *assumption of mine* shall be examined very thoroughly to find out the truth.

Scientists believe that all the celestial bodies are resting on the fabric of space-time, which holds them at their respective locations by pushing them up from beneath. However, in case any black hole really rips open a hole in this fabric, then nothing would be left beneath such a black hole to support its enormous mass. This possibility raises the question that ***"How do the black-holes keep hanging or*** *hovering in their respective locations; why don't they fall?"* This question points to yet another possibility that *"In case black-holes, which are several million times heavier (massive) than any star, can remain floating without the support of the fabric of space-time from the underneath, then all the other celestial bodies too shall be able to do so."* In that case, the very existence of space-time would become unnecessary. However, even under this possibility, different celestial bodies would still be able to move in their respective orbits under the influence of gravitational force (this possibility is discussed hereinafter in the forthcoming chapter-10, *"The puzzle of Gravitation,"*). This *speculation of mine* necessitates that *the feasibility of the existence of the fabric of space-time shall be examined very thoroughly.*

I feel *"Had gravitation not been a real force that is independent of the curvature of space-time, then gas-clouds could not have collapsed to create different stars, nor the stars would ever end up in supernova-type explosions; similarly, the black holes could never have been formed."* The existence of the gravitational force seems to be a must for the creation of different celestial bodies as well as for the proper functioning of the Universe (please see the forthcoming chapters 8 to 10 and 15). In the absence of any binding force, the whole universe would have scattered, thinned out, and disappeared in a vast emptiness. *Moreover, persistently increasing acceleration of different free-falling objects suggests "A certain force is definitely acting upon them persistently."*

Contradictions in the predictions of Relativity

Recapitulating the foregoing discussion, the following *predictions of Relativity* seem to contradict either reality or the basic ideas of the *Theory of Relativity* itself: -

1. No *direct* or *physical* proof of the very existence of *Space-Time* is available; even then, the *Theory of Relativity* is totally based on its existence.

2. The prediction *"The ticking speed of time changes due to the change in the strength of the gravity field and/or the speed of the moving objects" defies reality.* Had this presumption been correct, then the electrons orbiting the atomic nucleus right from the inception of the universe (for the past 13.7 billion years) would have moved ahead in the time in the reference frame of the atomic nucleus; accordingly, the structure of matter would have destroyed long back, but in reality, nothing like this has ever happened.

3. The prediction *"The ticking speed of time totally stops at the speed of light"* defies the reality; in case the ticking speed of time really stops at the speed of light, then the light, too, shall stop moving because *no activity can happen in the frozen time. On the other hand, light takes a time period of billions of years to reach us from the furthest galaxies; this means that time didn't stop for even light.* Further, the prediction that *"The objects moving with very high speeds, move ahead in time, i.e., they move into the future,"* is incorrect because those rays, which have not yet been emitted, cannot reach our eyes in advance; accordingly, all the fast-moving objects would have become invisible because light from a future event can't come to our eyes; however, this never happens in reality.

4. The fact that *"M.I.T, USA, in the year 2011, developed a camera, which can take the snapshots of even the moving light-waves"* indicates that the prediction that *"Nothing can move at speed faster than light"* is not correct. This conclusion is supported by the fact that *neutrinos were found in 2011 to move at speed faster than light speed.*

5. On the one hand, Relativity predicts: *"The far-ends of the fabric of space-time, and the galaxies that are situated at those locations, are expanding at the speed of light, or rather, at faster than light speed,"* on the other hand, it predicts *"Nothing except the mass-less energy waves can move at light-speed."* Both of these predictions contradict each other.

6. The aforesaid prediction regarding the expansion of the fabric of space-time (Universe) at faster than light speed contradicts another prediction "The mass of moving objects that are moving at the speed of light, becomes infinite," whereas all the celestial bodies in the expanding universe, whether big or small are though supposed to move faster than light-speed, have definite masses, thus the later prediction violets reality. The aforesaid fact clearly indicates that the Universe is either not at all expanding, or it is not expanding at the speed of light.

7. The prediction "The far ends of the fabric of space-time are expanding at faster than the speed of light" yet again contradicts the prediction "Light waves propagate rippling through this fabric" because in such a case, light, while propagating in the direction of the expansion of the universe would propagate at the double of its normal speed, whereas, in the direction opposite of the expansion, it won't at all be able to

propagate. This prediction contradicts reality: light propagates in all directions at the same speed.

8. The prediction that "The Universe, including all the galaxies and all the stars, is spread over the two-dimensional fabric of Space-Time" contradict reality because all the galaxies and all the stars are not spread only in a two-dimensional plane; they are, in fact, spread in all three dimensions. The fact "Light propagates in all three dimensions" clearly indicates that space-time is not spread like a two-dimensional fabric; instead, it is spread in all three dimensions. This fact also contradicts the prediction that fabric of space-time warps beneath massive objects because the three-dimensional space cannot warp like a two-dimensional fabric.

9. The prediction that "In the absence of any force, all the celestial objects move as if they are falling freely." clearly contradicts the law that "Nothing can move until and unless some force acts on it." Further, Relativity is unable to explain "How such objects, in the absence of any force, will move at a definite speed in a definite direction?"

10. The prediction that "Different planets orbit their respective stars because curved space pushes back them and thereby compels them to follow its curvature" contradicts the fact "No effect of the expansion of space-time is felt on the size of our solar system." This fact indicates that "Either the universe is not expanding; alternatively, space-time is not capable of pushing the planets, and thereby compel them to orbit their respective stars." In turn, this fact indicates that Gravitation is an independent Force.

11. The Fact "Different planets orbit the sun in almost the same plane" indicates that "Neither the planets move along the surface of curved space nor they push it; as such, curved space cannot compel all the planets to follow its curvature." This fact clearly indicates that "The Gravitational Force is not a consequence of the warping of Space-Time."

12. The presumption "Planets, while moving through space, lose their energy by way of emission of gravity waves; they shall, because of this reason, gradually spiral toward their stars" contradicts another presumption "The planets orbit their star because they follow the curvature of space." In case the second presumption is correct, then the planet shall always

move in a fixed orbit because the size of the indentation cast by their star will not change due to the depleting energies of the planets orbiting them. Therefore, the planets would orbit the respective star at gradually reducing speeds; eventually, when their energies would be exhausted completely, they would stop moving instead of spiraling inwards.

13. The prediction that "Gravity is resulted due to the warping of space-time" contradicts reality because in such a case, no micro-gravity condition shall exist anywhere within the interior of the indentation formed in space. Moreover, in that case, the gravitational force cannot have uniform strength everywhere on the surface of the Earth and other planets.

14. The high/low tides, which are produced on the seawater, as well as on the solid surfaces of the moons of Jupiter, clearly indicate that "Gravitational force doesn't result due to the warping of space" because any deep-well, which is made in space, can never cause the seawater or the solid surfaces of the moons of the planet Jupiter to rise and recede (fall back) alternately. This prediction clearly defies reality; "No such effect is possible in the absence of a force; merely an indentation formed in space, cannot produce such an effect."

15. The fact "Light deflects only from the close vicinity of the Sun" indicates that even the most-massive objects like stars can warp a very small portion of space, that too, at a very close distance from them, this means that the indentation made by the sun must be very small and shallow. Contrary to this conclusion, the fact "Planet and other distant objects like the Oort-cloud objects (please refer to the annexure-1) orbit the sun from a far-off distance"; indicates "The Sun must create a very huge and deep indentation in the fabric of space-time." This contradiction in these two predictions regarding "deflection of light" and "The size of the warped space that is supposed to produce the effect of gravity" clearly indicates, "Either both of these two predictions or at least one of them shall be wrong."

16. The fact that "Light propagates in all the three dimensions" in conjunction with another fact "Light does not follow each and every curvature of space" indicates that either light does not propagate rippling through the fabric of space-time or no such fabric does exist. Moreover, in case

the light really follows the curvature of space, then light, while moving straight toward a black hole, should deflect in the direction opposite of the gravity-center of any black hole instead of bending toward the black hole.

17. Our straight line of vision, along the Earth's surface, clearly indicates that either "The light does not follow the curvature of space-time formed around the Earth" or "The fabric of space-time doesn't warp due to the mass of the Earth; alternatively, no such fabric does exist."

18. The prediction that "Light-waves propagate through the medium of space-time" contradicts two other predictions of Relativity: 1) "The fabric of Space-time warps beneath massive celestial objects." & 2) "Gravity is not a force like other forces; it is merely a consequence of warping of space." This proposition of mine is supported by the following fact: In case "The light really ripples through space-time, as well as space-time warps beneath celestial objects," then the dark and opaque planets would not obstruct the path of light waves coming from the other side of such planets because the light-wave coming from the other side of the dark planets should ripple through the warped space and propagate from underneath such a planet without any restriction, and similarly, the light emitted from bright stars would never reach us because the light emitted from any bright star, would, because of the warping of space-time, not get any medium of space-time to ripple through, and reach us. This consequence of relativity clearly violates reality; this fact clearly tells that neither the light propagates through the medium of space-time nor any warped indentation is formed beneath any celestial object.

19. The fact that "We can very distinctly differentiate between the different kinds of *energy waves from each other such as the radio waves to the gamma-rays; moreover, we can easily identify a particular wave among enumerable waves propagating from all the directions simultaneously, through the same medium"* indicates very clearly that *light and all other sorts of energy waves do not propagate rippling through the fabric of space-time.* In case different kinds of energy radiations have been waves, then such differentiation would not have been possible. *This fact, in turn, indicates that neither the so-called energy waves propagate in the form of waves nor any medium is required for their propagation.*

20. Relativity predicts that different objects, in the absence of any force, move in space as if they are falling freely; however, different nebulae have different but distinct shapes; neither such nebulae move like free-falling objects, nor the gas and dust particles within them, are moving like freely falling objects. This fact clearly indicates that the aforesaid prediction of Relativity violates reality.

21. While a light-wave would propagate undulating up and down, through the two-dimensional fabric of space-time, then as illustrated below, it would certainly and continually displace this fabric at higher than light-speed to form the sine-waves of light. This possible displacement of the fabric of space-time, at faster than light speed, defies the prediction "Nothing can move faster than light speed."

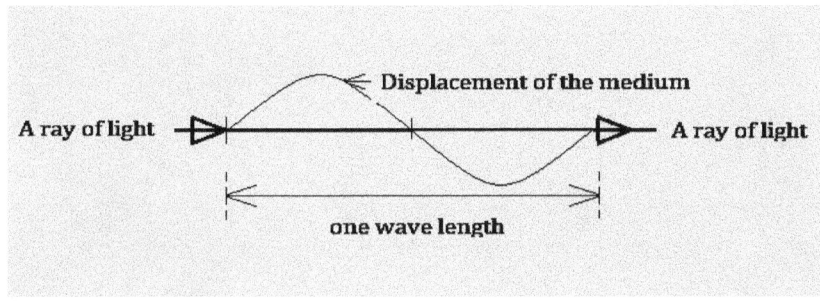

7: The String Theory

By the mid-1900s, scientists started to believe that everything, every phenomenon, as well as every functioning of the universe, could be understood and explained by *Relativity* and/or **Quantum Mechanics,** which was devised in conformity to *Relativity*. However, these two theories seem to differ on the following point:

Quantum Mechanics predicts that all the activities of the universe are governed by the natural or fundamental forces that are: *Gravitation, electromagnetism, Strong Nuclear-force, and Weak Nuclear-force*; scientists believe that all of these forces are activated by the emission and absorption of the respective force-carrier particles. On the other hand, *Relativity* predicts that *Gravity* is not a force like other forces; it is merely a consequence of the warping of space-time. So far as this point is concerned, both of the aforementioned theories are incompatible.

The aforesaid situation greatly perturbed physicists because they had to remember two independent sets of mathematics to describe the physics of the universe. Especially, they were unable to describe the physical processes in the situations when the fundamental forces had to battle for dominance, i.e., *at the time of the birth of the universe or in the center of the black holes*. They, therefore, started a search for a **Grand Unified Theory** based on a single set of mathematics that can describe all of the forces.

By that time, scientists did discover that while the heavy particles are smashed head-on, in a high-power particle-accelerator, then various fundamental particles such as *the* **Quarks, Leptons***, and force-carrying* **Bosons***, etc.*, are produced. This discovery provoked physicists to ask – **"With what stuff these fundamental particles are made of?"**

To resolve both the problems discussed above, the then physicists, based on very complicated mathematics, started to find out the relation between the vibration-pattern (wave-nature) of various particles (matter-waves), and their properties. This effort of the physicists led to the formation of the **String Theory**, which envisages that *"All sorts of the particles are actually the loops made of one-dimensional tiny string-like stuff; these strings, because of their intrinsic properties, always persist in vibrating in the* **multi-dimensions.***"*

A brief overview of the String-Theory:

This theory envisages that the whole universe, that is, all the known and unknown things, all particles, whether real or virtual, space-time, light and all other types of energy, etc., are believed to have been made of oscillating *string-like* objects, which do not have any dimensions other than the length. The universe is supposedly comprised of a web that is formed by these minuscule strings. Particles are now supposed to be **waves** that travel on these strings. Different types of particles are thought to be a collection of the nodes (points of zero amplitude) formed on these vibrating strings. According to this theory, these nodes are formed at such places, where two threads of the said web meet each other. These strings are truly tiny, many billions of times smaller than an individual proton that exists within an atomic nucleus; these strings vibrate at countless billions of times per second, probably in ten dimensions of space or maybe in eleven or more dimensions; there may be as many as twenty-six or even more dimensions.

The **String Theory** envisages "Instead of having different types of the elementary particles, we have a single type of one-dimensional object known as the fundamental or elementary string, which is the basic constituent of the matter as well as that of all other entities such as the light, all sorts of energies, and space-time, etc. *The very same string, when oscillates in a different manner and in a different dimension appears as a different particle.* In other words, different vibrational states of the same string appear to us as different elementary particles, or energy waves, etc." This is similar to a musical instrument in which a single string, when vibrates differently, produces different musical notes. Similarly, any change in the frequency of these strings results in the change of places of the nodes and also their count per unit length; as a result, the same string is converted into another kind of particle. Possibly, a very special vibrational state of the said string may have properties similar to the hypothetical *"Graviton."* This theory also envisages "Two pieces of the strings, due to their inherent properties can join together to form a single string, alternatively, a single string can divide or split into two different strings." This theory is though very tough and complicated for the common man to understand; scientists believe that anything can happen in the *world of particle physics*.

These theories now have 5 different versions in which the strings can perform different types of vibrations. Scientists envisage that the total numbers of the dimensions are not limited to only 3 or 4; instead, there may be as many as 9, 10, or many more dimensions; even as many as 25 to 26 dimensions are possible. Different scientists believe that possibly, the gravitational force while traveling through all these 26 dimensions, dilates to such an extent that it loses most of its strength; this might be the reason that the *gravitational force* becomes so weak.

In the big bang theory, which is described in the next chapter, it is not clear how the universe did begin from nothing or zero, string theory has tried to remove this anomaly also. Before the formulation of this theory, the universe was thought to be only one and single, but the string theory envisages *multiple universes,* having *multiple dimensions*. Such a constellation of different universes is called the *"multiverse."* These multiple universes are supposed to be like different slices of bread, which are known as *"membranes"* or simply *"branes."* Scientists now believe that the collision between two or more "branes" may result in explosions similar to the big bang. Such an explosion did not happen only once in history; such collisions might have been repeated several times, after regular intervals; explosions similar to the big bang are supposed to be regular activities of the multiverse.

Some Clarifications Needed on The String Theory

The String Theory has been formulated with hypothetical 9 or even more dimensions to maintain conformity with different theories such as *The Theory of Relativity, Quantum Mechanics, and the Standard Model of particle physics, etc.* Scientists believe that multiple dimensions might really exist; any single line, which has zero breadth and height, can be considered as a separate dimension; however, because of the very compact and near-zero size of such lines (dimensions), we fail to cognize or take notice of them.

It is very hard to understand "What these extra dimensions are?" All the distances in all the directions can be measured in the known three dimensions. Every country on this globe has different cities; every city has different buildings, and every building has its own set of measurements of length, breadth, and height. However, all these measurements, which can be taken at different places, fall within the known three dimensions; it is very hard for a common man to imagine any other dimension except that of the *time*. Scientists believe that perhaps the "Large Hadron Collider can discover these extra dimensions," however, so far, no extra dimension could be found.

Until the existence of these multi-dimensions is not proved, this theory will remain only an unproven hypothesis, not a fully established theory.

Scientists have many expectations from this theory because it can provide solutions to many such problems that have, up till now, remained unresolved. However, to enable the common man to understand this theory, the following points need a little more clarification:

1. *What these extra dimensions are? And why do they affect the gravitational force alone; why not they affect any other force as well,* or at least the electromagnetic force, which too, like the gravitational force, permeates up to very long distances?

2. *With what stuff are these strings made of? And where from these strings get the energy to keep vibrating continuously?* If they are made of energy, then how could the energy be confined in any particular place; why does it not dissipate or flow to some other place of lower intensity?

3. Since different types of the energy-waves vibrate at very high frequencies, these strings should have very high tensions. These strings might be of two types; the first type may have open ends, and the second type of the strings may be of closed-ended type. Now, a question arises *"Where the ends of these strings are tied to, and how such a high tension is produced in both the types of these strings?"*

4. In case a particle, due to the change in its frequency, may really be transformed into another particle, then *how the tensions of these strings are adjusted or varied to produce the new frequency?*

5. In case the very same string can be transformed into different particles, then how is it possible to transform their masses and different charges too; i.e., *how the electrical charge and the color charges, etc., are transformed into another kind of charge?* For example, electrons and protons, apart from the different masses, differ very widely from each other in their electrical charges as well. *In such a case, how is it possible to transform an electron into different particles, that is, sometime in quark, sometimes in a proton, and sometimes in an antielectron, etc., by merely altering its frequency?*

6. Different particles, depending on their energies, are supposed to have different masses. In case any particle is transformed into another particle of higher or lower mass, then its energy should either increase or decrease. This means that such particles would either gain energy or lose it. The law that *"Energy could neither be created nor be destroyed"* puts a question on this concept *"Wherefrom such particles gain extra mass, or what happens to the mass that they lose?"* This is contrary to the law of conservation of energy.

7. When two particle-beams are made to collide, then different types of new particles are produced. Some of such new particles immediately decay and produce some other elementary particles. The fact that new particles are produced only after the collision, not at the time when their energy was increased, indicates that *"It is the energy of the collision that produces new particles, not the higher energy imparted to such particles."* This point needs clarification.

8. It seems from the above probability that some of the energy of the collision might have momentarily transferred to the colliding particles. Due to this reason, *"The same string might start to vibrate in a different manner instead of vibrating at its natural frequency. As a result, the same string might behave like a new kind of particle; thereafter, when this extra energy is released, then the same string might transform back into the same or any other elementary particle."* This idea needs thorough verification.

9. In case, after such a collision, different particles of more than one kind are produced, then new strings should also, compulsorily, be produced. This possibility raises the question that *"Wherefrom these new strings are produced?"* This contradicts the law that *"Nothing can be produced from zero or nothing."* A similar question also arises, *"After the decay of any such particle, where the strings of the particles so decayed, disappear and how?"* This, too, is against the law of conservation of energy.

Scientists might have the answers to all of the above questions, however, with the aim that the common man too, may understand this theory, answers to the above questions shall also be included in this theory.

Some Alternate Possibilities:

Different scientists have repeatedly observed that when subatomic particles are made to collide at high energies, then different kinds of new particles are produced. The particles so produced, include some known elementary particles, as well as some rare particles like the muons, mesons, positrons, etc.; these newly created particles decay after a very, very small time period, or rather, almost instantaneously. In this context, the law that *"Particles can neither be created and nor destroyed"* poses a few questions *"Wherefrom such new particles did get produced?"* And *"How do they get destroyed?"* On the other hand, the string theory envisages that if a state of vibration of a string is altered, then such particle is converted into a new particle. I feel that in case this is true, *then the energy of the collisions might be momentarily absorbed by the colliding particles, due to which the natural order of fluctuation, that is, frequencies and amplitudes of the strings comprising these particles, might be altered. This change of frequency might, in turn, convert the colliding particle momentarily into an entirely different particle. A short while thereafter, when the energy of the collision departs in the form of a photon or any other particle such as neutrinos, etc., then these particles might be converted back into the original particles. Similarly, the energy of the collision might result in the splitting of the composite particles into their constituent sub-particles* (Chapter-6: under the **Rossi and Hall Experiment**). This possibility shall also be examined carefully.

There may be yet another possibility: As proposed by me earlier, in Chapter-3, the so-called elementary particles might also be made up of two or more constituent particles, which may be hundreds of times smaller than the so-called elementary particles; these tiny constituent particles might possess the charges of opposite nature. In such a case, the constituent particle having the smaller mass might orbit its heavier (more massive) partner because of their mutual attraction. In such a case, subject to the condition that this proposition of mine is correct, whenever such particles are made to collide in a particle collider; the speeds of the orbiting constituents of such particles may increase manifold due to the energy of the said collision; moreover, the normal orbit of the constituent orbiting the heavier one of such a particle may become elongated, rather than highly elongated and elliptical. In case this notion of mine is correct, then the mass of such an elementary particle, after the said collision, might ostensibly seem to increase in the proportion of the square of the increased speed of the orbiting member. As a result of such an increase in the speed of the revolving partner, the angular momentum of that particle would also increase; consequently, it would offer greater resistance to any effort to change the direction of its motion or to increase/change its acceleration. In such a case, though the quantity of matter (mass) in such a particle will remain unchanged due to that its weight produced by the force of gravity would also remain unchanged; however, it may ostensibly seem as if its mass has increased manifold; later on, when the speed of the revolving component of that particle and its orbital speed/path will become normal, then its behavior will also become normal.

It is necessary that this proposition of mine shall be verified very thoroughly.

PART– 3(A)

Mysteries of
The
Universe
&
The way it Functions

8: The Universe at Work

The meaning of the word "Universe" is probably known to everybody; however, some of the readers may not know how big it is? The entire ambit of the universe or its total span could probably not yet be seen. We have, by very powerful telescopes, seen the galaxies that are spread over distances ranging from 13 to 14 billion light-years. Up till now, about 200 to 400 billion galaxies have been seen in the known universe. Out of all these galaxies, almost every galaxy contains several hundred billion stars. Mankind, despite such a vast expansion of the universe, didn't know till 1924 that other galaxies/stars also exist beyond our home galaxy the Milky Way; before that period, the Milky Way was, naturally, considered the entire universe.

In ancient times, all the stars, except for the planets, were considered stationary. Afterward, people realized that while the earth moves ahead on its path, some of the stars, which are located comparatively nearer to us, appear to slightly change their positions relative to the distant ones. This is similar to the phenomenon of looking out of a moving train; when we peep out of the window of a running train, then the nearby objects such as the ground down below and the nearby trees, etc., appear as if they are moving with very high speed in the opposite direction, whereas the far-distant objects appear almost stationary. The aforesaid apparent displacement of the said moving stars enables scientists to measure their distances relative to us.

Our Milky Way is, in fact, a medium-sized galaxy wherein about 200 to 400 billion stars are distributed in the shape of a disk. This disk of stars measures about 100 to 125 thousand light-years across in diameter. On the other hand, the thickness of this disk is about one thousand light-years only, that is, about $1/100^{th}$ of its diameter, or even less. In the beginning, probably based on this fact, the universe was believed to be two-dimensional in shape; however, contrary to this belief, the universe is probably, spread in all three dimensions.

The universe comprises approximately 400 billion galaxies; many galaxies, out of them, are much bigger in comparison to the Milky Way. The biggest galaxy, seen up till now, is about 60,000 times bigger than the Milky Way. All the stars contained in the Milky Way are not distributed evenly in a totally flat disk, this disk bulges out in its center like a ball. All the stars in the Milky Way are not equal in size; some of them are bigger than our Sun, whereas some are smaller than it. The biggest known star is almost 2 to 3 thousand times bigger than the sun; even much bigger stars may exist in such a vast universe, in one or the other galaxy. As against the vastness of the universe, we, with the naked eyes, cannot see beyond a few hundred light-years. It may be inferred from this fact that "When we can't even see the entire Milky Way, then how is it possible to see the entire universe, which is billions of times bigger than the Milky Way."

Earlier, we have seen that none of the stars or the galaxies in the whole of the universe is stationary. The Milky Way, containing several hundred billion stars, is rotating very slowly around the center of the Milky Way, which is known as the galactic center. The stars located in the middle part of our galaxy, including our Sun, orbit the galactic center at an amazing speed of almost 250 kilometers/second. However, our galaxy is so big that despite such a high speed the Sun and all other stars take a very long time period of about 250 million years to go once around its axis. Besides the rotational speeds of different galaxies, the entire universe comprising all the galaxies is supposed to expand continuously at near-light speed. We, the living beings living on the Earth, are also moving at the same speed; however, we are unable to feel it. That's why we think that "We are stationary, that is, the earth is stationary, or it is at a position of absolute rest, though it is not."

Scientists all over the world have observed that stars located very close to the galactic center of any spiral galaxy, seem to rotate with much higher speeds as compared to the rotational speeds of the stars located near its outer edges; therefore, they seem to trail behind the stars located nearer the galactic center. Probably, this is the reason that it is predicted in **Relativity** that when distances of different stars would increase from the center of any spiral galaxy, the rotational speeds of the stars of that galaxy would gradually drop out. However, scientists, on actual measurements, have found that, contrary to the above prediction of **Relativity,** the rotational speeds of the stars follow almost a flat pattern; when the distances of the stars, from their respective galactic center, increase, they maintain almost the same speeds, rather than their speed increases slightly. Probably this is the reason that all the stars appear to maintain almost the same *relative locations with respect to each other*.

In case our Milky Way is seen from the above, from a distant point, then it would look like a spiral of stars, as shown in Fig-4 "A." An imaginary cross-section of a spiral galaxy is shown in Fig-4 "B," which shows that the stars, located very close to the galactic center, might orbit the galactic center with all the possible angles. Side-elevation of a spiral galaxy is also depicted in Fig-4 "C." However, since the spiral arms of any galaxy are much thicker and dense in the center as compared to their edges, their cross-section, if seen from one side only, might look like a wave, as is depicted in Fig.- 4D.

Fig. - 4A

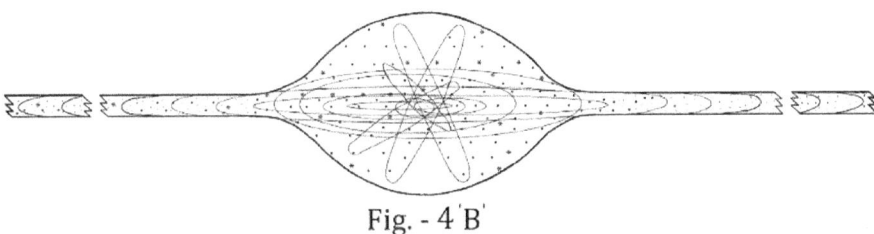

Fig. - 4 'B'

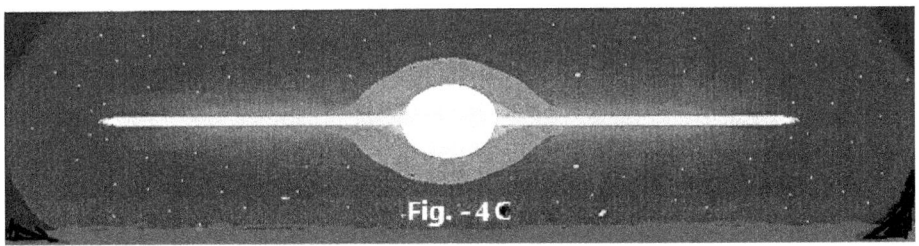

Fig. - 4 C

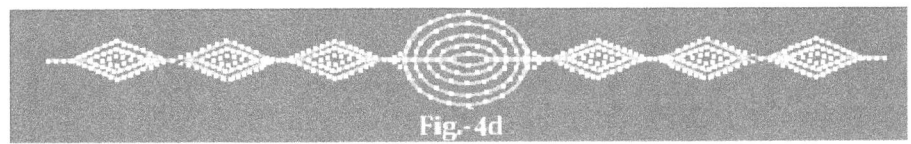

Fig.-4d

All the stars are so far away from us that they look like point-sized shining objects. Wherever we look at the night sky, it appears all the same in all directions; small bright spots are seen spread all over the sky. The galaxies are located at such far-off distances that it is not possible to even see them with naked eyes. Despite this limitation, some astronomers, who used to observe the sky with only the most primitive telescopes, predicted during the middle of the 18th century (around 1750) that all the visible stars might lie in a single band; this band is now known as "Milky Way." The then-scientists also predicted that this band should appear like a flat disk. However, it took almost 150 years to accept this idea fully. Since the different stars located at equal distances from the center of the galaxy rotate with almost equal speeds, it is difficult to make out the difference between their speeds; they seem to be stationary relative to each other, however, after long and patient observations, the stars located nearer to us may be seen to move in the background of the distant stars. Based on the amount of the said displacement of the nearby stars, astronomers work out the distances of the furthest stars.

The ancient philosophers believed that the universe always remains in a steady state; nothing ever changes in any corner of the universe. But, contrary to this belief, the universe is a very active place; on the one hand, numerous new stars are regularly created at different places, and on the other hand, different stars at different places end up their lives with very powerful, or rather violent explosions. Sometimes a massive star, or a black hole, swallows another star, and at another place, a star, after running out of its fuel, collapses into a small but much denser object. At yet another place, two galaxies may merge into each other as a regular feature of the universe. In fact, the entire universe is the workplace of Nature where numerous new events go on regularly, but at different places. This means that enormous changes take place everywhere in the universe at different places; the phenomena of *creation and destruction* go on side by side. However, every new event takes place in such a far-off place that we can't see or even imagine what is happening all around. Since we can't see or take notice of the changes that are going on at different places, the universe falsely seems identical in all directions. In fact, the universe is neither eternal nor immortal; these days, scientists believe that *"In the remote past, the universe started with a very powerful explosion known as "big-bang," and since then, it is expanding at a very high speed." Astronomers also believed that one day, the whole structure of the universe will shatter and defuse in space.* Different scientists during different times have devised different theories to describe the functioning of the universe. A broad outline of some of such theories is given below.

Background of the "Big bang" Theory

Probably, based on the fact that the universe appears identical from all the directions, *Russian scientist Friedmann,* in the year 1922, made two very simple assumptions: **(1)** *"The universe appears identical in whichever direction we look at."* & **(2)** *"This will also be true if we would observe the universe from anywhere else."* Both of these assumptions, in general, may seem to apply to the whole of the universe; however, in fact, they don't apply to any particular portion of the universe because *"Neither all the galaxies nor all the stars are identical; distances between them are also not identical."* Based on the fact that the universe ostensibly seems identical from every direction, it should not mean, *"The earth is at the center of the universe; scientists believe that since the universe is apparently identical from everywhere, it appears the same from the earth also;"* I, however, feel that there may be yet another reason for the same: *"Since we can see in all the directions only up to a limited distance, we always remain in the center of this limit of our vision; no matter from wherever we see the universe, we will always remain at the center of this limit of our visibility;"* however, we can't see *"How the universe looks beyond this limit, or how far it has actually expanded in different directions."* Despite this limitation, all the theories about the beginning and end of the universe are based on the aforesaid predictions made by *Friedmann*; therefore, such theories may not give a true or very accurate picture.

Our true understanding of the universe began in 1924, when astronomer, *Edwin Hubble*, focused a powerful telescope on an apparently vacant tract of space. To his astonishment, he found that the portion of space, which was thought to be empty, is actually full of a host of other galaxies. In 1929, Hubble made yet another discovery that almost all the galaxies are moving away from each other at very high speeds*. This discovery made by Hubble brought a revolution that divided the world of science into two separate camps. One of the groups of scientists interpreted that the universe could not be static and also that it must have had a beginning sometime in the far-remote past. The other group, including even **Einstein,** firmly believed in the steady state. This controversy continued for almost the next 40 years. Finally, in 1970, Roger Penrose and Stephen Hawking, working

---

* In Chater-15, "Dark Energy," a possibility is discussed that Universe might not be expanding, instead, the frequency of light coming from the distances of billions of light-years, reduces gradually, therefore, the light coming from such a galaxy exhibits red-shift, not due to the expansion of the Universe.

together, proved that the universe began in the remote past with a powerful bang; that is the reason why it is expanding continuously. This theory, eventually, brought the *end* to the concept of the *"Steady-State."*

×   ×   ×

A renowned American Physicist **George Gamow,** who was of Ukrainian origin, suggested in 1948, "The early universe must have been very hot and dense; it would have been glowing with white-hot light" (having a white-hot color temperature). Thereafter, after a time period of about 8-10 years, **Robert H. (Bob) Dicke** and **Jim Peebles,** based on the aforesaid speculation, suggested that apart from the scattering the matter-particles, which later on condensed into galaxies, a tremendous blast of radiation also must have been released by the *"big bang."* *They further argued: "The light from very distant parts of the universe, would just be reaching us now, i.e., in our present era;" we may still be able to see this radiation in the form of microwave radiation due to the cooling down of that glow.* The duo of Dick and Peebles, therefore, started preparations to find out any such residual or relic radiation.

Almost at the same time, i.e., in 1964-65, the duo of **Arno Penzias** and **Robert Woodrow Wilson,** who were unaware of the aforesaid prediction of **Dicke and Peebles,** accidentally discovered radiation of microwave that seemed to come from all the directions. This radiation remained almost unchanged day and night and even throughout the year; it didn't vary even as the earth was spinning on its axis and orbiting the sun. In case a very powerful telescope is focused beyond the visible stars, then a very faint glow could be seen. Scientists, based on this fact, believe that this glow is coming from the *surface of the last scattering of the big-bang.* The *"Background Microwave Radiation,"* discovered by *Penzias and Wilson,* was unanimously accepted as the proof of the hypothetic *"Big-Bang,"* and as well as that of the predictions made by **Friedmann,** that *"The universe is identical in all directions."* This microwave radiation, jointly predicted by *George Gamow, Dicke, and Peebles,* was believed to be the residual radiation of the big-bang.

From the very beginning of civilization, both religious people and philosophers did make different imaginations about the moment of the inception of the universe. However, none of them had any proof to support his idea; therefore, it was logically not possible to predict *"How the universe did begin?"* In the present era, scientists have, at least, two clues that are: *(1) All the galaxies are running away from each other & (2) The residual background microwave radiation (discovered by Penzias and Woodrow Wilson) that always remains unchanged.* The present-day hypotheses of the **"big-bang"** and *"expanding universe"* both are, therefore, based on at least a few grounds.

Almost at the same period, i.e., in the year 1965, the renowned scientist *Roger Penrose,* based on *The Theory of Relativity,* proved that "A massive star while collapsing under its own gravity might get trapped into a region of zero surface area, i.e., it would shrink to zero volume and become a *Black Hole.*" In 1970, scientists *Stephen Hawking* and *Penrose,* working together by reversing the direction of time in the above theorem, proved *"In case the theory of relativity is correct, then the universe would have begun with a big bang."* This simply means *"The correctness* of the concept of *"big-bang,"* totally depends on the *correctness* of the *unproved postulates of Relativity."*

In the above context, I personally feel that *"Whenever any process is reversed, then we should know for sure that up to what limit it could be reversed."* If this limit is not known, then we would never know at what stage this process should be stopped. This may be understood by the example of the relation between the volume of the gases and their temperatures; when the temperature of a gas is increased or decreased by $1°$ C, then its volume also increases or decreases in the ratio of $1/273$ of its original volume. If a graph is plotted showing the above relation between the volume and temperature of the gases, then it would show zero volume at $-273.15°C$ or $0°K$. In case the graph is extended below this temperature, it would even show negative values. However, gases liquefy below a certain temperature; therefore, the volume of a gas can never become zero or negative. *This fact indicates that we shall not reverse a process beyond the actual limit. If we go beyond this limit, then the results so obtained might not be correct or realistic; particularly when the possibility has never been verified "Whether or not the energy of light, i.e., its frequency, might continually attenuate over a time period of billions of light-years." (Kindly refer to the Chapter-15, "the Dark Energy")*

Based on the aforesaid fact, it may be deduced that "When we don't know that at what point in time, the universe did start to expand, then how it would be decided that how much back in time, we shall go?" However, it is not possible to imagine what exactly did happen in the beginning? Whether the universe, at the time of its birth, was only point-sized or much bigger? This means that the said theory of the beginning of the universe is based only on assumptions. *Although I am not very sure about reality, I understand that a somewhat similar description of the creation of the universe is given in the "Rig Veda" too. Anyhow,* since the scientists didn't have any other clue, other than the *"expanding universe"* and the *"background microwave radiation,"* scientists were unable to devise a better theory.

A brief overview of the *"Big Bang Theory"* is given below.

The "Big Bang" Theory

This theory envisages, "Before the said Big Bang, infinite energy was concentrated in a very small point, which was even smaller than the size of an atom; both **space** and **time** were also believed to be encapsulated in the said point wherein they had infinite curvature." This point was supposed to be infinitely dense and hot, even millions of billion times hotter than the core of a supergiant star. Scientists believe that both **space** and **time** too began with that Bang. Because all the laws of science and mathematics break down at infinity, it is not possible to find out "What did exist before the *Big Bang,* or what was the reason for the beginning of the universe?" Nobody knows when and where the "big bang" did occur or what was the cause of this explosion; however, it is estimated that the said explosion did occur approximately 13.79 billion years in the past. Some scientists believe "The said explosion was stronger and louder than the blast of billions of hydrogen bombs exploding together, and some of them believe that it was much milder and quieter." Anyhow, the energy that was released from the said explosion started to expand at higher than light speed. Scientists believe that the energy so released did spread all over in all three dimensions like an expanding sphere. All the matter particles, and subsequently the universe, etc., were created within this expanding energy; it is believed that the outer edges of the universe have been expanding since then at higher than light speed.

I envisage that the energy so liberated could have expanded in any one of the following three ways:

1. Had all the energy been released in one sudden burst, then all of that energy would have expanded like the surface of a bubble or a thin-walled hollow ball. In that case, it should enclose a vast (empty) space measuring billions of light-years. Further, the opposite portions of the surface of such a ball would move away from each other with a speed more than twice the speed of light. Therefore, it would not be possible to see or receive any radiation from the opposite surfaces of this expanding bubble.

2. Alternatively, that point might have emitted energy for a few billion years, and thereafter it gradually ceased to emit energy. In that case, the universe would have formed like a hollow sphere having a thick wall and two distinct surfaces, outer and inner. By now, these surfaces might have become very faint due to continuous expansion.

3. As a next possibility, energy might have continued to ooze out of that point. Accordingly, newer waves of energy would have kept following the earlier ones, due to that the universe might appear similar in every direction, like the interior of a sphere. In case energy is still oozing out from that point, then we shall probably be able to see its source. However, to the best of my knowledge, no such source of energy has ever been seen.

In all the aforesaid cases, no radiation from the opposite walls of the expanding universe would ever reach the other end because their relative speed would be **double the normal speed of light.**

It is not possible to find out "In what manner did the universe exactly expand?" However, scientists envisage that initially, the universe and space-time both did expand together at a speed much faster than that of the light. The energy expanding at such a high speed might have cooled down very rapidly; accordingly, in about the trillionth of a trillionth second, it cooled down to such an extent that the said expanding energy started to convert into matter particles. This expanding energy was still so hot that the particles formed at that particular point in time consisted only of quark-gluon plasma. The particles so created were moving randomly in the shape of countless eddies that were whirling and swirling in all the possible directions. Such particles, due to their random speeds, must have been colliding with each other. It is believed that in the beginning, there was only one single force known as the *"Super-Force,"* however, within one billionth second after the big bang, the said super force was divided into four fundamental forces: *"strong nuclear force," "weak nuclear force," "electromagnetic force"* and the *"gravitational force." However, the newly created particles did possess very high energies, and they were moving at such a high speed that none of these fundamental forces were able to bind them together and create atoms.* Therefore, the particles within the expanding universe continued their uncontrolled and random motions; as a result, the temperature continued to reduce very rapidly; no other activity was possible at that very point in time.

Almost after one second of the big bang, the temperature of the expanding universe fell to about ten billion degrees centigrade, which was still about a hundred thousand times hotter than what exists in the core of the sun. At this temperature, some of the energy would have converted into very light particles like the neutrinos, photons, electrons, and their antiparticles. The types of particles that did produce at that particular point in time would have depended on the rate of fall of temperature. These particles, because of their random motions, must repeatedly be colliding with each other; the collision of particles with antiparticles would have resulted in the annihilation of both of them. This would have resulted in the production of more photons. Scientists believe that due to some unknown reason, the production of normal particles would have marginally outnumbered the antiparticles; otherwise, the universe would not have survived. ***Contrary to the above belief, I feel that nature probably knows that what exactly it should do under any particular condition*** (Kindly refer to chapter-3, "Particles and Antiparticles"). ***The difference between the numbers of particles and antiparticles produced at that time was probably not merely a coincidence; they, following the well-set laws of nature, were destined to produce and subsequently decay in that manner only.***

After about 100 seconds of the Big Bang, the temperature would have fallen to one billion degrees centigrade, which persists in the cores of the hottest stars. At this temperature, some particles like quarks would not have been left with sufficient energy to escape the attraction of the strong nuclear force. As a result, they would have started to clump together to create protons and neutrons, which, in turn, combined to create nuclei of some lighter elements like Hydrogen and Helium. At that particular point in time, the nuclei of some light elements like Lithium and Beryllium, etc., would also have been produced in very small quantities; however, it was not possible to produce much heavier elements. Production of such bare nuclei of different matters would have stopped within 3-4 minutes of the "big bang," however, the newly created electrons, because of their very high energies and very high temperature, would have been moving at such high speeds that the electromagnetic attraction of the newly created nuclei was unable to capture them to form atoms. Therefore, while the universe continued to expand, electrons continued to move freely, eluding the attraction of the newly created bare nuclei of different elements.

The universe continued to expand thereafter, for several hundred thousand years, without any further activity. During that time period, the photons produced within the expanding energy might have been interacting with the electrons and bare nuclei. Scientists believe that in that duration, the photons were coupled with the bare nuclei; because of this reason, the infant universe did look like a ***white-hot opaque fog.*** During that period, the newly created universe was not emitting any light because the photons were captured by the atomic nuclei; this period is known as the ***"Dark Age of the Universe."*** After about 379,000 years of continued expansion, the temperature dropped down to a critical limit of, say, 3000°C. At this temperature, the electrons were not left with enough energy to overcome the electromagnetic attraction of the bare nuclei; as such, the atomic nuclei did instantaneously capture the electrons and produced the atoms of very simple elements, mostly Hydrogen and Helium. Scientists believe that at this temperature, the photons would have decoupled from the nuclei and spread out in the form of a sudden flash of bright light that spread spherically. Scientists also believe, "The radiation emitted from the very hot early universe might still be around today in the form of photons, which would have cooled down to only a few degrees above absolute zero." Accordingly, these photons would now be reaching us as microwave radiation known as the *"cosmic background microwave radiation."* This radiation, reaching us from all directions, is thought to be the signature of the *leftover* or *relic* radiation emitted for the *first time* from the said white-hot fog. This relic radiation is considered indisputable proof of the big bang. Presently, the cooler temperature of this microwave radiation has dropped down to almost 2.7°K.

Scientists believe that this microwave radiation was emitted from a set of points in the space, which were located at such a vast distance that these photons, emitted at the time of photon-decoupling, are being received now. These sets of points are known as the *"Surface of the last Scattering."* In other words, this radiation, which is supposed to have been emitted about 13.7 billion years ago from the said "white-hot fog," is being received now, i.e., after covering the vast distance between the said surface of the last scattering and the existing location of our Earth.

Thereafter, the particles within the expanding universe, because of their random speeds, continued to collide with other particles, which caused these particles to coalesce together. Soon more particles would have started to accumulate in different places, due to which countless localized dense accumulations of the matter-particles would have formed in different places. Probably for the same reason, numerous clouds of gas and dust did gradually produce all around. In due course of time, these clouds might have grown bigger. Those clouds that were located closer to each other would have developed gravitational bonds, due to which bigger clusters of clouds would have formed in different places. However, those clusters of clouds, which were separated by very larger distances, might have developed numerous other groups of clouds. All such clouds formed at that particular point in time might have fragmented in yet smaller but denser areas due to the random motion of the particles. In the next few million years, the localized denser regions of these clouds would have collapsed under their own gravity and would have become stars. The universe is supposed to have been created in the aforementioned way.

My own view on the Big-Bang Theory

In case the universe was expanding at faster-than-light speed, then it would not have been possible for the gravitational force, which is supposed to move at the speed of light, to reach out to the particles moving ahead at the higher speeds. However, the gravitational force of the particles, which were moving in the front line, must have reached out to the particles following them from behind because the gravitation force of the front-line particles and the particle coming from behind were moving toward each other. As a result of the attractive force acting between them, the speed of expansion of the matter particles might have slowed down gradually.

It is not known *when, and exactly at what point in time, matter-particles were first produced or at what exact speed the universe was expanding in the beginning?* Further, since the entire matter is supposed to have been produced in the first 3-4 minutes after the *Big-bang, all the matter particles should have been produced within a very small region (3 to 4 light-minute); accordingly, gravitational force should have immediately pulled back all these matter particles and squeezed them into a zero-sized point.* However, defying this possibility, the universe did persist in expanding; this seems impossible.

It is also not known that "At the time of decoupling of photons, what was the exact speed at which the universe was expanding?" However, space-time was also expanding then along with the universe at a faster-than-light speed. In that case, the flash of light would have traveled at the same speed at which the expanding space-time was moving; therefore, it was impossible for the photons (light) to flash out of the expanding universe. On the other hand, due to a gradual reduction in the speed of the newly created particles, it might have become possible for those photons (light) to move ahead of such particles. Even in that case also, the photons couldn't have moved ahead of the expanding fabric of "space-time," which was moving along with the universe at a speed faster-than-light. Moreover, since the photons are supposed to move in the medium of space-time, they couldn't move out of this medium and propagate ahead of this fabric; they were bound to keep moving with the expanding universe. On the other hand, the said flash of light might not be able to propagate in the direction opposite of the expansion because the expanding space-time would have dragged back such photons back to the point of their origin, this simply means that such photons can never be seen from behind because they would not have been able to travel in the direction opposite of expansion of the universe.

The Probable Loose-ends of the Big-Bang Theory

The fact that "Photons always seem to come from the place of their origin" puts the concept of the "relic radiation" under question because *"In case this radiation was originated at the time of decoupling of the photons, then why this radiation doesn't appear to come from the point of its origin, that is, from a small region, instead of appearing to come from all around?" This fact indicates that this radiation is not coming from the point of de-coupling, which clearly means that the same is not the relic of the "Big-Bang."*

Because of this anomaly, the probable loopholes, which might exist in this theory, are discussed below:

1. Unless the possibility of the attenuation of the energy of the light coming from the far-off distances, due to the lapse of very long time periods, is not studied, the authenticity of the ***Big-Bang Theory*** cannot be ascertained (please see Chapter-15 for the details).

2. In case the universe expanded like a thin-walled sphere, then the light, in all directions, shall appear to propagate on the outer wall of this sphere. Accordingly, the light moving from any point on its curved surface shall appear as if it is moving on a two-dimensional sheet. However, since the light propagates in all three dimensions, not on a two-dimensional sheet, the universe might not have expanded like a thin-walled sphere, though as shown below, it must have expanded from the point of ***Big Bang*** as a center. The Earth was created within the expanding Universe at a much later point in time after the Big Bang. In case, at the time of emission of the said first glow, the speed of expansion of the universe was lesser than the speed of light, then that glow might have moved ahead of the future location of the Earth before its creation. Since the photons always move ahead in straight lines, it shouldn't be possible for them to come back to the earth after its birth; therefore, we shall never be able to see that so-called first glow. On the other hand, in case the infant universe at that moment was moving at faster-than-light speed, then that glow would not be able to reach the end of the universe that was moving at a much higher speed. Afterward, when the universe would have slowed down, then, after some time, that glow would have

overtaken the speeding Earth. Only at that moment, the said first glow could have been seen, that too, for a very short moment. *However, in that case, it should appear to come from the point of its origin, not from all around.* Anyhow, after that moment, the said glow would have moved ahead of the Earth that was moving at a comparatively slower than light. Therefore, we should never be able to see it again; it just can't keep hovering on the earth forever. *In this scenario, the microwave radiation, coming from all directions, shall not be considered as the "relics of the first glow."*

3.   The probable way in which the Universe did expand is depicted in Fig-4E, below: -

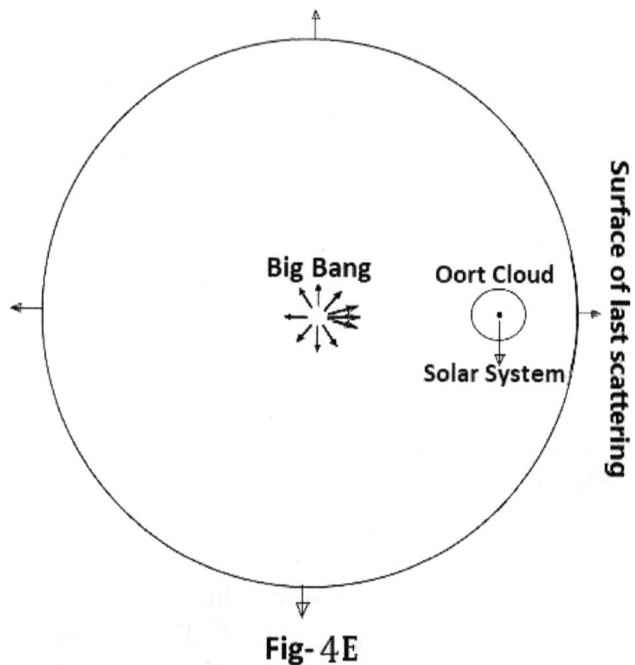

Fig- 4E

About 13.8 billion years in the past, the universe did spread out spherically from the point of the big-bang as its center; the Earth was created within the expanding universe after about 9 billion years thereafter (its present age being 4.6 billion years). Accordingly, as shown in Fig- 4E, the earth is not at the center of the *surfaces of the last scattering; accordingly, different parts of this surface must be moving away from the earth in different directions and at different speeds.* Since the said surface is supposed to expand at the speed of light, no radiation from this surface shall be able to reach the earth, especially from that portion of this surface that is moving in the direction opposite of that of the direction of earth's motion because whenever the light coming from this surface will reach the present location of the earth, the earth would move away with the expanding universe. *Accordingly, no radiation from that side of the said surface shall be able to reach us. As a result, "The said microwave radiation reaching us from different distances and directions should come with different intensities, different frequencies, and different amounts of redshifts;* **this radiation shall never appear identical from all the directions." This fact suggests that the said microwave radiation might be arriving from a surface that surrounds the solar system spherically from all around and whose temperature is about 2.7°K. Consequently, the microwave radiation coming from this surface from any direction shall always remain almost unchanged. The fact "While the Earth goes around the sun once, a variation of only 1/10,000 or less, is observed in the said background micro-wave radiation" suggests that the distance between the Sun and the source of this radiation must be at least 2×10,000 times greater or even more when compared to the distance between the Sun and the Earth. I feel that the Oort cloud or Opik-Oort cloud (see addendum-1), which is a surface of very cold gases, ice lumps, and leftover planetesimals, and which surrounds our solar system spherically from a distance of about 1½ light-years, might fulfill all of the above requirements.**

4. The so-called first glow was emitted almost about 0.5 billion years earlier than the light emitted from the oldest galaxies, i.e., about 13.2 billion light-years in the past. If the said light can still be seen as light waves, the said first-glow shall also reach us as light waves, not as microwaves.

5. At the time when the said first glow did flash out, that is, 13.8 billion years in the past, the earth didn't exist; the same (Earth) was created almost 9 billion years after the said emission of the first glow. By this period of 9 billion years, the said first glow emitted from the surface of the expanding matter would have diffused in space. However, the radiation emitted from the said expanding surface, at a right point in time, i.e., much before the birth of the earth, would have reached the early Earth while the same was growing up gradually; thence onward, the radiations emitted from the said surface thereafter would have persisted in reaching the early earth regularly. As such, the radiation now arriving at the Earth should not be considered the "leftover radiation of the first glow; the radiation reaching the earth now was actually emitted much later."

6. Since the said first glow was emitted when the temperature of the expanding matter had dropped down to around 3000°C, the said first glow could not have been "white-hot" as the scientists believe, instead it should have been of orange or dull red color. Thereafter, the temperature of the said surface of the last scattering would have continually dropped down with time. As a result, the frequencies and the cooler temperatures of the subsequent waves emitted from this surface at different points in time, would have persisted in reducing gradually.

7. Because the size of the early universe, at the time instant when the so-called first glow was emitted, would have been very small, around 3.8 hundred-thousand light-years, the fact "Any glow should appear to come from its origin" contradicts the belief that the microwave radiation coming from all directions is the relics of the said first glow. The said glow emitted 13.8 billion years in the past cannot come from the future location of its source, i.e., "The present location of the surface of the last scattering." Of course, by now, the matter scattered by the big bang must have moved to far-off distances, but *that expanding matter can't persist in emitting the said first glow forever.* Moreover, because the fabric of space-time is also expanding with almost the speed of light or even higher, the light should not be able to travel in the direction opposite of the direction of the expansion of the universe; any ray shall not reach us now from this direction. *The only condition that light may propagate in any direction at its normal speed is "Light shall be able to propagate without the medium of the expanding space-time."*

8. After the Big Bang, the fabric of space-time is presumably expanding persistently at the speed of light; accordingly, the continuum of space-time shall gradually thinned-out, and in turn, its density per unit area/volume and its strength both shall also diminish gradually until and unless this fabric is also not growing at the same rate. The growth of this fabric in all three dimensions, at the rate of the speed of light, means that the size of this fabric at its far-parts must be growing in the proportion of the 3rd power of the distance. This further means that whatever quantity of space-time was generated during the 1st second of its inception, 8 times of that quantity would have generated in the 2nd second, and 27 times in the 3rd second, and so on. Accordingly, after 13.8 billion years, the present rate of the growth of this continuum at its outer edges should be millions of trillion times higher than what it was during the 1st second

of the inception of the universe; this is absolutely impossible. This fact indicates that the continuous expansion of the fabric of space-time is not feasible.

9. The formula E = mc2 suggests that the energy, amounting to at least 90 billion times (approximate value of C2) greater than the entire mass of the universe, would have been released from the big-bang. The fact that "energy always flows out from its higher density to the lower density" attracts a few questions: 1) "Wherefrom such an enormous amount of energy did come?" 2) "How such a huge amount of energy could have accumulated in an atom-sized point, that too, against its nature to flow outwards." Such questions are not unnecessary; they point to the possibility that something must have happened before the "big bang." How could the entire universe be created out of nothing? What was special about that point where the entire energy, violating its nature, did accumulate? Why not anywhere else? What was cooking within that point that resulted in the said explosion of energy? What was the reason that the said point exploded exactly at that particular point in time? All these questions clearly indicate that something must have happened before the "big bang," which resulted in the creation or accumulation of energy in that particular point and finally resulted in the said Explosion. The fact "Time is the concept of the duration between two or more events or activities" clearly indicates "If some activity did *really happen before the Big Bang, then time surely did exist beforehand; it was not created by the Big Bang.*"

10. Scientists believe that nothing except that point, which contained space-time and the entire energy, did exist before the "big bang;" curvature of space-time before the said "bang" was said to be infinite. A question arises on this concept "In case nothing existed, then in what place did the universe expand?" In case the aforesaid notion that nothing existed before the big-bang is correct, then a vast void or limitless-emptiness would have been permeating all around. After the "big bang," the universe and space-time, etc., would have expanded within that very emptiness/void. The term "Space" merely defines the distance between two or more objects; it is nothing except the concept of distances or volume. In this scenario, another question arises "In what manner did the distances measured within that "primitive void" differ from those

measured in the present-day interstellar space? The universe is presently expanding within that very primitive void; day by day, it is growing bigger and thinner. Scientists believe that after a few billion years in the future, the universe would gradually dwindle and become so sparse that finally, it would defuse completely. However, this would not be the death of this void; it would continue to exist. Neither this void was born, nor would it die with the universe. That primitive empty-place is, in fact, "space;" it surely did exist before the beginning of the universe. Both space and time are not physical entities; none of them was born with the "big bang" they surely did exist even before the said "big-bang."

×   ×   ×

At this juncture, I am unable to withhold myself from raising an irrelevant issue that is given below; the same is not, at all, concerned with the subject matter of this book in any way:

The questions raised in points number 9 & 10 in the above paras are neither against any religion or community and nor do these paras, in any way, question or intervene the moment of creation. God has bestowed mankind with a bigger brain and unquenchable curiosity so that he may not just try to understand the unknown things; he may also successfully understand them. It is rightly believed that nothing could happen against God's wishes. If he hadn't granted permission, then such thoughts would not have come up in anybody's mind. This effort of mankind is as per the wishes of God. God, with a special purpose, has definitely blessed mankind with the extra curiosity and bigger brain; we shall not negate his wishes.

Anyhow, in case anyone has felt offended by the above thoughts, then I tender my apology for the same; I have no intention to offend anyone or any religion. I respect them all equally.

The Size, Shape, and the Age of the Universe

Scientists believe that our universe did begin about 13.8 billion years ago in the past; since then, the stars, moving at very high speed, have spread in all directions. Scientists also believe that the universe is expanding with an average speed that is almost equal to the speed of light. In the year 1929, Hubble observed that almost all the galaxies were running away from each other; he also noticed that the further away the galaxies were located, the faster they were moving away. Based on this fact, scientists did conclude that the speed of expansion of the universe is increasing continuously. Later, in 1970, the "big bang" theory envisaged that the universe, in the beginning, started to expand at greater than light speed. *Based on all the above facts, it could be deduced that in the beginning, the speed of expansion, which was greater than the speed of light, would have first slowed down and after some time, it again started to increase. This fact was later confirmed when in the 1990s, some scientists while observing an exploding star (supernova), concluded that billions of years earlier in the past, the stars were moving at a comparatively much lower speed. Now, based on the fact that though the speed of expansion of the universe is continuously increasing, the present speed of expansion is almost equal to the speed of light, it may be concluded, "The average speed of expansion of the universe, over the past 13.8 billion years, should have been much below the speed of light." Based on this conclusion, it could be deduced that the universe might have taken much longer than 13.7 billion years to spread through the distance of 13.7 light-years. Since this conclusion differs from the prevailing belief, it could not exactly be said what is the actual age of the universe or what its actual size is?*

There is no single opinion on the shape of the universe. Some scientists think that it has spread over a two-dimensional sheet; some people think it looks like a three-dimensional sphere. Yet another group thinks that its shape might be like a doughnut, a toroid, or an "O" ring.

Scientists believe that the universe is almost 13.79 billion years old, and the estimated age of our own galaxy, the Milky Way, is around 13.2 billion years. On the other hand, the furthest galaxies, too, are located around 13.2 billion light-years away from us. This means that our galaxy and the aforesaid furthest galaxies would have been created almost at the same point in time, that is, about ½ a billion (500 million) years after the "big bang" at close proximities to each other. It may also be predicted that these galaxies, at the time of their creation, would have been located very close to the point of the Big Bang, within a radius of ½ billion light-years. All of the galaxies created after the Big Bang would have moved in different directions along with the expanding universe but at almost identical speeds. Our Earth was created billions of years thereafter. In case the above prediction is correct, then a puzzle comes up *"How is it possible that the light rays, emitted from those so-called furthest galaxies, emitted much before the birth of the earth, are reaching us now?"* This is explained by the help of Fig–5 'A' depicted below:

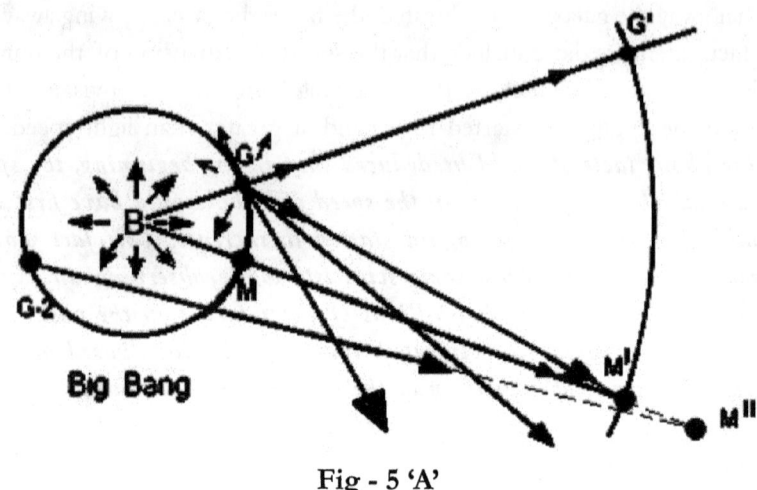

Fig - 5 'A'

The point of the "big bang" is denoted in Fig–5 'A' by the letter "B." The locations of the Milky Way and that of the said galaxy that was created at the same point in time are both respectively shown by the letters M and G. The present locations of these two galaxies are respectively shown by M' and G'. These two galaxies, immediately after their creation, would have started to move radially outward in different directions, that is, toward their present locations; while moving on their respective paths, they would have continually emitted light rays in all directions. Both these galaxies were moving with the expanding universe at almost the same speeds; therefore, by the time the Milky Way would arrive at its present location M', the said galaxy G, after traveling by almost equal distance, would also have moved from its original location to its present location at G'. As shown in Fig–5 'A', the galaxy G, at the time of its creation, would have emitted light rays in all directions including the future location of the Milky Way, that is, toward M' also. With the passing of time, the rays of light propagating toward M', the present location of the Milky Way, as well as the Milky Way itself, would have continued to move toward the said point M'. When these rays would reach the earth at the said point M', the said galaxy would appear at its original location G, from where the incoming rays were first emitted, not at its present location G'.

Likewise, the rays emitted from yet another galaxy G-2, which was created at the same time, but on the other side of the point of the "big bang," would not have yet reached the earth. The rays emitted by the said galaxy G-2, in the direction of the future location of the Milky Way, shown by M", would probably reach the earth in the next few billion years. This means that we, at present, cannot see all the existing galaxies or the entire universe.

All the galaxies, which were created at the same time period after the "big bang," would have immediately started to move in different directions with equal speeds. Therefore, since then, they should have traveled through equal distances, and by now, they would be located at equal distances from the point of the "big bang." Accordingly, the shape and size of the universe would depend on the fact that whether space-time did spread on a flat, two-dimensional sheet of space-time or spherically in all three dimensions. In case all the galaxies are located on the surface of a hollow sphere, then due to the limitation of our vision, they may appear to lie on a two-dimensional sheet. In that case, we may be able to see either a very small flat portion of the total surface of such a sphere; alternatively, the galaxies located within the range of our sight may seem spread on a spherical cap. However, we probably, won't be able to see the entire surface of such a sphere.

Irrespective of the above fact, whether the galaxies are spread in the two dimensions or in three dimensions within a spherical ball (not on its outer surface alone), the surface of the last scattering (that was created in the first instance) would be moving in the front line; the same would be followed by different surfaces of galaxies, which would lag behind one after another, in the order of different time periods of their creations. In that case, some galaxies would be moving ahead of the Milky Way, and some galaxies might be following it from behind. A schematic diagram of the universe expanding in this manner is depicted below in Fig- 5 'B'.

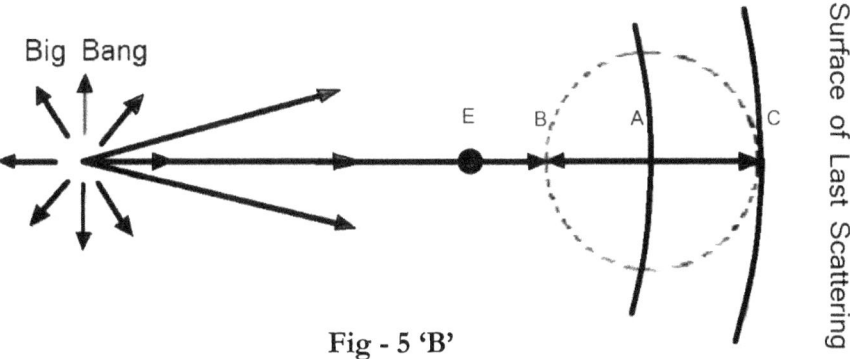

Fig - 5 'B'

A surface of galaxies, which is moving ahead of the Milky Way, but behind the surface of the last scattering, is shown in this figure as "A." Since the beginning, i.e., even before the creation of our Earth (E), the said surface of galaxies "A" would have continually emitted light rays in all directions. Now, consider a ray of light that was emitted from the surface "A" at the right point in time toward the future location of the earth, which is shown by the letter "B." At the same time-instant, the earth, immediately after its inception, would also have started to move at the speed of expansion of the universe and in the same direction in which the universe is moving. Now, by the time the earth would reach its present location "B," the rays emitted from the surface "A" would also reach the same point "B." As a result, the surface "A" would appear from the Earth as if the same is still located at "A" from where the incoming rays were originated. However, during this span of time, the said surface "A" would also have moved from "A" to its present location "C." This example clearly reveals that "all other galaxies and stars must have moved to comparatively greater distances as compared to the distances at which they seem to exist in the present."

Based on the above example, it may be envisaged that by now, the universe might have expanded even beyond its estimated ambit of 13.7 billion light-years, where the furthest galaxies ostensibly appear to exist; accordingly, if in the present era, we can see the oldest galaxies located at a distance of 13.2 billion light-years, then subject to the condition that such galaxies do still exist, they would have moved further away from the locations where they presently seem to exist. Therefore, we cannot find out the exact speed of the expansion of the universe; only the apparent speed of different galaxies relative to the earth can be calculated. In case this presumption is correct, then the actual expansion of the universe, and its actual age, both might be much greater than what we have estimated; however, it is not possible to find out these values.

9: Creation & Life Cycle Of The Stars

The idea of *"the creation and destruction of the stars"* is not very old. About 250 years ago, nobody could have believed that even the stars too have to undergo the cycle of birth and death; the stars were considered eternal. However, in the year 1755, **Immanuel Kant** and **Pierre-Simon Laplace** proposed that different stars and planets were created by the gradual collapse of the interstellar *gaseous clouds, which are* known as *Nebulae;* this theory is known as the *"**Nebular Theory.**"* The modern theory explaining *the formation of stars* was later on formulated in the year 1970 by the Soviet astronomer **Victor Safronov**.

In those days, no equipment that was capable of observing the process of the *formation of stars* or verifying the aforesaid theory was available. Later on, in the year 1990, ***the Hubble telescope*** was installed in the Earth's orbit; this telescope is capable of seeing even very feeble light coming from very distant stars, even the stars located at as far-off distances as 12 to 13 billion light-years. The said telescope helped us to acquire valuable knowledge in the field of astrophysics. We can now look back into the past, like a "Time Machine," and see various cosmological events with our own eyes as if we are watching a very slow movie.

Although the creation of the stars is a regular feature in this vast universe, even then, it is not possible to observe this process; even a small event at the cosmological level may take several million years for its completion. This means that the phenomenon of *star-formation* can't be observed within the period of even a few hundred thousand years; moreover, the stars, during their creation, are shrouded with very thick blankets of *interstellar dust and very dark clouds of gas*. Even the Hubble telescope is unable to see the spectacle of the *star formation* because the visible light waves cannot penetrate through the thick blanket of gas & dust; only the ***infrared waves*** *a*re capable of penetrating through such a blanket. Keeping this fact in view, the *infrared-telescope* named *"**Spitzer**"* was established in the earth's orbit in the year 2003 to solve this problem. **Spitzer** helped us to study numerous stars under different stages of their formations.

A broad outline of the theory of star-formation developed based on these observations is given hereunder:

The Nebular Theory of Star formation

Apart from the visible stars, massive clouds of gas and dust, known as *"Nebulae,"* also exist within all the galaxies, as well as in interstellar space too. Such clouds might spread over very large areas measuring up to hundreds rather than thousands of light-years. These clouds, depending upon their sizes, may contain several hundreds of thousand times more mass than that of our Sun. These clouds mostly consist of molecules of different gases like Hydrogen, Carbon-mono-oxide, Helium, etc., and therefore, they are also known as "Molecular Clouds." These clouds also contain very fine grains and dust particles, solid objects like pieces of ice lumps, heavy elements in small quantities, and various other compounds; so far, a total of 118 types of molecules of carbonic and inorganic substances have been seen within such clouds. Since the stars are continuously formed within these clouds, they are also known as the *"stellar nursery."*

Huge quantities of the molecules of *gas, dust, and several other elements/ compounds* are spread all over within different nebulae, but in a very sparse and exiguous manner. These *molecules of gas and dust, etc.,* due to their mutual attraction, gradually accumulate in different places; as a result, several dense gas clumps/ clouds are formed within a nebula. The larger clumps of gas gradually shrink or collapse within themselves, first due to accretion and thereafter due to gravitational force; thereby, numerous dense cores are formed randomly within these clouds at different places. The turbulence within these clouds, and the mutual attraction between different dense cores, cause these clouds to fragment into yet smaller dense regions or pockets. The mass of different dense pockets might range from less than the mass of the Sun to several times its mass. The sizes of such pockets may vary from 1 to 5 light-years or even more. Scientists believe **"Due to some unknown reason, the gas surrounding these cores starts to rotate slowly in the shape of a flat disk;"** however, they couldn't establish the reason **"How and why this rotating motion is induced."** Anyhow, gas and dust from these rotating clouds, which are known as *"protoplanetary disks,"* gradually start to accrete onto their central regions. As a result, the gravitational force of these central regions starts to increase gradually; *this increasing gravitational force* causes them to gradually squeeze within themselves at much higher rates. In the next few million years, these cores, because of gradually increasing friction and pressure within their cores, grow immensely hot and start to oppose any further compression; finally, balance is created between the gravitational crunch and the outward pressure. Several hundreds of such hot cores might have been created within a large cloud almost simultaneously. With time, these hot cores turn into the seeds of different stars; the mass and size of each of these seeds would have depended on the mass of their parent gas clumps.

Scientists believe that later on, these *rotating gas clouds* flattened out due to the centrifugal force generated within them. This is considered analogous to making the raw pizza bases, or "the Rumali Roties (a kind of very thin Asian bread)," which are made by rotating and tossing the flat dough cakes in the air. Scientists, on similar lines, believe "gas clouds also flatten out because of their rotatory motion." The size of any *protoplanetary disk* depends on the mass and/or the amounts of gas contained in its mother clouds. The sizes of different *protoplanetary disks* may range from ½ to 100 light-years in diameter or even bigger. When any such rotating gas disk surrounding the dense core collapses inward, its rotational speed gradually increases due to *the conservation of its angular momentum.* This increase in the rotational speeds results in the slowing down of the process of direct accretion of gas onto the central core; or rather, the *gas and dust* contained in the central part of such gas envelopes are forced to spread outward. As a result, this process of accretion is retarded. Despite this retardation in the process of accretion, the core gradually grows bigger in mass, and in due course, it becomes a young hot star known as *"Protostar."* The persistent compression of the gas causes the temperature of the core to increase gradually; when its temperature reaches up to about 3000°C and above, the core, from time to time, emanates bipolar jets along its rotational axis. This is illustrated in Fig-6 below. Magnetic activities are also found associated with these jets. Scientists believe that these cores shed off their excessive angular momentum through these jets, this reduction in the angular momentum results in the reduction of their rotational speeds.

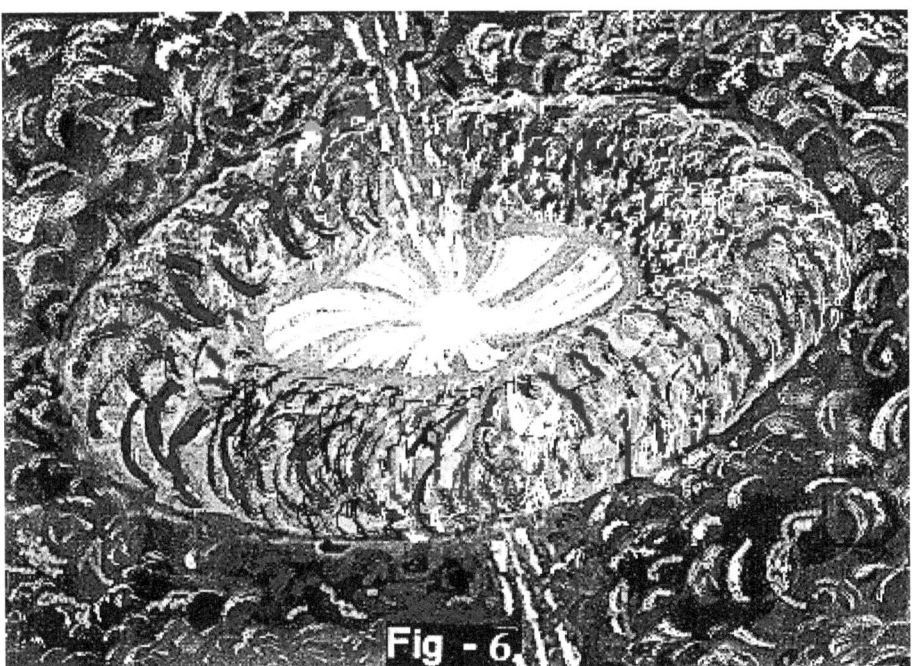

Fig - 6.

These rotating gas envelopes, or the Protoplanetary Disks as they are called, gradually become thin and less opaque due to continual accretion of the gas and dust onto their cores; eventually, these envelopes disappear. Consequently, the young star, after about 1 million years of its creation, becomes visible. Such a young star, in this stage of its creation, is known as *"T. Tauri Star,"* which, for about the next 10 million years or so, continues to grow bigger and more massive. During this stage of its growth, it continues to eject the bipolar jets. Gradually, its core is heated up to a temperature of about 8 to 10 million degrees; at the right temperature, the process of nuclear fusion of the Hydrogen atoms is triggered in its core. Consequently, such a *T, Tauri star* becomes a fully functional star. In due course of time, when the amount of gas falling onto the core reduces, the bipolar jets gradually disappear.

Formation of the Planets

Scientists believe that the rotation of gas in the protoplanetary disks results in mutual friction of gas and dust particles, which heats them. These particles, on heating up, tend to stick together; radiation of heat from the core accelerates this process. By the time the star reaches the *T. Tauri stage*, the gas enveloping the disk becomes thin; consequently, because of reduced friction, the gas envelope starts to cool down. This results in condensation of the less-volatile materials, which, in turn, results in the formation of small grains near the central part of the disk. Within the next few hundred thousand years, these dust particles coagulate to form about 1 kilometer-sized objects called "planetesimals." Gas and dust start to accumulate on these planetesimals at a much higher rate. As a result, the planetesimals, within the next 2-3 hundred-thousand years, grow up as big as the Moon to Mars-sized objects. These bigger bodies act as the seeds or embryos of the future planets. Thereafter, these seeds, due to their increasing gravity, start to suck large quantities of gas and dust from their near surroundings and grow at runaway speeds. Generally, rocky and metallic planets are formed in the regions lying in the near vicinity to the hot core because large portions of the available *gas and volatile matter, etc.,* are sucked away by the star itself. The gas giants like Jupiter and Saturn are normally formed at larger distances from the star, beyond the frost line, where the star is not able to suck much gas and where the temperature remains low. It is estimated that it takes a time period of a few hundred million years for the formation of Sun-like stars and their planetary systems.

Probable Loopholes in The Nebular Theory

Though this theory enjoys very wide acceptance worldwide, even then, it doesn't clarify the following points, at least, to my satisfaction:

1. There is no explanation of the points that (a) *"Why and how did the clouds start to rotate before the formation of the stars?"* (b) *"From where they get the energy to rotate, and how is it decided that in which direction will they rotate?"* There must be a hard and fast rule for this purpose.

2. There is a very vast difference between the *flattening of the pizza base* and that of *the protoplanetary disks*. **No force attracts the dough cake towards its center; on the other hand, the gravitational force of the dense cores always attracts the protoplanetary discs toward their centers.** Thus, the conventional explanation of the *flattening of the protoplanetary discs* doesn't seem satisfactory, at least to me.

3. Since the molecules of gas & dust in a protoplanetary disc rotate at almost the same speeds, the possibility of friction between them leading to the formation of planets seems very remote.

4. If asteroids were formed due to the coagulation of dust particles over each other, then all the asteroids should have almost the same composition. Contrary to this possibility, some of the asteroids are made of carbonaceous material, some asteroids are made of silicates, and others are made of the metals like Nickel and iron; there is no explanation for the same.

5. In case planets are formed due to the coagulation of dust particles, then the internal structure of the comets and asteroids, etc., should be granular like sandstone. A Japanese spacecraft "Hayabusa-2" has already collected rock samples from an asteroid named "Ryugu;" in case these samples are found of igneous origin, then this theory would need a thorough review.

6. The explanation that "the excessive angular momentum of the rotating stars is released or vented out through the bipolar jets" is unable to fully satisfy me. This seems purely an assumption; no proof is available in its support.

7. No proper justification for the formation of bipolar jets is available. (Please see Chapter- 12.)

8. It is not clear at which particular place the planets would form and why, that is, how and why dust particles would start to coagulate at any particular place, not anywhere else?

To *remove the above-listed doubts, I have worked out an alternative solution to the process of "the formation of stars." I wish that this speculation of mine* shall be thoroughly examined at the highest level. However, before we discuss this speculation, it seems necessary to plug the loophole listed in serial No. 1; for this, we shall solve the mystery of *"how the whirlpool is formed in the washbasin simply by opening its drain plug"*? I feel that this *very puzzle* holds the key to the mystery under question. In fact, the whirlpool is formed because of the suction created at the drain opening due to the force of Gravity; *a force is essentially required to generate any motion.*

The Puzzle of The Whirlpool

In nature, the whirling motion is produced almost everywhere; electrons revolve around the atomic nucleus, Moon orbits our Earth, the Earth, in turn, orbits the Sun, whereas the sun, along with the family of all of its planets, asteroids, and comets, etc., revolves around the Galactic Centre of the Milky Way. This series of whirling motions doesn't end here. Probably, a few people might know that the Milky Way, accompanied by other nearby galaxies, which are in substantial numbers, orbits a bigger galaxy named Andromeda. Such a group of galaxies forms a *local group of galaxies,* which are also known as *"local clusters of galaxies."* The *local clusters of galaxies,* along with a few other nearby *local groups of galaxies,* orbit a much bigger local cluster. As a result, countless *super-clusters of galaxies* consisting of hundreds of *local groups of galaxies* are formed. This series of *whirls formed within another whirl* is an enigma that probably conceals a very complicated and deep mystery of the universe. Solving this puzzle of the *whirls swirling within another whirl* is an open challenge for every curious critter. As far as I know, this puzzle has not yet been solved.

The whirlpool formed within a washbasin is the best-known example in nature of such a whirl. Washbasins generally have plugs or lids by which the outlet of their drainpipes can be closed or opened at will. If the washbasin is filled with water and its drain plug is removed, then the water flows out through the drain in a spiral path. The whirlpool, so formed, resembles a little with the spiraling galaxies. It might mean that the whirlpool can enable us to solve the puzzle of the formation of spiral galaxies. However, it seems to me that no serious effort has ever been made to unveil this mystery.

In case we study the process of formation of a whirlpool within the washbasin, then it would become obvious that the quantity of the water flowing out of the drain, as well as its speed, both depend jointly on the size of the drain and the height (head) of the water surface above the drain. In case we carefully observe the above phenomenon and think over this process a little, then it would become evident that the whirlpool isn't created immediately after the opening of the drain; it is formed after a very small-time gap. The higher is the water head and/or the bigger is the size of the drain opening, the more vigorous would be the whirl so formed. In case the size of the drain is very small, then the size of the whirlpool so formed would also be very small. In such a case, the water, though continuously flows out from even beyond the outer edges of the whirlpool, the surface of the water beyond the outer edges of such a whirlpool might seem standstill. It may thus be concluded that the *rate of flow of water* and *the size of the drain* both jointly control the process of the creation of the whirling motion and also that of the size of the whirl so formed. On the other hand, when the size of the drain is so big that the entire water flows out in one gush without any restriction, then the whirl is not at all created.

Now, in case some granules of a chemical named Potassium-Permanganate or that of any other soluble color are sprinkled in the basin, then it would be seen that the whirl is not created immediately after the opening of the drain; instead, the water from all directions, first, starts to flow radially, i.e., in a straight line toward the drain; a very short while thereafter, different water currents, coming from all directions, adopt independent, but identical curved paths; the curvature of these paths gradually increases until the whirl doesn't acquire a stable speed and shape. This observation signifies that the whirl is not formed immediately also that the curvature of different water currents increases gradually until the whirlpool does not stabilize. Further, in case the water contains some suspended particles, alternatively, a fine powder is sprinkled over its surface, then after almost all the water is drained out, i.e., when a very thin film of water is left in the basin, then the whirl loses all of its flurries, it ceases to rotate in a whirl; the remaining water coming from all directions flows out radially in straight lines. The above observations suggest **"when the drain-opening doesn't offer any resistance to the water flowing out or the quantity of the water flowing out decreases substantially the whirl couldn't be created."**

All the observations described above indicate, *"Though the vigor of the whirl depends on the head of the water and the size of the drain; the whirlpool can only be created when the quantity of water flowing out through the drain is more than its capacity."* This conclusion is supported by the fact that *"When the size of the drain opening is far more than the* quantity of water flowing out, i.e., when the entire water flows out abruptly, without any resistance, then the whirlpool is not created." Thus, we can conclude that *the size of the drain opening and the resistance created therein that restricts the flow of water through it both are equally important, and of course,* **"The force of gravity is of utmost importance because it induces the water to flow through the drain; nothing can flow in the absence of a force."**

The above conclusion seems contrary to the theory of Relativity, wherein gravitation is not considered a force; the smaller celestial objects are thought to orbit the massive stars due to the curvature of space-time, not due to the pull of gravity (please see chapter-5). Suppose the rotary motion is really created obeying this theory. In that case, the variations in the sizes of the drain or the head of the water shouldn't have any influence on the size or speed of the whirl because such variations, in any way, would not affect the size and the depth of the curved portion of the washbasin or that of space-time that is warped due to the mass of a celestial body. This fact clearly indicates "the whirling motion in the gas-clouds, around the newborn stars, is not induced due to the curvature of space-time; a force acting from all-around that acts toward the central core, is definitely required for this purpose.

Anybody can observe that the whirl so formed always rotates in a definite direction; a few scientists believe that this is linked with the spin of the Earth. The Earth always spins on its axis from west to east, and so does the seawater; however, no whirling motion is induced in the seawater due to this reason. Had both the direction of the rotation and speed of the whirlpool has been controlled by the speed of the Earth's rotation, then the speed of rotation of the whirls shall never be affected either by the size of the drain-opening or by the head of the water; in that case, the whirl shall always rotate at a constant speed. This fact indicates that "The direction of the rotation of the whirl does not depend on the direction or speed of the earth's spin."

I understand that scientists believe that *such a spiral motion in the whirls is probably induced by the gravity waves.* However, a question arises on this concept *"Why is the whirl not created immediately after the drain is opened, and why does it die off during the last moments when the flow of the outgoing water becomes very weak?"* Had the aforesaid belief of the scientists been true, then the creation of the whirl and its rotational speed, both shall have no relation to either the size of the drain or with the quantity and speed of the water flowing out. In case the gravity waves are capable of swirling the water so vigorously, then measuring the energy of such waves shall definitely be possible; however, up till now, it has not become possible to feel or detect the gravity waves (*Please also see Chapter-11 for the latest discoveries relating to Gravity Waves*). Here, one more question arises: "How do the gravity waves *create the whirl?*"

An explanation of this enigma of the formation of the whirlpool, as speculated by me, is given below, which is based on the most fundamental rules of science: -

Immediately after the opening of the drain, the ***gravitational force*** pulls down the water molecules through the drain. Therefore, the water starts to flow toward the drain in the form of enumerable water streams from all around; however, this force doesn't act on all the water molecules directly and equally. The value *of the suction so produced goes on reducing gradually with an increase in both the distance from the drain and the increase in the surface area of the water; the further away the water molecules are located from the drain, the lesser is the value of the pull acting on them.* Accordingly, the water molecules located just above the drain are immediately sucked in by the gravitational force, but ***inertia*** prevents the rest of the water molecules from flowing towards the drain instantaneously; the distant molecules from all around start to flow after a bit of time lag. *During this initial instant,* only that much water from all around rushes at a time toward the drain that can flow out instantaneously. *Since the water flows from a bigger area toward a gradually reducing area, its speed increases gradually in the inverse proportion of the reducing area.* As a result, *the speed of the water molecules heading towards the drain, in the front line, would increase at a much faster rate than that of the molecules coming from behind.* ***Accordingly, during a very small initial moment, the water molecules from all around would rush freely and radially toward the drain, i.e., in a straight line without any restriction.***

*The above scene won't last long; the gravitational force would cause a continual increase in the speed and quantity of the water flowing toward the drain. However, up till the quantity of water flowing out remains lesser than the capacity of the drain opening, the water will continue to flow radially. However, w*hen the quantity of the outflowing water tends to surpass the capacity of the drain, both the quantity of the water flowing out and its speed would become saturated. As a result, *different water currents rushing from all around would obstruct the free flow of each other; this will produce congestion at the drain, the value of which will be greater near the drain.* ***As a result, the pull of gravity would fail to increase the speed of the out-flowing water any further;*** but this limitation won't apply at the banks of the basin, *where the speed of water molecules would still tend to increase* ***because the effect of the congestion will be very meager at this point***. Accordingly, *t*he congestion *produced at the drain will reverse the pattern of the* **rate of increase of the acceleration** *of the water currents rushing toward the drain;* ***the speed of flow would though persist in increasing, the rate of increase of its acceleration would gradually reduce as if a*** *brake has been applied on different water currents.* As a result of the said congestion, the molecules of water coming from upstream, i.e., from behind, despite their momentum and the higher accelerations, would not be able to flow freely because the molecules that are flowing ahead of them are gradually slowing down. **Accordingly, such molecules would not get any room to flow radially, i.e., straight toward the drain, though their momentum and the force of gravity, which will still be acting on these molecules, will produce a tendency in the molecules coming from behind to keep moving at higher speeds.**

Under the above scenario, each molecule coming from behind because of its momentum and a comparatively higher rate of acceleration would develop a tendency to move ahead of the molecule that is flowing just ahead of it but with a gradually reducing rate of acceleration. **Since the molecules coming from behind with higher accelerations won't get any room to move straight toward the drain, they, in their endeavor to keep moving with higher acceleration, would move sideways, i.e., in a direction perpendicular to the line of their final goal, the drain** (this phenomenon is similar to that what happens when an obstacle comes in the way of a current of the flowing water). **The congestion produced at the drain would thus produce an altogether new component of acceleration in the water molecules that will act transversely, i.e., in a direction perpendicular to the acceleration acting radially toward the drain; accordingly, the water molecules will flow in the direction of the resultant of these two accelerations.**

All the water molecules that will begin to flow from all around the bank of the basin would start with zero speeds. Therefore, the rate of increase of the speeds of all these streams **would be infinite times higher in comparison to their initial "zero" speeds.** Now, since this acceleration of water molecules, acting in the transverse direction, is generated due to the difference between the rate of increase of acceleration of two consecutive molecules flowing toward the drain one after another, the value of the said transverse component of acceleration would also be the highest at the starting point. Next, when these water molecules, under the influence of the gravitational force, would move ahead toward the drain, the speed of such molecules would though gradually increase, the rate of acceleration, i.e., the ratios of increase in the speeds of the molecules flowing one after the other would go on reducing due to the congestion produced at the drain. **Accordingly, the value of the transverse acceleration would also go on reducing gradually,** and at the same time, **the value of the radial acceleration of the water molecules would continuously increase due to the increasing value of the suction that would inversely depend on the square of the distance from the drain; the value of the acceleration in this direction would reach its peak at the end of its journey.** Accordingly, **the different water streams, depending upon the varying values of these two accelerations, would flow in the direction of the resultant of these two accelerations. As a result,** each of these water streams would adopt an identical *spiral path; this increase in the speed of the spiral flow would also be controlled by the gradually increasing centrifugal force, which will oppose the suction.* This phenomenon is further explained below.

The whirlpool formed in a washbasin consists of countless individual streams of water, all of them rushing toward the drain from all around. Out of all such streams, the point of origin of a single stream that originates at the banks of the washbasin is depicted in Fig-7 by the coordinates "0-0". The stream of water, which initially started to flow radially toward the drain, is depicted by a solid straight line that joins the starting point "0-0" and the mouth of the drain, which is depicted by the letter "D." *When the water molecules located at the bank of the basin are sucked a little in the radial direction, they would be simultaneously dragged away tangentially to a much greater distance by the comparatively stronger tangential component of the acceleration generated at this point. While this stream of water would move further ahead in the direction of the resultant of these two components of accelerations, the transverse speed of the water stream would go on decreasing, and its speed in the radial direction would gradually increase. As a result, the transverse displacement of the water molecules would continually reduce, and the displacement of these molecules toward the drain would continually increase. Accordingly, after almost halfway to the drain, where suction created by the drain becomes stronger than the transverse component of the acceleration, the said stream would gradually bend toward the drain instead of going away and approach the drain through a curved path whose curvature would continue to increase.*

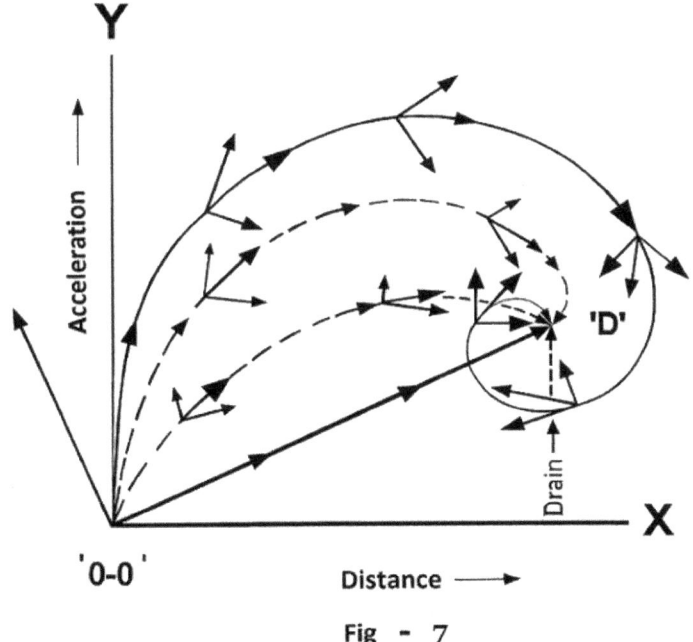

Fig - 7

The different water streams that were generated during the transient period (before the whirl could grow fully and stabilize) are depicted in the said drawing by different dotted lines. The approximate magnitudes and directions of different components of the accelerations acting on different water molecules are also shown in the said Figure-7. These *streams would approach the drain through a curved path whose curvature would gradually increase according to the varying values of the resultant velocity. Accordingly, the centrifugal force generated in the water molecules would also increase gradually. This increasing centrifugal force would try to counterbalance the suction of the drain; however, as these molecules would reach closer to the drain, the strength of the force of suction would increase at a rate much higher than that of the rate of increase of the centrifugal force.* Accordingly, the force of suction, generated at a distance very close to the drain, would suddenly overpower the centrifugal force; as a result, the water molecules, which reach very close to the drain, would be sucked in abruptly into the drain opening. ***Analogous to the above phenomenon, the spiral arms of the barred galaxies, just before entering the respective galactic centers, bend abruptly and very sharply toward the said galactic center.***

The fact that ***"It is not possible to create such a whirlpool in the microgravity conditions"*** *clearly reveals that* ***"The suction created at the center by the gravitational force*** is a must for the creation of such a whirlpool." *Further, since the maximum transverse acceleration is generated at the outer edges of the whirl, the maximum rotational speed is also generated at that point; even though this rotational speed goes on decreasing due to the congestion produced at the drain, the same seems to increase ostensibly because of the increasing curvature of these water streams. Spiral galaxies, too, exhibit similar patterns of rotational speeds; the stars located at the outer edges of the galaxy have a slightly higher rotational speed as compared to that of the stars orbiting the galactic centers from close quarters.* ***This similarity in the rotational speeds of the whirlpool and that of the spiral galaxies indicates that "The galaxies, too, were formed by the gravitational force acting toward their centers."*** *However, such a pattern of rotational speeds of spiral galaxies is* the *opposite of the predictions of* **Relativity.**

It is easy to understand the above phenomenon in another way. Although the gravitational force produces the persistent increase in the speeds of all the incoming water streams, the congestion created at the drain compels these streams to take a comparatively longer time to reach the drain. Therefore, all the water streams travel through much longer distances than the shortest (radial) distance between the bank of the washbasin and the drain. Analogous to the fact that "A much longer rod needs to be coiled to accommodate the same into a smaller space," the said longer streams also necessarily need to adopt different but identical spiral paths to reach the drain.

The mystery of the creation of the whirlpool doesn't end here; this riddle has one more unresolved twist *"such a whirl never rotates in an arbitrary direction,"* in the *Northern hemisphere,* the whirlpool always rotates in the clockwise direction, whereas, in the *Southern hemispheres,* it always rotates in the anticlockwise direction. Such a behavior of the whirlpool suggests that "a natural force, which acts in the opposite directions in both of the hemispheres," decides the direction of the rotation of the whirlpool. Out of all the natural forces, only *the magnetic force of the Earth's magnet* acts in the opposite direction in both the hemisphere. As per the prevailing conventions, the lines of the magnetic force come out of the Earth's surface in the *Southern Hemisphere,* whereas, in the *Northern Hemisphere,* they enter into the earth's surface. Since the whirlpool rotates in the opposite directions in these two hemispheres, the direction of the magnetic field remains unaltered relative to the direction of the rotation of the whirl. This could be understood by observing a timepiece from both sides; in both cases, the clock though always runs in the same direction, when the same is viewed from behind, it appears to run in the opposite direction.

Anyone may wonder, *"How the magnetic force may affect the nonmagnetic fluid like water?"* However, all the substances contain some free electrons; moreover, water possesses the property of diamagnetism too. Therefore, while the water molecules flow toward the drain, the Earth's magnetic field might compel them to rotate in a pre-fixed direction. This speculation of mine may be tested by reversing the direction of the magnetic field around the whirlpool swirling in a washbasin by placing a strong permanent magnet or an electro-magnet below the washbasin before opening the drain plug. In this context, please refer to the forthcoming section *of this chapter, namely-* *"The alternate Speculation on the formation of the Stars."*

It may also be noticed that "The gases, which damage the Ozonosphere, are mostly produced in the Northern Hemisphere;" even then, these gases cause damage to the said layer in the Southern Hemisphere. This is a matter to ponder whether this damage could be linked to the direction of the earth's magnetic field? This appears to be a matter of serious research.

I have no idea what the scientists think on the aforesaid matter? Anyhow, the magnetic field certainly affects subatomic particles very prominently. In the year 1956, American physicist Chien Shiung Wu, who was of Chinese origin, did notice that the nuclei of a radioactive substance, namely, "Cobalt-60," start to spin when lined up in a magnetic field, whereas when the polarity of the applied magnetic field was reversed, then the direction of rotation of these nuclei, was also reversed. This experiment suggests that the direction of rotation of the charged particles can be controlled by the polarity of the magnetic field. In other words, nature knows that under what circumstance a particle would rotate in what direction or in what direction the same should be permitted to rotate. This fact indicates that the direction of the magnetic field might have some relation to the direction of the spin of the water-whirl too. The direction of rotation of the galaxies, stars, planets, etc., might also be decided by the Magnetic Force. This might also be a subject of research.

As brought out in chapter-3, "Quantum Mechanics," under the subheading "Some Least-known Properties of the Light," the magnetic force can arrest even the photons or the so-called light waves. Lately, I gathered from an unauthentic source that the light coming from the distant galaxies had been observed to exhibit redshift in the Northern Hemisphere of the Earth, whereas, in the Southern Hemisphere, such light has been observed to exhibit blueshift. I am not very sure about this discovery; however, I have learned that some of the scientists have inferred from the above discovery that light can travel at different speeds in different places. *In case this information is correct, then it might also be possible that* **"Instead of the speed, the frequency of the light might be affected by the polarity of the Earth's magnetic field; it might decrease in the Northern Hemisphere and increase in the Southern Hemisphere."** At least some possibilities of such an effect might exist. This possibility also appears to be a matter of research.

As brought out earlier, the whirling motion is not created in the water alone; several examples of the whirling motion can be seen in the universe. This fact indicates that such a series of whirls, in which smaller whirls are formed within the bigger whirls, might hold the key to a deep mystery of the universe. This appears to be a matter of research whether there is or isn't any relation between the directions of the rotation of the celestial objects with the direction of the magnetic fields of the same stellar object or the nearby massive objects like the stars? In the present era, Mars doesn't have any magnetic field, but it certainly had one a few billion years ago in the past. This fact suggests that the riddle of the definite direction of rotation of the whirls must necessarily be solved. If this could be done, we might be able to understand various unresolved mysteries of the universe in a much better way. This problem should not be taken lightly.

An important question remains unresolved: "How do the stellar objects orbit the massive ones?" We have seen that the water-whirls are formed due to the congestion created around the central opening. On the same line, the nucleus of an atom attracts the electron, but the weak interaction or the quantum repulsion prevents it from coming very near to the nucleus; this is probably why the electron keeps revolving around the atomic nucleus. On the other hand, "When a celestial object under the influence of the attractive force of a massive object, starts moving from its "zero" initial speed, then *the rate of the increase of its speed at the starting point would be infinite times of its initial zero speed, which will be similar to the variations in the speed and the acceleration of the water molecules moving in the whirl,* while this object would move ahead, its speed would though continuously increase, the ratio of the increase in its acceleration would continually decrease. However, this similarity is not adequate to make that object move in orbit; it will most probably, directly collide with the massive object. In the forthcoming section of this book, an effort has been made to solve this problem.

An Alternative Speculation on Formation of the Stars

Scientists believe that the process of star-formation was probably triggered by the supernovae explosions that compress the molecules of the gaseous clouds. However, primitive stars were born millions of years before any supernova did explode. This fact signifies that supernovae did not necessarily trigger star formation; of course, *such explosions must have provided the raw materials like loose gas and the solid fragments of the dying star cores; The star debris might have facilitated the formation of the next generation stars. The stars of the next generation and their planets, etc., might have formed due to the accretion of gas and dust on these solid fragments of different sizes, not at any arbitrary place. Immediately after such an explosion, the fragments of star-core of random sizes would have been scattered to different distances in a disorderly manner.* The following fact supports this prediction of mine:

On 12th Feb.'2001, when NASA's spacecraft named **Near Shoemaker** did land on the asteroid named **433 Eros**, it was found that this asteroid is a large chunk of rock that is composed of silicates, basaltic materials, and a few metals like Aluminum, Nickel, Iron, Magnesium, Gold, and Platinum, etc. Such a composition of this rock, especially the presence of **gold**, indicates that "All the asteroids/comets were probably created out of the debris of the star-core scattered by a Supernova or Hypernova explosion because the heavy metals are not supposed to have been produced from the Big Bang." Scientists believe that comets/asteroids might have been created under the extreme heat that exists in the near vicinity of the Sun; later-on such objects might have migrated to their present locations. However, in the absence of a force that is much stronger than the Sun's gravity, such a migration against the powerful gravitation of the Sun seems impossible.

In fact, several asteroids have either their own moons or binary companions that orbit each other; this fact reveals that the smaller objects, too, have their own gravity fields, which are resulted due to either the warping of space-time or due to the *attractive force* acting between such objects, there can be no other reason for the same.

My speculation described below is purely based on the two facts given above.

The earliest stars were supposedly created about a few million years after the "big bang." By that time, the expanding gas, which did possess *diamagnetic properties,* would have cooled down to nearly absolute zero degrees and become superconductive. This gas would have expanded in a very uneven and inhomogeneous manner, due to which different dense gas clumps having different quantities of gas would have been produced at numerous places; *later on, such clumps or pockets of dense gas would have been created due to the accretion of gas and dust over different solid fragments that were produced as a result of different supernova explosions.* These dense gas clumps, which were spread all over at different distances from each other, were surrounded by very thin and sparsely distributed gases. The cores of these gas pockets, due to continued accretion of gas/dust, would have persisted to gain mass; subsequently, the increasing gravitational force of these cores would also have caused them to gradually and continually collapse inward and thereby become more compact.

Initially, the gas falling into the cores of different gas pockets would have, first, flown radially in numerous straight lines, due to which several concentric spheres of gas of varying diameters would have formed around the central cores. Next, analogous to the formation of the whirlpool in a washbasin, the incoming gas currents, due to the creation of congestion around the cores, would have adopted different curved paths, as shown in fig-7A and fig-7B.

The gas clumps, which were formed during that point in time, didn't have either regular shapes or homogeneous gas distribution. Therefore, different gas streams would have come from different distances and directions; each one would have been carrying different amounts of gas. These gas streams would have been moving at different speeds, which were proportional to the respective distances traveled by each of them. Accordingly, that stream of the superconductive gas, which was flowing at a greater speed, and carrying a greater amount of gas, would have created the strongest magnetic field, stronger than the field produced by all other streams. This strongest magnetic field would have created its mirror image (See chapter-2 "Superconductivity") in the molecules of all the other gas-streams rushing toward the central core. This strongest magnetic field would gradually push all other gas streams, from both sides, toward its magnetic equator; finally, it would establish all of them in this plane **in the shape of a two-dimensional whirling disc,** the magnetic field of the said disc will pass through its axis of rotation. These swirling gas streams would cause the collapsing core to rotate in the same direction as an independent body; **its rotational speeds would depend on the distances from where the gas (from within the gas pocket) was being sucked.** The increasing gravitational force would cause these cores to gradually adopt spherical shapes.

It is thus clear that the **gravitational pull** did induce the rotational motion in all the in-falling gas currents; the direction of the swirling of these in-falling gasses was, as explained below, was decided by the basic rules of **Electromagnetism and Electro-dynamics**. The curvature of space did neither cause the gas to flow toward the center nor did it cause the gas to swirl in a particular direction.

The whirling of the supercooled gases, which also contain free electrons, should have produced an electric current that will flow in the direction opposite to the flow of the free electrons. Accordingly, the electric current, which is generated in these gas streams, would flow outward from the core. When any of these gas streams would be gripped by the **Right-hand** in such a manner that the thumb points toward the direction of the electric current, i.e., away from the core, then, according to the **Right-Hand Grip Rule**, the lines of magnetic force produced by the flow of the electric current, would point toward the fingertips. Accordingly, the lines of the magnetic force produced jointly by all such gas currents will enter the spinning-axis of the core from its top, i.e., from its geographic North, and would emerge out from the underneath (geographic South) of the core. **In other words, the whirling superconductive gases would produce a magnetic field at their axis of rotation; the direction of this field will point downward, i.e., from the geographic North of the whirl to its South. This direction of magnetization of the core is very important because almost all the celestial objects, including our Earth, too, have a similar direction of magnetization.**

In the case described above, the electric current flows through all the gas streams in the direction away from the core, and the magnetic field acts directly downwards. In a star under making, the direction of the rotation of the gas streams is controlled by **Fleming's Left-Hand Rule of the electro-dynamics,** which states, "In case the central finger, the forefinger, and the thumb of the left hand, are held at the right angle to each other, in such a way that the forefinger points toward the magnetic field and the central finger points toward the direction of the current, **then the magnetic field of the core will exert a thrust on every individual gas stream toward the direction of the thumb;"** that is, toward the **left of an observer who is looking in the direction of the flow of the electric current, i.e., looking away from the core. Accordingly, when viewed from the geographic North of a star under the process of its creation, each gas stream,** *while approaching its fixed goal, the central core, would be first deflected toward the right hand of the observer, then it will bend toward the core. Accordingly, all such gas streams, when viewed from the top, i.e., from the geographic North of the star, would approach the core through a curved path swirling anticlockwise, that is, from the Geographic West of the star under the process of making, toward its Geographic East. This direction of rotation of the gas-whirl so formed is also of the utmost importance because almost all the celestial objects not only rotate in this manner relative to their magnetic poles, they also orbit their respective stars/gravity-centres in the same direction.*

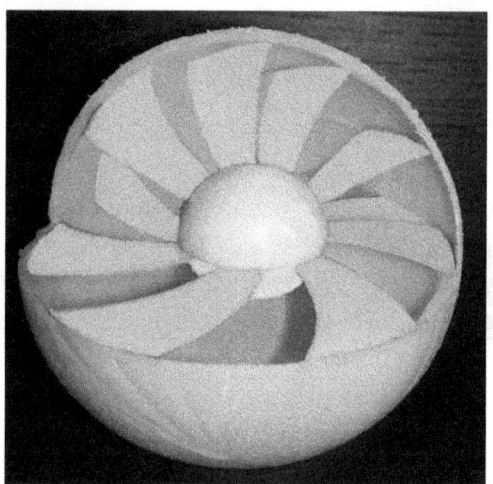

Fig- 7A

Fig- 7B

Due to the continuous ingress of the gas, the mass of the core as well as the pressure within it would increase continually; its temperature would also increase gradually and continually. In due course of time, the gravitational crunch would turn these rotating cores into different red-hot spherical balls, i.e., the embryos of the future stars. When these embryos would get heated up adequately, they would start to emit infrared radiation of adequate strength; before this point in time, it would not be possible for anybody to see the early stages of the process of the star-formation, i.e., the formation of different swirling cores. Before the creation of a star embryo, a zone of hot gases having no superconductivity would have been created around the core; this hot region would have been separated from the very cold region of superconductive gases by a region known as the frost line. When the magnetic field of the core, due to its gradually increasing strength, would reach out to this cold zone, the mirror image of this field would be gradually created in the molecule of the infalling superconductive gasses. Thus, right from the initial stage of the star formation, the infalling matter, due to its magnetic property, would have been organized in a two-dimensional rotating disk. The speed of the rotation of this disk, i.e., the gas envelope, as it is called, would directly depend on the distance from where the gas was drawn and inversely proportional to the mass and size of the rotating gas; the bigger and massive would be the rotating body, the slower it will rotate. The gradually increasing centrifugal force generated in the rotating gas envelope would start to resist the process of accretion of the gas and dust onto the core. Eventually, the rotating core would be separated from the gas envelope and become the protoplanetary disk.

*Within any nebula, several rotating cores, rotating in the above manner, would gradually grow side by side and gradually become more massive; as a consequence, their growing gravitational force would start to suck gas from the gradually increasing distances, even from beyond the frost-line, where the gas would be very cool and superconductive. As a result, the size, as well as the speed of the in-falling gas streams, and in turn, that of the gas envelope, would gradually increase. At the same time, the core would also continue to grow bigger. As a result, the gravitational field of the core would continue to suck gas from still greater distances. However, this process can't continue indefinitely. Eventually, a point will come when either the stock of the gas in that gas pocket will exhaust or the strength of its magnetic field will cease to extend any further. As a result, the size of the **protoplanetary disk** will also cease to grow. The more would be the quantity of gas rushing from this envelope towards the core, the less opaque the protoplanetary disk would gradually become. In due course of time, the gas envelope will gradually thin out and finally disappear.* In the meantime, the star embryo will continue to suck the gas and become more massive.

A similar process would have begun simultaneously within that nebula all-around at numerous other places, too.

It could thus be seen that all of such rotating cores, as well as their rotating gas envelops, would have been created by the incoming gas currents that were coming from different distances. Different gas currents would have started to flow from different distances. Accordingly, depending upon the distances from where the gas started to rush toward the outer edges of the respective cores, as well as their gas envelopes, both of these rotating structures would have different rotational speeds and different angular momentum as well.

A similar process would have started simultaneously within different stellar nebulas at different places. As a result, several hundred to millions of such star embryos of varying sizes and masses would have formed almost at the same time within different nebulas. *Since all such embryos and their rotating gas envelopes were created by a similar process, magnetic polarities and directions of rotations, etc., of all of them would also have been similar.* These embryos, depending upon their mutual distances, might repel each other with a very feeble force due to their similar magnetic polarities. Initially, such cores, at the time of their creations, would have simply continued to accrete mass from their respective gas envelopes; no other activity, because of their inadequate gravitational strengths, might have been possible.

In the due course of time, different cores, due to continually increasing mass and increasing compactness, would have been converted into numerous whirling maelstroms of red-hot gases of different sizes and masses. Similarly, different gas envelopes would have become protoplanetary disks. *The gravitational force would have further compacted these maelstroms of hot gas and thereby turned them into **swirling fireballs of spherical shape, i.e., into the embryos of the future stars**.* The bigger cores would have continued to collapse until the nuclear fusion didn't start within them; thereby, they would have become the newborn stars; the smaller cores would have become either the failed stars or planets. These young stars so formed would have started to emit the star wind at a very high speed of about 45,000 Kilometers per second, or even at a much higher speed than that.

After a few million years after the "big bang," different protostars, or the *spinning fireballs*, would have become so massive that their gravitational force would have gradually extended up-to-the-far-off distances. Depending upon the strength of the gravitational force of such fireballs, they would have started to draw gases even from the surroundings of the smaller fireballs located in their close vicinities. Next, amongst a local group of such star-embryos of different masses and sizes, the gravitational force of the biggest amongst all such fireballs would have gradually become so strong that apart from the loose gases, it would have started to attract even different smaller fireballs too, that were located within its gravity field. As a result, different smaller gas pockets created in the near vicinity to each other would have been drawn, one after another, towards the bigger one. These gas pockets, along with the star embryos, which are developing within their cores, would have started to move toward the most massive fireball moving on the same path on which the loose gas was flowing. However, the magnetic repulsion acting between them, and the pressure of the star-wind as well, might have resisted the inward moments of these embryos and thereby put them, too, on different curved paths; the angular moment of these whirling bodies might also have compelled them to adopt curved paths. The speeds of the in-falling gas, as well as that of different fireballs, would have increased with time; *the centrifugal force produced in these smaller fireballs, due to their mass and the increasing curvature of their paths, would have finally produced a balance between the gravitational pull of the central star-embryo and the centrifugal force of all such star-embryos spiraling toward the most massive star-embryo. As a result, these smaller fireballs would have established into different stable orbits around the big one.* The smaller fireballs, while orbiting the bigger one, would also have continued to suck gas from their near vicinity. In this manner, a supergiant maelstrom of gas would have been formed around the most massive fireball; the smaller spinning fireballs would have been its integral part. Such spinning fireballs would have continued to suck the loose gasses from their surroundings. Thus, a system of several thousands of smaller whirls spinning within a giant whirl would have been formed. The gravitational force of the biggest protostar, being much stronger than that of the smaller ones, would have sucked more gas from larger distances, including the areas lying around the smaller fireballs too. This would have probably starved the smaller ones, and as a result, either the rate of growth of the smaller fireballs would have retarded, or the same might have totally stopped at some stage of the formation of this giant whirl. In the absence of a continuous supply of gas, much smaller fireballs would have cooled down immaturely; accordingly, depending on their masses, they might have been turned into either smaller stars, failed stars, or planets of different sizes. All of the newly created stars and the failed stars might have been surrounded by a few planets and their moons, too. In the due course of time, smaller protostars situated outside of the said giant maelstrom might also have been captured by the central giant-protostar of such a giant maelstrom.

Such a process of star-formation, if spread over a very large area, then in the early stage of the creation, then such a process might have resulted in the formation of a whirling configuration of the first-generation star-embryos, in which several proto-stars along with their families of different planets, etc., or at least the gas-pockets comprising a gas-whirl, would have formed a giant whirl within which different smaller whirls of star-embryos would have been orbiting a massive central proto-star in the manner as shown in Fig- 7C. Such a configuration or congregation of the stars may be called the primitive, mini, or dwarf galaxy. Such dwarf galaxies would have persisted in growing by accreting more mass. In due course of time, bigger galaxies would have captured smaller ones and grew still bigger. The giant-sized central stars of such mini galaxies might have, in due course of time, become black holes directly. Explosions of different dying stars might have scattered the stardust and star debris in the form of numerous solid cores of different sizes. In the due course of time, the gas ejected by such explosions would have accreted on different solid cores; this would have resulted in the creation of the second generation of stars and their planets, which would have been the integral parts of different primitive or dwarf galaxies. Such a process would have started from the beginning, and the same is still going on.

Fig- 7C

The bipolar jets emanated by the young stars, under the process of their creation, are very much similar to the jets emanated by different "Quasars," (See Chapter-12 for details) and also to the jets that are produced just before the death of different super-massive stars; the death of such supergiant stars, results in Hypernovae explosions. However, as of today, no clear explanation of

234

the formation of such jets is available. At present, it is believed that stars and black holes, shed off the excess material swallowed by them through these jets; however, this explanation doesn't seem satisfactory to me at least. *Accordingly, entirely different speculation on the formation of these jets, is given in this book in the forthcoming Chapter–12, "The Quasars."*

The Death of the stars

The law of nature that "Anyone, or anything, that was born, is sure to die one day or the other" is irrevocable. Even the stars and the universe, too, are not exempted from this law.

Immense energy is produced deep within the cores of different stars by the fusion of the Hydrogen atoms, which are converted into the Helium atoms. In this process, a very small amount of matter is destroyed and converted into energy. Any star about the size of our Sun needs to burn hundreds of million tons of hydrogen within a small time period of one second so that it may sustain its existence. Accordingly, all the stars, in order to shine continuously, must necessarily accumulate a very large quantity of gas to meet their energy requirement for their entire lifetime. Normally, average-sized stars, during their estimated lifetime of 10 to 14 billion years, continuously burn hydrogen at the above rate. More massive stars in order to produce much higher temperatures to maintain their shapes against gravitational collapse, need to burn the fuel at much higher rates. However, massive stars, due to the much higher rate at which they consume their stock of the fuel, run out of fuel in comparatively much shorter time periods. The life span of a star depends on the total stock of fuel it has accumulated, as well as on the rate at which it consumes its Hydrogen stock. This is an irony that the bigger is the star, the shorter is its lifespan. The bigger stars complete their lives within a very short period of even ½ billion years to 1 billion years, as against the life span of 10 billion years of the average-sized stars of the size of our Sun. The stars, which are smaller than the sun, may have much longer lives.

It is the huge stock of fuel that provides life to the stars. However, the stars, whether large or small, one-day run out of their fuel. Thereafter, the very gravitational force that created all the stars by compacting the Hydrogen-clouds becomes the cause of their deaths, too. However, stars of different sizes and masses end up their lives in different ways. Although such stars are said to have died, even then, some activities still continue within their remnant cores, even after their so-called deaths. Life never accepts defeat so easily; the gas envelope separated out from the core of the dying stars provides the raw material for the creation of the second generation of stars; in addition, *the seeds of life were* also *produced* from the chemicals that were produced by the explosion of different dying stars.

A brief description of the different stages from which the stars of different masses pass through before their deaths is given below:

The Brown Dwarfs

The non-shining celestial bodies, which are marginally smaller than the luminous stars, are called the "Brown Dwarfs." Such bodies are, in fact, *"failed stars,"* which failed to trigger the nuclear fusion within their cores. The luminous stars must possess at least 80 times as much mass as that of the planet "Jupiter" so that they may support Hydrogen-fusion. Celestial bodies, which are 3 to 70-80 times as massive as Jupiter is, develop as failed stars. Such failed stars, during the process of their formations, would have become very hot due to the heat generated in this process; they, therefore, glow like stars or at least with a dull red light. Nuclear fusion would also have started to some extent in the cores of some of such bigger bodies; however, the process of the gravitational collapse could not produce the required temperature and pressure to sustain nuclear fusion. Whatever amount of energy is produced within these stars radiates out almost the whole of it. As a result, such bodies gradually cool down and become almost invisible. They retain their shapes due to natural repulsion between the atoms. It would be wrong to say that they died immaturely; in fact, they couldn't grow as stars due to the lack of the required quantities of gas.

The surface temperature of such stars normally ranges from 5 to 6 hundred degrees to 1000 degrees. At this temperature, some radiation, in the form of infrared rays and dull-red light, is radiated out from these failed stars; charged particles and X-rays, etc., are also emitted from these failed stars from time to time. Such radiations indicate that they are not completely dead bodies; instead, some nuclear activity is going on in their cores. These dwarfs also have magnetic fields of their own and may have a family of some planets and their moons too. Such a system of failed stars and its planetary system could be said to be the "Mini Solar System." Since the heat in adequate quantity is not generated within these dwarfs, they, in the next few billion years, may become very cold, due to which activities within their cores, if any, may stop totally. Even after this fate, their gravitational force would continue to exist. If such stars acquire some extra mass from any collision with other celestial bodies or in some other manner, they too may turn into small stars. However, if this doesn't happen, then they will continue as dwarfs forever.

The Red Dwarfs

Smaller stars having masses of up to half the mass of our Sun, or a little lesser, are also capable of supporting nuclear fusion, but at a much slower rate. The lesser the heat generated, the duller and redder the light emitted out from such stars. And therefore, they are known as "Red Dwarfs." These stars can also have families of a few planets and their moons. In case these stars don't get any further mass, then they would continue to burn their fuel for very long periods of time, at a very slow rate. The surface temperatures of these stars remain around 3 to 4 thousand degrees, due to which they shine with luminosities of around $1/10^{th}$ of our Sun, or even 1/10000 times lesser than that. Such stars might be far-more in number as compared to the visible stars; however, because of their faint light, it is very difficult to observe or find them. The luminosities of such stars would go on diminishing with depleting fuel, and because of that, it would become even harder to see them. Since such stars burn their fuel at a very slow rate, they have a much longer life as compared to the sun-sized stars. These stars, because of their much longer lives, might be capable of living even after the end of the universe. However, even such stars are also not immortal; some day or other, they also would run out of fuel; thereafter, they would cool down slowly.

The Red Giants

When the bigger stars, which have masses ranging from ½ solar mass up to 10 solar masses, consume about 2/3rd of their total fuel, then nuclear fusion within their cores gets slowed down. This, in turn, reduces the production of heat; as a result, when the outward pressure created by the heat of their cores gradually diminishes, gravity becomes dominant. This causes the gas envelopes of their cores (mainly comprising Hydrogen and Helium) to suddenly contract and collapse. This sudden collapse has two consequences, which are, 1) "More hydrogen enters into the hot zone" & 2) "Temperature of the core again shoots up, which in turn, reignites nuclear fusion in the Hydrogen layer that immediately surrounds the Helium layer formed deep within their cores." The heat generated by nuclear fusion causes the star to expand again. And as a result, the process of nuclear fusion spreads in a bigger region. Such a process causes such a star to expand rapidly, and become much bigger, say up to even 10 to 1000 times its original size; its luminosity may also increase by 1000 to 10,000 times. Any Sun-sized star may swell up to 100 to 200 times its original size. At present, the diameter of our Sun is about 15 hundred-thousand kilometers; however, in the next 4 to 5 billion years, when it will consume most of its fuel and consequently inflate by about 100 times its present size, it may reach up to the earth's orbit, or perhaps it may swell further, to an even greater extent and may reach up to the orbit of the planet Mars. Expansion of these stars, to such a great extent, causes their heat energy to spread over a much bigger volume. This results in the dropping of their surface temperatures to about 3 to 5 thousand degrees. Such stars, at this temperature, appear to be red or orange-red; as such, they are known as the "Red Giants."

When the process of nuclear fusion reaches the outer layers of the stars, then their gas-shells swell to such a great extent that the grip of gravity, on their outer gas-layers, becomes very weak. As a result, the outer gas layers of the red giants are blown or thrown away by the radiation pressure and the shock-waves produced by the nuclear activities going on within their cores. In about the next 1 million years or so, the entire gas shell of such a star is disposed of, leaving behind the bare core, which is very dense and hot. The gas so liberated accumulates around the star like a hollow shell or ring, which is known as the "Planetary Nebula." Such rings are short-lived, and therefore, in about the next 10 to 50 thousand years, they gradually expand and thin out to such a great extent that they become completely invisible.

When any massive star, much more massive than the sun, uses up its fuel completely, its core is squeezed to such a great extent that its core temperature shoots up to several hundred million degrees. Such a high temperature facilitates the nucleosynthesis of much heavier elements. Such stars, after running out of their fuel, swell many times bigger than the red giants. Such massive stars, depending on their sizes, are called the "Super-Giants" or the "Hyper-Giants." Being much massive and heavier than the red giants, they end their lives in different fashions, which are briefed hereinafter.

The White Dwarf

What could happen to a star when it runs out of fuel and how its cores could support themselves against gravity was proposed for the first time by an Indian student **Subramanian Chandrasekhar** in the year 1928. According to his calculations, any dying star, which has about 1.4 times* the mass of the sun or lesser, after using up all of its fuel, and disposing-off its gas envelope, would eventually

---

* This limit of mass is known as the "Chandrasekhar Limit"

stop shrinking (due to its own gravity) and would settle down as a white dwarf; this would become possible because of the force of repulsion acting between the electrons, this force would finally counterbalance the force of gravity. The compact material formed due to the contraction of the cores, of such stars is called the "Electron Degenerate Matter." The small but very dense cores, leftover after the end of these Red-Giants, are known as the "White Dwarfs."

In the beginning, the dense core, left behind after the death of a Red-Giant, glows with very bright white light due to its residual heat energy. It may glow with a luminosity of about 250 times greater than that of the sun. However, in the absence of any nuclear activity, it gradually cools-down; its glow would gradually fade-out due to this gradual cooling. The white dwarfs that are formed after the deaths of different stars of different masses would differ in their sizes. The white dwarf left over after the death of a star as massive as our Sun, would be almost equal to the Earth in size, that is, about 13,000 kilometers in diameter. White dwarfs, after a few billion years of their formations, would cool-down to such an extent that they would not emit any light. Such cores are known as "Black Dwarfs." A black dwarf is, probably, made of pure diamond that measures several thousand kilometers in size; however, because of its mass and gravitational force, we can neither bring it to our earth nor start mining diamonds on its surface.

In fact, the leftover core of a so-called dead star is not at all a dead celestial body, that is, a totally inactive body. In case any other star or big planet collides with it, or it sucks-in sufficient matter from any other nearby star or its own planet, then the process of nuclear fusion may restart within its core. In case, its mass persists to grow, then after a certain critical limit of mass, it would shrink further and, as explained hereinafter, convert into a "Neutron Star," with a very powerful explosion.

Supernovae and the Neutron Stars

The most intense and luminous interstellar glow is known as "Super Nova." The word "nova" means new or a new star; that is, the word supernova means a very bright new star. In fact, the glow of a supernova is produced by the explosion of a dying star (a supergiant), which emits a very energetic and extremely luminous glow that lasts for a very small time period on the stellar scale. However, both, its remnant core and the gas envelope that separates out from it, may continue to shine for a period of a few weeks to a few months. Such explosions may emit as much energy in their brief life span as the sun could do in its entire lifetime of about 10 billion years.

The process of nucleosynthesis in the cores of bigger stars is carried out at a far-rapid rate; therefore, the cores of the bigger stars, having about 3 to 10 solar masses, generate temperatures of hundreds of million degrees. Such a high temperature enables these stars to start fusing the Helium atoms too. The fusion of Helium atoms produces some heavier elements such as **Selenium, Carbon, Nitrogen, Oxygen, etc.** However, the nucleosynthesis of Helium atoms does not produce much heat; therefore, the star begins to contract under its own gravity. As a result, nuclear fusion spreads, for a small time period, in the inner layers of its gas shell and heats it up. This process, being short-lived, causes repetitive swelling and contraction of the core. The star eventually inflates very rapidly and grows much bigger, even much bigger than the red giants.

The process of nucleosynthesis, within the cores of the much bigger stars, generates temperatures as high as eight (8) billion degrees; this temperature is about 60 times higher than the temperature of the core of our Sun. At this temperature, the element "Iron" is also produced, which is comparatively heavier than the elements produced in the smaller stars. The gravitational force tries to fuse these iron atoms too, but it fails to do so because a much higher temperature is required for this purpose. The element Iron is said to be the enemy of the stars because it absorbs almost their entire heat energy. As a result, the process of atomic fusion comes to a sudden and almost spontaneous halt; as a result, the generation of heat is also suspended. Consequently, in the absence of any resistance in the form of the outward pressure, the entire matter contained in the cores of such stars, is suddenly squeezed into a very small point. The crushing force in such a big star is so strong that even the mutual repulsion between the electrons fails to counter it; as a result, all the electrons enter into the nucleus and convert all the protons into neutrons. Finally, the inward rush of the matter is halted by the mutual repulsion among the neutrons. This sudden halt of the in-rushing matter creates a powerful shock wave. The matter rushing inward is not just halted; it is rebounded back with almost equal speed. Consequently, the in-rushing matter bounces back at a very high speed. The shock wave, so produced, shatters the gas shell of the dying star and sweeps the same off, up to very large distances. Some very heavy elements, in very small quantities, are also produced during such explosions due to a very high temperature that is produced in this process. The remnant core left behind after the total destruction of such a massive star is a very small and dense core having a radius of about 20 to 30 kilometers. Almost all the mass of such a massive star squeezes into this dense core. This compact matter is known as "Neutron Degenerate Matter," and the remnant core is called "Neutron Star."

The Binary Stars and their Deaths

Some stars, during the process of their formation, are created alone as a single star, such as our sun, whereas some of them might have taken birth in pairs of two or even in a group of more than two stars. In the beginning, the seeds of these stars, because of their low masses, would not be able to exert any appreciable force on each other. However, as they grow bigger, their mutual attraction would also grow stronger. Ultimately, because of almost equal masses, they would have started to orbit each other; more precisely, they would orbit the point of their common mass. In these pairs, the star having more mass is known as the "Primary Star," and the other one as the "Secondary Star." The heavier one of these two stars runs out of fuel comparatively much earlier and becomes a white dwarf; due to that, it becomes almost invisible. The secondary star, which has relatively lesser mass, becomes a red giant. This red giant, therefore, appears to orbit an invisible object. Sometimes, another star, having a family of several planets, might also orbit a binary pair of stars. Such a congregation of stars is known as a "Triplet of stars."

During the year 1963-64, scientists observed a strange phenomenon: *"A giant star was not just orbiting an invisible point, it was a matter of great surprise for the scientists observing this phenomenon because gas, from the gas-shell of that star, was gradually spiraling toward that invisible point where the infalling gas did persistently disappear without leaving any trace whatsoever."* This invisible point is known as **"CygunusX-1,"** which was later accepted as a ***black hole,*** perhaps, of medium size. Astronomers first thought that a primary star in a pair of binary stars must have collapsed and transformed into a white dwarf, whereas the other one, orbiting the dwarf, might be a red-giant. And as such, because of the very small distance between the two, the white dwarf is gradually snatching away gas from the gas shell of its companion.

In such a pair of stars, when any invisible white dwarf accumulates adequate mass, then nuclear action might once again start within its core. Such a dwarf, after acquiring adequate mass, might end up in a supernova and become a neutron star. Such an explosion may throw away the gas-cover of its companion red-giant, or even its remnant core too. Scientists have observed a few stars moving at very high speed outside our galaxy; it is believed that, perhaps, such stars have been thrown out of the galaxy due to the explosion of the primary star of a binary pair.

Hyper-nova

Massive explosions causing deaths of extremely massive stars, say 10 to 30 solar masses or even bigger, are known as "Hyper-novae." These stars use up their fuel within half a billion to 2-3 billion years. Thereafter, they end their lives with explosions like a supernova that might be hundreds to thousands of times more energetic and vigorous than any supernova explosion could be. Such exploding stars, just before their deaths (explosion), emanate very powerful bipolar jets of charged particles and gamma-ray radiation, etc., from their poles (please also see chapter-12, Quasars, for further details).

Extremely high temperature, matching the energy of the hyper-nova explosion, is generated at the time of such explosions. During that extremely small fraction of time, very heavy elements like Silver, Gold, Platinum, Uranium, etc., are also produced at such a high temperature. Since these elements are created within very limited time periods, they are so rare in nature.

Fig- 8

An imaginary depiction of a "hyper-nova" is illustrated in Fig-8 above. Probably, the gas shell and debris of the dying star are thrown mainly in two different directions, 1) "Charged particles and energy radiation is mostly swept away in the directions of both of its poles," and 2) "The gas shell and the star-debris, etc., are thrown away in the direction of its magnetic-equator, this debris might spread like a doughnut or a toroid." Such a direction of disposal of the star debris might be related to the strength and direction of the star's magnetic field. What actually happens could only be established by astronomers and scientists.

Quark Novae

A very powerful explosion, much bigger than the hyper-nova, was observed about 13 to 14 years earlier from now, i.e., sometime in 2006-07. The energy of this explosion suggested that probably still a bigger star had exploded. The gravitational force of such a massive star would be so strong that even the mutual repulsion between neutrons would fail to stop the gravitational collapse, which would be finally halted by the mutual repulsion of the quarks. Such an explosion is known as "Quark Nova," which is much more energetic and stronger than the hyper-nova. The dense core left behind such explosions is known as the "Quark Star," which is made of the "Quark Degenerate Matter." Such stars, on the name of the strange quark, are also known as the "Strange Stars."

Electroweak Stars

It is speculated that in case a still bigger star explodes, then even the repulsion between the quark degenerate matter would fail to support their cores. It is believed that gravity would squeeze quarks completely and burn them down, thereby converting them into Leptons. An extremely great temperature would be generated in this process, so high that the electromagnetic force and the weak nuclear force would merge into each other, thereby, they would become a single force. Such stars support themselves by mutual repulsion between the leptons. Such hypothetical dense cores are called "Electroweak Stars."

Black-Holes

The next stage of the star-collapse is the *"Black Hole,"* which is discussed separately in the forthcoming chapter -11.

The Spark of Life

In the beginning, a little after the "big bang," only very light elements such as Hydrogen and Helium were created; no heavy element could be created at that time. However, some heavy elements have also been seen in the oldest galaxies. This means that extremely massive stars were formed in the infant universe, which might have very short lives of, say, about ½ billion years. The heavy elements and their different compounds would have been created within the cores of the first-generation stars and also during their explosions. Subsequently, these chemicals might have spread all over the universe due to explosions that would have been similar to a hyper-novae or supernovae. The second and third-generation stars and their planets might have formed from the star-dust and the debris that was produced after the death of these first-generation stars. Apart from Hydrogen, Helium, and other light elements, different heavier elements like Nitrogen, Oxygen, Silicon, Calcium, Carbon, Phosphorous, and their compounds, such as water vapor, Carbon-di-oxide, Methane, Ammonia, and different oxides, nitrides, and carbides of different elements, etc., were also formed during these explosions. All these compounds are the raw materials from which life did evolve* after a lapse of an eon of tens of billion years after such explosions. In brief, life was created by the death of the stars. The death of the stars also resulted in the creation of the next generation of stars and their planets on which various life forms originated in due course of time. The deaths of stars have provided us home, such as Earth-like planets, to live and flourish. It is a cruel reality of nature that the "death of one is the life for others."

We knew beforehand that the entire universe and all living beings are made of different matter particles. Now, as brought out in this chapter, i.e., 'Life Cycle of the Stars,' we have seen that the force of "Gravitation" is the main tool of Nature for the creation of the stars and planets; ironically, the same force causes their destruction too. The raw materials for the origin of life were also created indirectly by the same tool that caused these raw materials to spread all over the universe. However, gravitation is merely a property of the matter that was created

---

* Kindly refer to chapter-16

during the "Big Bang." Thus, matter and all of its intrinsic forces (See chapter-3) are responsible for the creation of the universe and eventually for the origin and evolution of life.

By now, we, mankind, have find-out the answers to some of the eternal questions, except for the question of wherefrom the energy was created that caused the "Big Bang" and how the different properties of the matter were decided?" The answer to these questions is still eluding mankind; it is probably, beyond his capability to solve this riddle.

PART – 3 (B)

FEW

UNRESOLVED

MYSTERIES
Of the
Universe

10: The Puzzle of gravitational Force

Although **Relativity** does not consider gravity a real force, even then, scientists of the later generations have, perhaps, different ideas; the Standard Model of Particle Physics envisages that only a single "super-force" existed before the big bang, which later on split into **Gravitational Force** and *three other fundamental forces,* namely: **The Strong Nuclear Force, The Weak Nuclear Force, and The Electromagnetic Force.** All the four fundamental forces are supposed to work in a similar way, that is, by emitting and absorbing different kinds of force carrier particles. However, contemplating a little, it can be seen that gravitational force is much different from the rest of the three natural forces because the properties and strengths of the remaining three forces start to change at high temperatures, whereas the temperature has no effect on the strength of the gravitational force, which works with the same vigor at temperatures ranging from absolute zero to hundreds of billions of degrees K. All the other forces repel the like forces, whereas, gravitation always attracts, it never repels any object. Normally, all the other forms of energy always flow out from the points of their higher densities to the lower densities. Contrarily, the gravitational force always attracts everything toward the center of gravity, where it has the highest strength, which never flows out or dissipates, *i.e., it doesn't follow the rule applicable to other sorts of energies.* Every massive body, under the influence of gravity, tends to collapse and shrink in size. The gravitational force can crush the matter and compact it into a point of almost zero size, where this force becomes strongest amongst all other natural forces. The gravitational force of such a zero-sized massive body can capture even all other sorts of energies. As discussed earlier in chapter-6, in *"Analysis of Relativity,"* gravitational force might not be merely a consequence of the curvature of the space-time; it might be a real force having a separate and independent identity. This force must be independent of the curved space. Probably, Gravitation is not a property of any particular kind of particle; it is the common property of all sorts of matter particles. The strength of the gravitational force of any object, because of the virtue of this fact, always bears a definite ratio to the combined mass of the total number of the matter-particles contained within that object.

Scientists, while contemplating over the problem of *the beginning of the universe*, are perturbed by the far-weaker strength of the gravitational force; the problem that worries them the most is the strength, rather than the lack of strength of gravity. They expect that "Since all the four fundamental forces were created by the splitting of a single super-force, all of them should have equal strengths." However, gravitation is much weaker than the remaining three forces, almost billions of times weaker. Probably, everyone has seen that even a small magnet can lift a small pin against the gravitational force of the earth. This fact, though, appears most ordinary to the commoners; the same is a matter of great concern to the scientists because, at a glance, it might seem that even a small magnet can overpower the gravitational force of the entire earth. However, a series of questions arise at this conclusion- 1) whether the entire gravitational force of the earth was centered only on that pin? 2) Whether the same magnet can lift comparatively heavier objects? 3) Would it lift the same pin from a much longer distance? 4) The Earth is holding the Moon from a distance of 3.8 hundred-thousand kilometers; this fact attracts the question, "can any magnet also do the same?" 5) If a small magnet is stronger than the earth's gravitational force, then can it also control the orbits of the moon or any other celestial body made of pure iron?

The answer to all the above questions would certainly, be negative. In case the distance between the pin and the magnet is increased a little, then it would certainly fail to lift the same pin. On the other hand, if the whole mass of the earth could be compressed into a small point, its gravitational force would become so powerful that not just a small magnet, even the most powerful crane, would not be able to lift the same pin.

It is clear from the above delineation that before jumping to the conclusion that *even a small magnet is stronger than Earth's gravitational force,* everybody has probably persistently ignored an important fact **"The gravitational force has the longest *range-of-effectiveness* amongst all the other Natural Forces."** Although magnetic fields of different celestial bodies, too, permeate very-very long distances, the range of effectiveness of the other two natural forces, i.e., strong and weak interactions respectively, is limited to atomic diameters only. Similarly, the effectiveness of the electromagnetic force, acting between the nucleus and the electrons, is also limited to a very small region. On the other hand, gravitation is capable of exerting a force of attraction on different celestial bodies that are spread over distances of several thousand light-years or even beyond. However, similar to the rays of light, the intensity of this force acting between two objects decreases very sharply, when the distance between these objects is increased.

In the above scenario, it could be seen that the earth is exerting a gravitational force on the said pin from its center of gravity, that is, from a distance of almost 6,550 kilometers. Whereas the magnet can lift the said pin from a distance of only a few millimeters, which is about one billion times smaller as compared to the distance from which the earth exerts a gravitational force on the same pin. *Therefore, it would not be proper to say that even a small magnet is several billion times stronger than the entire gravitational force of the earth.*

As brought-out earlier, ***the gravitational force*** acts over very-very long distances, even at such far-off distances where all the other forces become ineffective. However, its effectiveness depends on the mass of different objects and the distance among them. This could be very well understood by the following example: Our Moon is located about 380,000 kilometers away from us. We may not feel the effect of the Moon's gravity from such a long distance; even then, its attractive force produces the high tide in the huge mass of seawater. Similarly, the sun, too, from a distance of about 150 million kilometers, produces a tidal effect in the seawater, but no such effect is observed on the sand particles lying on the beach or on the smaller quantity of the water stored in various other reservoirs. *The fact that the tidal effect is produced only at that surface of the sea that faces the Sun or the Moon indicates that such an effect is definitely produced by some attractive force exerted by these distant celestial objects this conclusion is supported by the fact/ logic 1)* ***any deep-well made in space, cannot cause the seawater to rise in the opposite direction of the Earth's gravity-center; this fact indicates that the tidal effect is not produced by the warping of space-time. 2) The size of the curvature made in space would be the same for the seawater as well as for the sand particles lying at the same distance from the Sun or the Moon, whether they lie on the surface facing the sun/moon or elsewhere, but no tidal effect is produced on the sand.***

In the above case, *the sun and the moon respectively exert attractive force in a similar manner on both the seawater and the sand, both of which lie respectively at the same distance from them; even then,* ***the tidal effect is produced in the seawater alone.*** *Reason for the same being: sand fails to act as a single massive body because of a lack of cohesion between the sand particles.* ***This fact indicates that the more massive the object is, the greater the gravitational force acting on it.***

Although the earth and the moon are located at about equal distance from the sun, even then, the moon has been captured by the earth's gravitation. Evidently, the Sun could not capture the Moon because of its lesser mass as compared to that of the Earth and also because it is much nearer to the Earth. The Sun, because of this reason, is unable to exert as much force on the Moon as the Earth does. This further means that besides the mass, the distance between various objects too has an important role; a combination of both of these entities, i.e., mass and distance, decide the strength of the force acting between any two objects. The structure of the entire universe depends on this property of gravitational force. Although the gravitational force is considered billions of times weaker than the other natural forces, it carries out its role properly and effectively. Had the nature of gravitational force been a little different from what it actually is, then immediately after the **Big-bang,** either the universe would have collapsed to a single point, or it would have expanded at an infinite rate and defused in the endless *space*. Whatever duties nature has assigned to the individual forces, it also has bestowed matching properties upon them so that they may perform the assigned work efficiently. Or another way around, it may also be said that the creation of the universe was done exactly according to the properties of the natural forces. And therefore, there is no point in comparing their strengths; Mother Nature has bestowed all the four forces with different properties and different strengths so that they may discharge their assigned duties properly.

According to Sir Isaac Newton, the value of the gravitational force acting between two objects can be calculated by the formula: $F_g = G. (m_1. m_2) /d^2$ where "m_1" and "m_2" represent their respective masses of these objects, and "d" is the distance between them. The distance between the center of gravity of two or more objects is, therefore, extremely important. The denser object, out of two different objects of the same mass, would be smaller in size, due to which the gravitational force would become more intense at the surface of the denser ones. When any star collapses due to its own gravity and becomes a black hole, then the intensity of its gravitational force increases a lot. As a result, any black hole, in comparison to the original star, exerts a much stronger force on the objects that are located at equal distances from their outer surfaces. An optical lens can focus the light rays on a very small area and thereby causes a piece of paper to catch fire. In the same way, when any massive object is compacted to a very small size, then the intensity of its gravitational force increases to such an extent that even the atoms of that object are crushed and compressed into a zero-sized point.

Scientists believe that the strength of the gravitational force acting between two objects decreases in the inverse proportion of the square of the distance between them. *However, gravitational force doesn't act on a two-dimensional surface alone; this force permeates spherically in all three dimensions.* **As such, the strength of this force should decrease in the inverse proportion of the volume, that is, in the inverse proportion of the third power of the distance (d^3) between two objects; accordingly, the strength of the gravitational force shall decrease in the proportion of M/d^3.** The value of the force exerted by two objects on each other is calculated by the formula- $(m_1.m_2)/d^2$; this would mean that both of these objects would exert equal force on each other. This formula attracts a question, **"How is it possible that any smaller object can attract a much massive object with the same force** *at which the massive object attracts the smaller one?" These two forces shall not have the same value.* In fact, the value of the total force acting between two objects should be the sum of the attractive forces that the individual objects exert on each other. However, very small objects shall, normally, not be able to exert any appreciable attractive force on the bigger objects from the far-off distances because their gravity fields fail to reach out to far-off distances. Normally, only the bigger objects attract the smaller ones from extremely large distances. In case both of the objects have considerable masses, and the distance between them is comparatively shorter, *then the total force acting between the two shall be the sum of the forces exerted by the individual objects on the other one.* This idea needs thorough verification.

The way in which gravitational force performs its function could be understood by the example of the light. Any distant star can be seen from a very long distance of hundreds of thousand light-years, whereas it is not possible to see the light of a small candle from the same distance. This doesn't mean that the rays emitted by a candle won't reach us from far-off distances. In fact, different light sources, depending on their energies, i.e., the intensity of light emitted by them, could be seen from different distances. Candlelight could not be seen from far-off distances because its intensity, after traveling through even a small distance, defuses to a great extent due to the continuous expansion of light in all three dimensions. In other words, the number of the photons per unit area, that is, the flux of light goes on diminishing to such an extent that it becomes difficult to see the source, i.e., the candle; *it is clear that it is the intensity of light, i.e., flux per unit area, that matters.* Any source of light emits a flood of photons, or successive light-waves, which continually propagate one after another. Therefore, the intensity of light, that is, *the flux of photons per unit area* at any particular distance from the source, normally remains almost constant. As a result, when an object comes out of the shadow of any other object, then the same would be instantaneously flooded with the light of almost the same intensity. It is clear from this example that *it doesn't matter how far away that object is located from the source of light because the photons did exist at that place beforehand; this phenomenon has no relation to the speed of light because such objects do not have to wait for the fresh light-waves to come from their source to illuminate them.*

Probably, the *Gravitational force* also works in a similar fashion as described above. Everybody might be conversant of the fact that any telescope having bigger lenses instead of smaller ones can see a faint light in a much better way because such a telescope catches more photons. Similarly, a battery of photocells can produce more power than a single cell. Analogous to this fact, any massive star would exert a stronger attractive force on a massive object than what it would exert on a smaller object that is located at the same distance or even closer.

I believe that the gravity field of any celestial body, unlike the deep gravity wells, as was envisaged by Einstein, permeates spherically, like an invisible spherical halo that surrounds an object from all around; this is similar to the magnetic field that surrounds the poles of a magnet. The ambit, or the reach of such a gravitational field of any object, like that of the light-rays, might extend to infinite distances; however, beyond a certain distance, its intensity would gradually become so weak that it might not be able to exert any appreciable force on other objects. Accordingly, any small object located within the gravitational field of any massive body may or may not exert any appreciable force on the said massive object or on any other small object that is located in the gravity field of the said massive body. However, the said massive body, depending upon the distances and masses of different smaller objects that are situated at different distances from it, would exert a much stronger force on the said smaller objects when compared to the force any smaller object would exert on the said massive object. Now, in case any small object moves within the limit of the effectiveness of the pre-existing gravitational field of any massive body, then the strength of the gravitational force exerted by the said massive body on the said moving object would be adjusted immediately and automatically in accordance with the varying distances between both of them; however, nothing like this will happen beyond the limit of the effectiveness of the gravitational field of any object. *Analogous to the light waves, the gravitational force, too, would not need to travel repeatedly between these two bodies.* This *would mean that* **within the ambit of the effectiveness of any gravitational field, the gravitational force acting among two or more bodies would immediately be adjusted according to their varying distances, without depending on the speed of gravity.** *However, any celestial object, say object-1, would not exert any appreciable force on any smaller celestial body (object-2) that is located beyond the limit of the effectiveness of its gravity field. However, the same object-1 may exert stronger force on a much more massive object (object-3) situated at a comparatively greater distance. This limit of the effectiveness of the Gravitational Field of different objects would depend on their respective masses; the more would be the mass of an object, the greater would be the range of its gravity field. However, this range of effectiveness would vary for different objects depending upon their respective masses.*

Before the formation of the *general relativity theory,* it was believed that in case any celestial body moves from its place, then the force acting amongst all other bodies, located even at infinite distances, would change instantaneously. This concept is only partially correct that too within short distances only, **it would not hold good at infinite distances.** As discussed in the previous paragraph,

the strength of the gravity field of any object can neither remain constant nor effective beyond the limit of its effectiveness, i.e., at infinite distances. *The distances between the neighboring stars located within any galaxy are far too long than the effectiveness of their gravity fields, which depends on their individual masses. As a result, even the neighboring stars, because of inadequate strengths of their respective gravity fields at such a large distance, are unable to have any influence on the speeds, locations, or respective orbits of each other or that of their respective planets. On the other hand, the gravitational force of the million times heavier supermassive black holes, which are located at the galactic centers of other galaxies, permeates over hundreds of thousands of light-years, i.e., up to the outer edges of their host galaxy. In fact, the gravitational field of a Sun-sized star, depending upon its mass,* **can suffuse over a distance of 1 to 1½ light-years only,** *whereas the distance between two stars normally exceeds 4 light-years or so. On the other hand, the gravity field of any supermassive black-hole can spread over a distance of several hundred thousands of light-years or even far beyond.* Accordingly, all the stars within any galaxy are held in their respective orbits around the central black hole without disturbing other stars. *In the above context, it is a matter to ponder that if the gravitational force of any star, depending upon its mass and the distance from other stars, fails to reach out to even the nearest star, then how the same, as was believed earlier, can reach out to all the stars in the entire universe?*

Although the *Gravity-field* of any supermassive black hole definitely permeates beyond the outer edges of the host galaxy, its gravity field, at such a large distance, becomes too feeble for the normal-sized stars located beyond that limit; accordingly, its gravity may not be able to capture the stars from other galaxy or exert any appreciable force on them. Albeit, the same black hole, in spite of very long distances, might exert a very powerful force on the supermassive black holes located in the centers of the other nearby galaxies. Even in such a case, the same central black hole of the said galaxy would not exert any force on the stars that are located within any other nearby galaxy. Therefore, the supermassive black holes of the other galaxies would have no influence on the movements of those stars. For this reason, other galaxies will not disturb the rotational movements of any star or its planet or any other celestial body like different moons, steroids, or comets, etc., orbiting their respective stars or planets located in the nearby galaxies.

*In the whole of the universe, numerous galaxies of different sizes and masses are spread all over, which are located at different distances from each other. As brought out in the previous paragraph, the gravity fields (****not the deep gravity well****) of their central black holes permeate up to different distances. Accordingly, the gravity fields of all such black holes, because of very large distances, become so weak that* **only the most massive black hole** *among other black holes of any group of nearby galaxies attracts and captures comparatively smaller central black* *holes of the nearby smaller galaxies located within the ambit of its gravitational field. Therefore, analogous to the formation of a whirlpool, the smaller black holes, along with their respective disks of stars, would orbit that very galaxy that has the most massive black hole in its center;* **no dark matter would be required for this purpose.** Such whirling structures of galaxies are known as ***Local clusters of galaxies*** or ***Local groups of galaxies***. All the stars of the respective galaxies that form such a group of galaxies would continue to move in their respective orbits, without being affected, in any way, by the gravitational force of different black-holes of the other galaxies of that group.

The distances between different groups of the ***local-clusters of galaxies*** are comparatively much greater as compared to the distances that exist among the black-holes of different galaxies situated within the same local cluster of the galaxies. The gravity field of only an extremely big black-hole would be able to attract the central black-holes of different local clusters of galaxies situated within that group. Accordingly, the gravitational force of that particular central black hole of that particular cluster of galaxies would compel all other nearby *smaller clusters of galaxies* whose central black holes are comparatively less massive to orbit that very local cluster of galaxies that have the most massive supermassive black hole in its center. In this way, still bigger clusters of galaxies would be formed in which different *local clusters of galaxies* orbit a comparatively bigger cluster of galaxies. Such giant clusters of galaxies are known as *"Super-Clusters of galaxies."*

The distances between different **super-clusters of galaxies** are so vast that even the most massive black holes of the other nearby *superclusters of galaxies* fail to exert any appreciable force on other supermassive black-holes that are located in the centers of different nearby *super-clusters*. Due to this reason, different super-clusters, in the absence of any strong attractive force acting between them, are floating almost freely in the space. Therefore, the formation of still bigger whirling configurations of *super-clusters of galaxies* becomes impossible. Such super-clusters are not connected directly to each other; instead, they are connected to each other like the links of any chain in which one cluster is connected to others, one after another, by very weak gravitational force that acts between the central black holes of each supercluster. Countless groups of such interlinked superclusters of galaxies are spread all over the vast universe. Because of this type of bondage, the superclusters are joined together like the spider-web. Such a chain of superclusters is known as "Filaments." These Filaments can form very long chains measuring billions of light-years. Webs of such chains are spread all around in the entire universe. In this expanding universe, only these superclusters are, perhaps, drifting apart, not the other whirling structures that are held together by the strong bondage of gravity.

Probably, only those *rotating structures* are stable in shapes, which are held together by the strong gravitational bondages.

Whenever two galaxies happen to come very close to each other, the attractive force acting between their respective black holes might increase exponentially. Accordingly, they might start to come still closer, spiraling toward each other; they may finally collide and merge into each other. The process of their merger might take a few million years to complete; thereby, a much bigger single galaxy is formed. After such a collision, the black holes of both the galaxies start orbiting each other to form a pair of binary black holes. Eventually, the smaller black hole might fall into its bigger companion. Likewise, if a much smaller galaxy comes very close to a massive galaxy, then the bigger one may stretch the smaller one like spaghetti, which after some time may add up to the bigger galaxy like an extra arm. In this case, the smaller black hole would start orbiting the massive one from quite a distance. This smaller black hole would gradually approach the bigger one through a spiral path. In due course of time, after reaching very close to the bigger one, the same would finally either merge with it or, alternatively, it might be thrown out of such merging galaxies to a far-off distance. Such supermassive black holes, which have been thrown out, would gradually accrete stars from their surroundings, and thereby they would, in due course of time, build new galaxies. All such activities depend only on the distances and masses of different galaxies and also the mass of their respective black holes, i.e., on the value of the gravitational force acting amongst them.

11: The Black Holes

A speculative image of the region surrounding a black-hole, from its near vicinity, is depicted below:

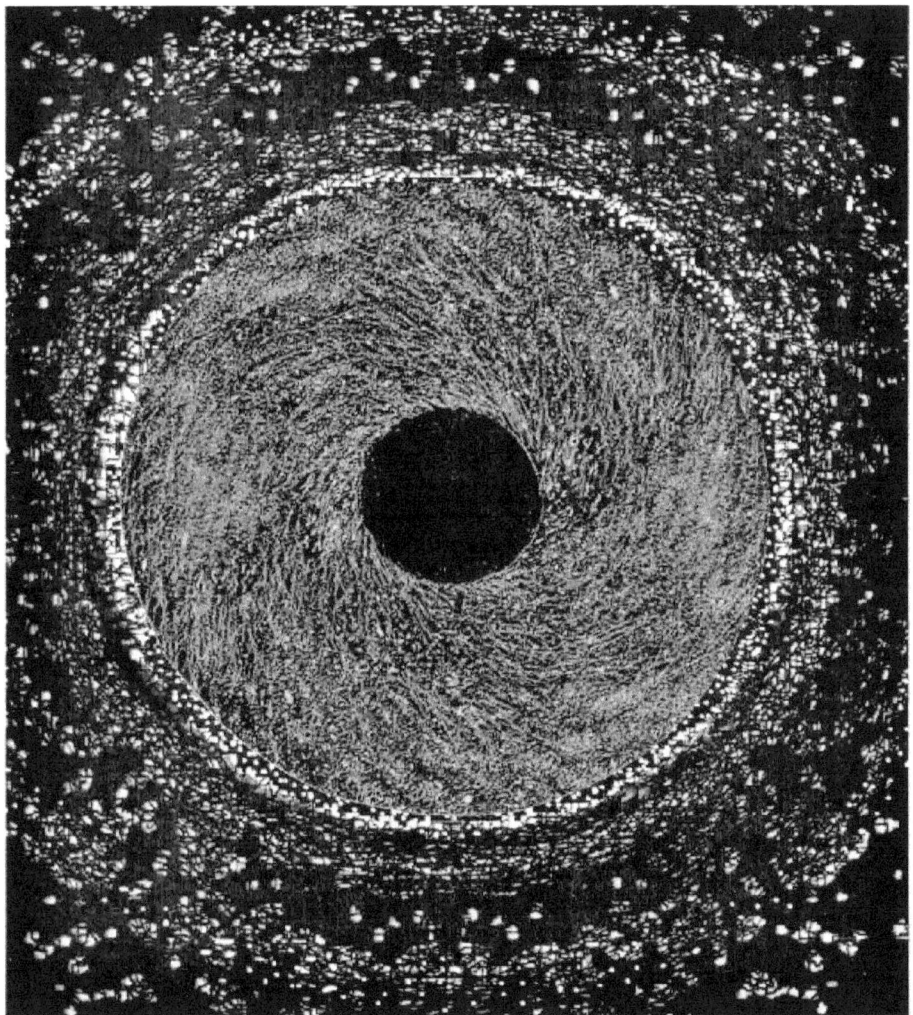

The Black holes are supposed to be point-sized, most compact, and totally invisible celestial objects, rather regions, having immense masses in different quantities. Black holes, because of their almost infinite gravitational force, can swallow anything from their vicinity; even light rays cannot escape their gravity. *Although the existence of the black holes was envisaged as early as 1783, this idea was not accepted for almost the next 200 years until Roger Penrose, in 1965, proved their existence. Although Blackholes do certainly exist, it is not at all possible to see any black hole. However, the effect of their gravitational force can be felt from very far-off distances.* Stars located very close to the galactic center of our home galaxy, the Milky way, do orbit an invisible object at very high speeds. This fact indicates the presence of a very strong gravitational field at the galactic center. Apart from this, *gases from the gas-shells of several stars located within the milky way have been observed to spiral into some invisible point, where the infalling gas disappears forever. Sometimes a dark and point-sized small object can be seen in the background of bright light or behind the gas falling into such an object. These invisible objects are most likely the black holes.* Infalling gas that spirals toward any massive black hole becomes so hot that it periodically emits X-rays, radio rays and infrared rays, etc. Accordingly, possible locations of different black holes can be predicted by locating the sources of such radiations.

When the fuel of a massive star is completely exhausted, then such a star collapses into a small point; consequently, that star becomes a black hole. *Scientists believe that the process of gravitational collapse is so rapid that the star would start to spin at a very high speed.* **Relativity** *envisages that at such a high speed, the entire energy of that star would be instantaneously carried away by* **gravitational waves**. *As a result, the star would shed off all of its energy within that very instant and settle down in a stationary state; it will also cool-down to 0°K. Thus, the star would turn into an* **extremely small object of zero activity and zero energy.** Therefore, it won't be possible for anybody to see such an object because it would not emit any sort of radiation. Such a black hole, because of its immense gravitational force, would continue to suck gas and other objects from its vicinity; therefore, its core would again start to rotate due to the very high velocity of the infalling gases.

The aforesaid idea of the formation of Blackholes is though based on Relativity; the same violates the interconvertibility of the matter and energy because the mass is considered the measure of energy; accordingly, any object with immense mass cannot have zero energy.

The Gravity Waves

Before we discuss the anomaly pointed out above, let's find out *"What the* **gravity waves** *are, and how a moving/spinning object can dissipate its energy by these so-called gravity waves?"*

Waves are normally produced by periodic vibration of the particles of any medium, about an average position that is the position of its rest. Einstein predicted that massive objects, while moving through space, produce gravity waves; these waves, or ripples, propagate through the fabric of space-time with the speed of light. *It was predicted that, similar to any other energy wave, the gravity waves too, carry away the energy from those fast-moving objects that produce these waves.* Sophisticated equipment of the modern era has found that the earth is very gradually moving closer to the sun, and accordingly, the earth might fall into the sun within the next 4 to 5 billion years. However, one example goes against this prediction: The moon is gradually moving away from the earth instead of coming closer to it. In case the prediction of the scientists is correct, then both Earth and Moon shall follow the same rule. However, they practically do not follow this rule in a similar manner. Therefore, it is very difficult to establish the fact that whether both, the earth and the moon, are really losing their energies by way of the emission of gravity waves? *Alternatively, it is also possible that the Sun, due to the gradual depletion of its stock of fuel, has gradually and very slowly started to swell, and as a consequence, it seems to us that the Earth is moving closer to the Sun.*

This book has emphasized in chapter-6, and at many other places, that no continuum like space-time does exist. *Anyhow, if gravitational waves are produced in space-time due to the movement of the celestial bodies, then such waves would complete their one frequency after the body producing such waves completes its one orbit.* Celestial bodies take very long time periods to complete their orbits. The earth takes a time period of one year to complete its orbit, and the Sun takes about 250 million years for this purpose; any galaxy may take many times more time to orbit a bigger galaxy. Accordingly, *if the gravity waves were produced by the movement of the celestial bodies, then the frequency periods of the waves so produced would be so large that it might not be possible for our equipment to take any notice of any such wave.*

The intensity of the gravitational force of the celestial bodies depends upon their mass and compactness. Therefore, periodic changes in the value of these two entities of any celestial body may result in periodic fluctuations in the strength of its gravitational force. Accordingly, a sudden change in the size or mass of a massive body due to the explosions like supernovae, or a collision between two black-holes, may create an illusion that such a body is emitting powerful gravitational waves.

Of late, the scientists have envisaged that "*Supermassive black holes spin at the speed of light, or even higher than that.*" It is also believed that "*The energy of such spinning black-holes is carried away by gravity waves.*" *The latter presumption contradicts the fact that* "***The energy in the form of the mass, can never be carried away due to simply its speed of the spin.***" *This presumption violates the presumption of* **Relativity** *that the mass of any object increases to infinity at the speed of the light.*

In order to detect gravity waves, scientists, since the long past, are very minutely observing several pairs of binary black holes. In the year 2010, scientists are said to have succeeded in detecting gravitational waves coming from a far-off pulsar, which is spinning at a very high speed. However, a question arises "why other pulsars don't emanate gravity waves?" Properties of gravity waves remaining unknown, aforesaid lone evidence doesn't seem adequate to establish this concept. The rotating magnetic field of the said pulsar might be producing some kinds of waves, which might have been mistaken as gravity waves. *Supermassive black holes that are far more massive than the Pulsars, and which are spinning at much higher speeds. The nearest black-hole located in the galactic center of the Milky Way might be a much better candidate for detecting the existence of the Gravity waves; however, the fact this black-hole has never been found to emit Gravity Waves, puts the concept of such waves under the question mark.*

The energy of any object is normally carried away by the waves having similar properties. For example, sound waves, or light waves, cannot carry away the heat energy of an object; this could be done by infrared waves only. Accordingly, the loss of heat and kinetic energies due to the emission of gravity waves doesn't seem logical. In case the Gravitational energy of any object is carried away, then *such an object should become a body without any gravity field. In case, the energy of moving objects is carried away by gravity waves, then the black holes must shed off their entire masses too because mass is considered to be the measure of energy,* however, this never happens in practice. *This fact also puts the concept of gravity waves under the question mark.*

The energy possessed by any object is of two types: *1) intrinsic energy due to its own mass, and 2) heat and kinetic energy, etc., which it acquires from some other source or from the high-energy surroundings; energies of this category, are not intrinsic to any object.* The heat energy of any hot object is continuously dissipated by radiation, thereby its temperature falls gradually; whereas, *its kinetic energy is reduced by friction, not by the loss of the heat energy.* The fastest rotating pulsar (a kind of neutron star) rotates on its axes at the speed of around 800 revolutions per second. At this spin speed, a pulsar of about 20 miles (32.28 km) in diameter, *would have a linear speed of about 80,000 kilometers per second, i.e., almost ¼ᵗʰ of the speed of light.* No moving object in space loses its energy due to friction, therefore, the kinetic energy of the said pulsar, following the law of conservation of energy, would remain intact, it can't be carried away by its speed. *The gravitational collapse would cause any object to just contract in size; the quantity of matter contained within that object, i.e., its mass, as well as its kinetic energy and heat energy, both would remain unaltered even at extremely high rotational speeds.* Likewise, *the entire mass and energy* of any rotating blackhole *would be concentrated in zero-sized volume,* which is evident from its *immense gravitational force.* Accordingly, *its rotational speed, and temperature, both shall increase to infinity, because the kinetic energy of any compacted body would be stored within it, in the form of its angular momentum that would be equal to the product of its rotational speed and its mass.*

In case, the aforementioned pulsar contracts further diameter-wise, and becomes ½ of its original size, then its volume would reduce in direct proportion of d3, i.e., by 8 times. *As a result, its rotational speed shall increase by 8 times, that is, double the speed of light or at least equal to it, if not greater than that.* Very large inertia would be generated in such a massive body which would be proportional to the product of its mass & speed. *Since the mass of any object can not be carried away by simply increasing its rotational speed, it must retain its entire angular momentum.* Therefore, in order to stop the rotational speed of such a massive body, and bring it to a total halt, a very strong force would be required instantaneously to break its inertia; *such a rotating object cannot come to rest suddenly, all by itself, without applying a force.*

Black holes are formed due to the collapse of extremely massive stars, which are comparatively far more massive than any pulsar or a neutron star; *naturally, they possess comparatively greater mass and greater inertia. Any star, after becoming a black-hole retains its mass; this is evident from its gravitational force. The famous formula E=mc2 clearly shows that the energy of any object can't be zeroed down until its mass becomes zero. This simply means that defying Relativity, entire energy, that is, the mass of a collapsing star, cannot be carried away by the gravity-waves; clearly, this prediction seems wrong. Neither can any object exist without mass, nor any object without mass would be able to retain the property of gravitation.*

If the mass of any black hole remains intact, its inertia will also remain intact. Therefore, a massive body like a black-hole cannot be brought to rest instantaneously without applying a very strong brake. *On the other hand, it is impossible to produce any such force from nowhere to bring the spin of such a huge rotating mass to an instantaneous halt.* It is impossible to produce such a big force all by itself. *Therefore, I feel that the rotational speed of a black-hole cannot, in any case, automatically come to an instantaneous halt. According to the law of conservation of energy, its speed should increase in inverse proportion to its volume.* Had it been possible to shed off energy by means of gravity waves, *then the spinning black-hole should also shed off its entire mass too. However, this never happens in reality. In view of this anomaly, this presumption of Relativity seems merely baseless speculation without any proof (it was repeatedly brought out logically in chapter-6, that the very existence of space-time is doubtful, rather impossible).* **Moreover, how the gravity waves could be selective, that is, how can they take away the energy of other kinds other than the mass, that too instantaneously, not gradually?** Contrarily, it appears to me that due to the concentration of the entire energy in a very small point-sized place, intensities of heat and kinetic energy of any blackhole might become **infinite,** not zero. **Since the gravitational force of a black-hole doesn't allow any energy wave to escape from its interior, its entire energy shall also remain confined within its core.** *In case of a direct collision of two black-holes, a large portion of the energy of collision would be absorbed by their combined gravitational force; even then, a small portion of this energy might escape from their magnetic poles in the form of a very powerful shower of almost massless particles of very high frequencies,* **not in the form of any disturbance or ripple created in space-time.**

Till March 2015, when the first edition of this book was published, gravity waves could not be detected. However, on September 14, 2015, so-called gravitational waves created by the collision and merger of two black-holes, at a distance of about 1.2 billion light-years, were said to have been detected by *The Laser Interferometer Gravitational-Wave Observatory of America.* Such waves were again observed in 2016 and even thereafter too.

In order to detect such waves, a laser beam is split into two beams by a half-polished mirror; both of these beams are then sent on two different paths, at an angle of 90 degrees to each other; each path is 4-kilometer long. Both of these beams are reflected back and are made to re-combine at a screen. Since both of these paths are exactly equal in length, no interference shall be observed. *However, it is believed that if space-time is affected in one of the arms by any gravity-wave, then the beam traveling through that arm would take a longer time to reach the screen; therefore, interference fringes will be observed.* However, in case any gravity-wave was really produced by the collision of two black-holes, then such a wave would be produced during a very short instant of collision. *In view of this fact, I feel that such a wave shall pass through the said 4Km long path within a very short time interval of $1.33 \times 10-5$ seconds, which would be very difficult to detect.*

The concentric wavefronts of the gravity waves shall follow each other like the outer surface of a hollow ball. These wavefronts, coming from a distance of billions of light-years, can't approach the Earth in a straight line like a ray of light; such wavefronts shall approach the Earth in the form of a flat surface or a sheet, separated from each other by one wavelength each, that is, such wave-fronts would approach the earth in the form of flat sheets coming parallel to the surface of the Earth. Accordingly, such wave-front will affect both the arms of the said apparatus simultaneously, not only one of its arms. Therefore, the interference fringes observed at the *aforementioned observatory* would have been produced by some other reason, not due to *Gravity-waves affecting one arm only.* However, *as envisaged earlier, in case the fabric of space-time doesn't really exist, then no ripples of the Gravity - wave could be created in the said non-existing fabric.*

$$\times \qquad\qquad \times \qquad\qquad \times$$

All the natural forces, that is, the "Strong Nuclear Force," "Weak Nuclear Force," "Electromagnetic Force," and the "Gravitational Force," are intrinsic properties of different matter particles. Therefore, energy, that is, *the capacity to do work*, could be derived from any of the above forces. This means that "Energy," in the form of these natural forces, is confined within the matter itself. The famous formula E = mc2 indicates that the energy of any piece of matter can never be zeroed down until its mass; that is, the value of "m" in the above formula doesn't become zero. *In case the above formula, as well as the belief that "mass is the measure of energy," are both correct, then the entire energy of any celestial body, because of its mass, could never depart from it.* This means that till such time the matter-particles are intact, their intrinsic energy would also remain intact. For example, the black hole located in the center of our galaxy has been exerting gravitational force for the last 13.2 billion years on each and every star of this galaxy; even then, its strength has not at all attenuated a bit. This is evident from the fact that our galaxy is gradually growing bigger in size by continual accretion of mass (by swallowing matter from outside). In view of this fact, *it can be said that* whenever *the gravitational attraction of any massive body compels any smaller body to move, then none of these two bodies would lose even a small portion of their mass, i.e., their intrinsic energy; simply, some of the potential energy, produced in the smaller object due to gravitational attraction, would be converted into kinetic energy.*

The above idea could be understood by the example of permanent magnets: These days, permanent magnets are made from various types of alloys and mixtures, such as Neodymium-Iron-Boron ($Nd_2Fe_{14}B$), Yttrium-Cobalt (YCo_5), Samarium-Cobalt ($SmCo_5$ & Sm_2Co_{17}), Alnico, Ceramic and Ferrite magnets, etc. Such magnets have very long self-lives. Pole strengths of such magnets don't diminish even a bit by repelling or attracting other magnets or other magnetic objects repeatedly millions of times. This becomes possible because of the special structure of their molecules, more particularly, because of the manner in which electrons are arranged within their molecules and atoms. Electrons within molecules of such magnets are arranged in such a way that their magnetic fields always remain aligned. *Molecular structure, rather the manner in which their electrons are arranged within their atoms/ molecules, doesn't change by performing any amount of work.* As a result, the strength of their poles doesn't diminish a bit; ***the magnetic force is an inseparable property of the matter-particles.*** *Electrons, for the last 13.7 billion years, have been orbiting the atomic nucleus due to their mutual attraction;* ***even then, their energies have not depleted even a bit, even after continually exerting force on each other for such a long time.*** *This became possible because the force-carrying particles could be exchanged in any number (Please refer to Chapter-3).* ***Thus, it can be concluded that energy cannot be separated from particles; it is intrinsic to them.*** *Accordingly, the energy of the universe would always remain proportional to the total mass of the matter existing within it, not a bit more or less.* ***Therefore, it is not possible to carry away the intrinsic energy of any object, regardless of the fact that how fast it is spinning or how much force it is exerting on the other objects.***

Any object, when heated, doesn't start moving simply because its energy has increased. Similarly, the speed of any moving hot celestial body doesn't reduce because it gradually cools down, that is, due to the loss of its heat energy. ***This ordinary fact indicates that celestial bodies don't use their intrinsic energies to move through space; a force is essentially required to move them in orbits, i.e., change their directions continually.*** The gravitational force of different stars and their planets depends on their mass, i.e., the amount of matter stored within them. The *speed in any celestial body is produced by the* **Gravitational force** *exerted by a bigger body that compels it to move. None of such bodies consume their intrinsic energies in this process; simply, their static energy produced by the gravitational force is converted into kinetic energy when such objects move.* ***When any planet, while moving in a curved path, speeds up, then the centrifugal force, so produced, holds it into a stable orbit.*** Inertia probably resists instantaneous speeding up of any planet; due to this reason, centrifugal force fails to instantaneously counterbalance the gravitational pull. ***Therefore, the pull of gravity would always enjoy an upper hand in this tug-of-war between these two forces by a very slight margin.*** *Different planets, As a result of the above phenomenon, might gradually spiral inward, bit by bit. On the other hand, due to the gravitational pull exerted by the sun, the moon is probably spiraling away very slowly from the earth.*

We know that hundreds of million tons of matter in the form of meteors and interplanetary dust, etc., are added every year to each planet and star, which results in a gradual increase in their respective masses. On the other hand, the stars continually burn hundreds of million tons of their fuel every year, resulting in a gradual reduction in their masses and, in turn, a gradual reduction in the strength of their gravitational force. It is not clear what would be the effect of a combination of both of the above phenomena or in what manner such a change will affect the orbital paths of different planets? *However, until the existence of the fabric of space-time is not proved, neither the existence of gravity-waves could be ascertained, nor it could be proved that the energies of the celestial objects are carried away by the so-called gravitational waves.*

The Conventional Theory of the Formation Of Black Holes

As brought out earlier, in chapter-6, the existing theory explaining the creation of Black-Holes envisages that when any massive star collapses under its own gravitational force, then the curvature of space-time around such a dying star increases gradually due to the continual shrinking of its mass into a zero-sized point. *To me, this presumption contradicts the very idea that gravitation is not a force because no object, in the absence of any force, can collapse against the repulsion of its own particles.* Anyhow, it is believed in the existing theory that light rays, while passing in close vicinity to a star that is collapsing under its own gravity, would follow the increasing curvature of space-time. Accordingly, such rays will gradually bend toward the center of gravity of the said collapsing star. Thus, the angle of deflection of those light rays would gradually increase; redshift, too, in such rays, would gradually increase. As a result, the light-source would appear to grow redder and dimmer. *Soon strength of the gravity field of that star would increase to such an extent that the entire matter of the said collapsing star would be squeezed into almost a zero-sized point; this will cause the curvature of the space-time around that zero-sized point to become infinite. As a result, that point-sized core having an infinite density would be surrounded by infinitely curved space-time of the zero surface area, such a state of matter and space-time is* known as **gravitational singularity** or simply **"singularity."** In this process, the gravitational force of the collapsing body becomes so intense that a hole of infinite curvature around that infinitely dense core is ripped open in the fabric of space-time. The persistent deflection of light rays within this hole would compel them to move along the edges of this infinite curvature forever, in a closed circular path; such rays would fail to escape from this curvature. According to the theory of relativity, nothing can travel faster than light; in case light can't escape, then neither can anything else. Therefore, the black hole would never be seen. However, its gravitational force would still be felt from outside of the edges of this region of infinite curvature. The gravitational force of a black hole becomes so intense that it drags every nearby thing into its point-sized core and merges the same with itself.

One can see only what happens outside of the edges of the infinite curvature, not from within this limit. Since no radiation can come out from inside of this infinite curvature, any information *about the events happening within* could not reach us or any other observer. Such a region of **infinite curvature**, from where the light rays can't escape, is called *"Event Horizon." When the black hole would grow bigger, it would catch light rays from increasing distances. This simply means that* **the size (radius) of this region of infinite curvature shall continually increase.** *Scientists further believe that no two rays would run on the same path; such rays should move on parallel paths, keeping a certain distance from each other. However, since* there is no rule to establish in what direction the light would deflect, the light rays would be deflected in either of the directions; this will create chaos.

   The aforesaid belief of the scientists poses a few problems **1) Increase in the size of the event horizon would mean that its curvature would not remain infinite; it will decrease very sharply with any increase in the size of the event horizon. This possibility defies the existing theory, which necessitates: "the curvature of space at the event horizon shall become infinite so that light could be trapped within the infinite curvature." 2) As a consequence of the increase in the diameter of the event horizon, no medium of space-time would be left within it for the rays that were moving on smaller paths earlier. In this scenario, a question arises "where, or in what medium, do the rays that were earlier moving along the zero-sized circles would now move, and how the light waves would move undulating?**

   Contrary to the hypothetical zero-size of the warped-space of infinite curvature, the black-holes, in reality, capture the light from a distance of several hundred-thousand km., i.e., their event-horizons measure thousands of kilometers in size, where space-time, in any case, can't have infinite curvature. This very fact alone can indisputably prove that zeroed surface-area, having infinite curvature, does not form around any black-hole. This possibility poses the question, *"How the black-holes are formed, and how do the black holes capture the light?"*

   Apart from the aforementioned anomalies, there is one more anomaly that not only contradicts reality it also defies the basic requirement that an infinite curvature must form in space. If not, then the light could not be trapped. This un-noticed anomaly existing in this theory is: *"Formation of the infinite curvature means, both, size and the radius of such a region, shall become zero; however, zero size means such a region does not exist at all, nor any other thing can exist within a zero-sized region."* Further, as soon as the diameter of a zero-sized region would increase, then the curvature of that region would not remain infinite; because the curvature of any region having a non-zero radius cannot have infinite value.

In fact, *the black-holes of different masses capture light from distances ranging from hundreds of kilometers to several hundred million kilometers. Therefore, the curvature of space at the outer edge of the event horizon that has such a large radius cannot have infinite value. This fact clearly indicates that the prevailing theory for the formation of black holes is not capable of solving the problem of "How any event-horizon can have such a big size or how can it catch the light in the absence of infinite curvature*

Different scientists are not sure whether the black-holes rotate or they don't. They have no single opinion: "Whether a necked singularity is formed or the black hole is surrounded by the event horizon." Our existing theories envisage *"Black-holes are formed when a massive object collapses into a zero-sized core of infinite density; in turn, the mass of that core causes space-time surrounding it to warp into a zero-surface area of infinite curvature."* The force of gravity is supposedly a consequence of the warping of space-time; therefore, *the long reach of the gravitational force of the black-hole shall mean that the size of the deep-well created below the black-hole shall, as depicted below, extend up to the distances where the effect of its gravitational force is felt, i.e., up to several million L-Y. On the other hand, the formation of singularity means that the event horizon of zeroed surface-area shall be created either at the bottom of the huge deep-well created by it or far-above the black-hole. On the other hand, in reality, the event horizon that surrounds the black-hole always lies in the mean galactic plane, not in any other plane, at a far-off distance.*

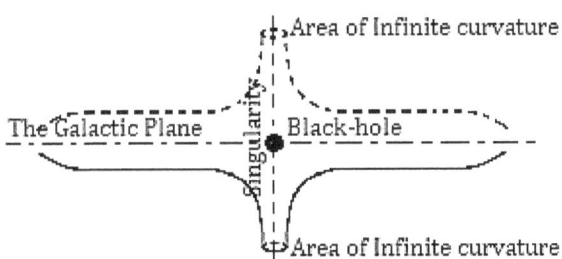

An Alternative Angle on the Formation Of Blockholes

All the anomalies highlighted above, and in chapter-6, put the process of the formation of the Blake-holes under question; the doubts raised above are clarified below:

The existing theory envisages that black-holes are formed due to the infinite curvature created in space-time. Similarly, the bending of the light-waves near the *Blackholes* is also explained by the formation of curvature in the space. *However, the actual size of the event-horizons formed around different black-holes defies the existing theory; infinite curvature cannot be created in a very vast area from where the Blackholes capture the light in reality. Further, as envisaged in this book at several places in Chapters- 4 to 10 (kindly refer to Chapter-6), the very existence of the fabric of space-time seems doubtful; in case the existence of space-time could not be proved, then the existing theory of the formation of Blackholes, too, could not be explained or proved, though they certainly do exist.*

In order to solve the aforementioned problem, I am giving below my own speculation on the formation of black holes; this speculation is totally independent of the Theory of Relativity.

The roots of the belief that the force of gravitation cannot cause light waves to bend or deviate from their straight paths go back to about 150 years in the past *when physicist **James Clerk Maxwell** established that **light is a member of the electromagnetic wave family**. It was thought that in case light could be influenced or attracted by gravity, then similar to the cannonballs, light-waves too should also come back to the earth after some time.* **The speed of light was probably not considered while conceiving the above belief.** *Contrary to the aforesaid belief, not only the light, different spaceships, too, because of their very high speeds, are able to escape Earth's gravitational field.* Anyhow, as per the prevailing theory, black-holes are formed because light waves are trapped in the infinite curvature of space-time; **this means that the gravitational force performs this work indirectly, not directly.**

According to the modern concept, the photons, because of their momentum, can travel infinite distances. In case this belief is true, then there is a strong likelihood that photons, too, have some mass because, in order that any particle could have momentum, it is necessary that it should have at least a little mass, may it be negligible. We have earlier seen that photons, too, can be trapped in the solution of Nano-sized magnetic particles (Chapter-3, Some Least-known Properties of the light). This fact suggests that analogous to the magnetic trapping, the gravitational force, in the capacity of a real and independent force, might also attract or even capture the fast-moving photons, provided they have at least a little mass. The value of the gravitational force acting between two objects depends on their respective masses as well as on the distance between their centers of gravity. The photons emitted by the distant luminous sources, while on their way to the earth, normally pass by the sun or other stars from a considerable distance; therefore, the paths of such photons are not affected by their gravity. However, when photons on their way to Earth pass from a very close distance from the surface of any star, where the gravitational force is very intense, then despite their negligible masses and high velocities, the gravitational force might divert such photons toward the center of gravity of that star, due to its very strong gravity. *Johann Georg Van Soldner* came up with a similar idea in 1801. **The gravitational force of any black hole being several million times more intense than that of any star; its gravitational force might exert a very powerful force on photons that pass from a close distance from any black hole. Consequently, photons might bend toward the black holes or even fall abruptly into its core through a spiral path;** in case this is possible, then the belief: *"The Event Horizon is created due to the infinite curvature of the space-time"* would go wrong.

In such a case, black-holes would be formed due to the inward collapse of extremely massive bodies caused by their own **Gravitational Force, not due to the formation of infinite curvature in space-time**. In that case, the very existence of the wormholes would become doubtful, rather impossible, because no direct proof of the existence of space-time is available. Moreover, it is logically pointed out in several places in this book that the force of gravitation can't result from warped space and also that medium of space-time is not necessary for the propagation of light. In case the fabric of space-time is really non-existent, **then the existence of wormholes too won't, at all, be possible.**

Present-day scientists believe that 1) *Supermassive black holes rotate at higher than light-speed; therefore, space-time adjacent to such black holes also rotate at the same speed. 2) when an object reaches very close to the event horizon, then the time for such an object stops to lapse; as a result, that object shall remain visible forever.* **However, I feel that there seems to be no consistency in the conventional prediction mentioned above because 1) In case such a black-hole loses all of its energy in the form of radiation of Gravity Waves, then the said black-hole should become stand-still; moreover, the whole of its mass, in the form of its energy, should have deserted it. However, nothing like this happens in reality; its mass remains intact. 2) Did the time really freeze, then an activity like the rotation of the black hole or space-time, etc., would not have been possible; no activity is possible in the frozen time 3) Nothing is supposed to move at higher than light speed. 4) In case any object falling into the black hole can be seen forever, then all other objects that fell into it during the past several billions of years shall also be seen. However, nothing like this happens in reality.** All the above-mentioned facts indicate that *the said concepts of the theoretical emission of Gravity-Waves, and freezing of time at the Event-Horizon, etc.,* are incorrect because such concepts contradict each other.

It is clear from the above discussion that the photons, instead of being trapped in the infinite curvature of space, are captured by the gravitational force of the black hole from its near vicinity; this is the reason we can't see beyond the so-called event horizon. *Yet another possibility might also exist, though the same is very remote: When the matter, falling into the black holes, reaches the event horizon, its speed might reach the limit of the speed of light, and thereafter, it may surpass even this limit also; this might also be a reason that nothing could be seen beyond this boundary.*

The Myth of the Wormholes

Relativity predicts that a conical **Worm-hole** necessarily exists beneath each and every black-hole. This concept is based on the prediction that an infinite curvature is necessarily created in space due to the formation of the black-holes. The **wormholes** have been fascinating the scientists as well as the fiction writers with equal intensity. It is believed that *"In case any astronaut enters a wormhole, then if he is lucky, he will emerge in a new time-zone, unhurt."* Two facts given here are capable of busting this myth: 1) It is not possible for any object having a finite shape and size to enter into a zone of infinite curvature, that is, into a zero-sized zone, until and unless such object is either squeezed into a zero size or divided into billions of very minute atom-sized particles. 2) Probably, some people might remember that sometime during the decade of 1960-70, a comet named "Shoemaker-Levy 9" was captured by the planet Jupiter, which compelled the same to orbit itself. Later, in June'92, this comet was divided into several fragments by the gravitational force of this planet. Two years thereafter, i.e., in June'94, all these fragments fell near the South-pole of the planet Jupiter. These facts tell very clearly that *nothing can escape a black hole unhurt.*

The Size of the blackholes

German scientist *Schwarzschild,* in 1915, proposed that stars having three times solar mass or above can become black-holes. However, even some of the neutron stars might be more massive than this limit. Quark-stars/strange stars and electroweak stars, etc., might be even many times more massive than the aforesaid limit. This possibility indicates that any black hole must be many times more massive than this limit.

Scientists have predicted that all the matter contained in any black hole is squeezed into an atom-sized or even smaller point. In the absence of any information coming from within the black holes, it is not possible to find out in what form or state, does matter exists within this point or how big its core could be in size. *It is also believed that magnetic force in any substance, is created due to the circular motion of the electrons rotating around the atomic nucleus. However, in the case of the neutron-stars, this concept doesn't seem to hold good because it is believed that during the process of the formation of such stars, all the electrons enter into the atomic nucleus and convert all the protons into the neutrons. In case this concept is correct, then no magnetic field could be generated in the absence of the electrons moving in a circular path.* *Contrary to this concept, all the neutron stars have very strong magnetic fields.* Likewise, the most compact stars, that is, the black holes, besides gravitational force, have very strong magnetic fields, too. *This fact indicates that the matter particles contained within the cores of the black-holes, even after having been squeezed and crushed to the last limit, retain at least the aforementioned two properties of the normal matter, i.e., Gravitation and Magnetism.* *This fact indicates that matter might exist within their cores in some compact form or in any other state in which the matter particles don't lose both of the aforesaid properties of the matter. It is believed that the density of the core of any black hole becomes infinite; may the black-holes of the same size (zero size) have a mass 3 times, 3 thousand times, or even 300 million times of that of the Sun. However, this seems impossible because the density of different black holes cannot have* **different degrees of infinity**.

We know that *even a small quantity of matter contains trillions of atoms, whereas a supermassive black-hole might contain several million times the mass of the sun.* It can be concluded from this fact that atoms in infinite numbers might exist in the point-sized core of any black-hole. However, scientists believe that all of these atoms in the form of the matter-waves might occupy the same place that measures even lesser than the size of an atom. Theoretically, this seems possible when all the matter-waves do exist in the same phase; in case some of the waves go out of phase, then the pairs of the in-phase waves and the out of phase waves will annihilate each other; this will result in the reduction of total mass, and as a result, the star would not turn into a black-hole. On the other hand, **trillions of trillion of atoms in the form of matter particles might not be able to occupy the same place; in this case, also, the core of the black-hole won't be able to shrink to the zero-size. Accordingly, the black-holes of different masses might have very small but finite sizes that might differ from each other.**

Actually, the size of the core of a black hole would depend on the fact that *to what extent the matter could be compacted.* A bigger black hole, formed by the merger of two identical black holes, would contain twice as much matter as compared to the original ones; its gravitational force would also be doubled. It would then squeeze the total matter with twice the force. In this scenario, there could be at least three possibilities: *1) In case it would be possible to squeeze the matter further without any limit, then, as believed, the core of such a black-hole would again be compressed to the zero size. 2) Had it been possible to compress the matter only up to a definite extent, then, although the final size of the new black-hole would be reduced in size, the same couldn't be compressed back to zero size, and 3) in case, the matter was already compressed to the last limit, then it would not, at all, be possible to reduce its size any further; it too would become double in size, volume-wise.*

The aforesaid discussion suggests that there is, at least, a slight possibility that even the core of any black hole might have a definite size; at least, the core of the supermassive black holes might have a finite size **because the infinite heat energy trapped within the black-holes, would resist compression of the matter particles into a zero-sized point.** At present, no *means exist to verify this prediction of mine; it seems that this is beyond the capacity of mankind.* However, a natural phenomenon supports this possibility: **"Whenever any black-hole devours a large quantity of gas from its surroundings, then sometimes, a very small black object is seen in the background of the infalling gasses. Had such an object been really a blackhole of the size lesser than an atom, then it would not have been possible to see anything in that background, that too from such a large distance. This fact indicates that different black-holes may have finite sizes."**

The Size Of the Event Horizon

Scientists believe that Event Horizon is the region of infinite curvature, having a zero surface-area; accordingly, its size, too, must be zero. **Ignoring this possibility, they further envisaged that the size of the event horizon shall continually increase with the increasing mass of the black-hole.** *In case the aforesaid prediction of the renowned scientists is true, then neither its size would remain zero, nor the curvature of the event horizon would remain infinite; in such a case, the event horizon, in the absence of the infinite curvature, won't be able to capture the light. Contrarily, black-holes, in reality, capture the light from very large distances of several hundred-thousand kilometers;* this fact indicates that the size of the event horizon of any black hole must depend on the strength of its gravitational force; logically, the more massive is a black hole, the bigger would be the size of its event horizon. *It is a fact that the gravitational force does not act only in two dimensions. Instead, it reaches out to all the objects located in all three dimensions. This means that this force acts or spreads out spherically, in three dimensions, not only in two dimensions. Therefore, the intensity of the gravitational force of the black holes and other objects, at any distance "d," measured from the center of gravity, should decrease in* *the inverse ratio of the cube of that distance, that is, in the proportion of the value of $1/d^3$ (*chapter-10, "Puzzle of Gravitation"*)*, as against the conventional value $1/d2$ (in the inverse ratio of the square of the distance). *The mass of any black hole, that is, the total quantity of the matter stored within its core, must bear a definite ratio to the size of its event horizon.* **Since black holes of different masses capture light waves from different distances, the outer edges of the event horizon of any black hole should be formed at such a distance from its center, where the intensity of its gravitational force (that should be proportional to the value of the ratio m/d^3), shall be adequate to capture the light rays.** *Accordingly, the sizes of the event horizons of different black holes, having different masses, would differ; however,* **the intensities of the gravitational force of all of the black holes at the outer edges of their respective event horizons shall always be equal.**

It may be envisaged on account of the above possibility that in case *the mass of any black hole becomes double of its original value, then the size of its event horizon would increase by 21/3 times only, that is, 1.27 times of its original size. Similarly, the size of its event horizon could only be doubled in size when its mass increases by 8 times, that is, by 23 times its original mass.* It may thus be seen that only a ratio could be established between the mass of the black hole and the size of its event horizon; it is not possible, by any means, to estimate the exact sizes, either of the event horizon or that of the core of any black hole.

The Relation Between The Mass Of Any Galaxies And that of the Black Hole located at its centre

Till to date, after the creation of the universe, hundreds of billion stars, after running out of their fuel, might have turned into black holes or into other hard-to-see compact objects, such as White/Black dwarfs and Neutron stars, etc. Accordingly, the total number of smaller black holes of different sizes and masses, as well as that of the other hard to see objects, might be greater than that of the visible stars. Therefore, the total mass of any galaxy cannot be estimated correctly on the basis of the total mass of the visible stars alone, as was attempted in the 1930s (Please see Chapter-13, Dark Matter).

Almost 99% mass of the entire solar system is concentrated within the Sun alone. All the planets, asteroids, comets, and all other smaller bodies, etc., are orbiting the sun because its mass is several thousand times more than the combined mass of all the other members of the solar system. The disk of stars comprising any galaxy, too, orbits a *supermassive black hole.* Analogous to the solar system, the mass of any of the galactic centres must invariably be many times greater than that of the combined mass of all the visible and invisible stars that comprise that galaxy. In case the mass of the supermassive black hole is lesser than the combined mass of the objects orbiting it, then it will be beyond its capability to compel such objects to orbit it. Our milky-way comprises about 200 to 400 billion stars; out of that, a good number of stars/invisible objects would be much more massive than our Sun. Cosmologists have estimated that the galactic centre of our galaxy contains only 4 million times mass as compared to that of our Sun. However, looking at the total number of the visible stars alone, and the combined mass of all such stars, the estimated mass of the galactic centre doesn't seem to be adequate to compel an entire lot of the stars to form a whirl, being far below than the combined mass of all the stars going around it. This could be understood from the example of the sport of "hammer throw," where any sportsman can swing the hammer, subject to the condition that his body weight and strength are both greater than the weight of the hammer, the inverse of this is not possible; only those celestial bodies that have lesser mass can orbit the heavier ones, the heavier (more massive) body can never orbit the lighter (less massive) one. *This fact suggests that the mass of the galactic centre should be many times greater than the estimated mass of the galactic centre of the Milky-way.* Our present estimation of the mass of the galactic centre is probably, calculated on the basis of the formula, $F = G. M_1 M_2 / d^2$. In case the square of the distance is replaced in the above formula by the cube of the distance "d" (please see Chapter-10), then the mass of the galactic centre of our galaxy might work out many times greater than what we estimate at the present.

Primordial Black Holes

Scientists believe that the shock waves produced by the *Big Bang* would have compressed matter so produced to almost zero (0) size. As a result, numerous low-mass black holes having a mass of a few thousand million tons might have been produced at that time. Such *hypothetical* low mass black holes are known as the "Primordial Black Holes." Generally, it is believed that a similar phenomenon might happen due to massive explosions like that of a hypernova or a hydrogen bomb, etc. However, such so-called low mass black holes would have produced only in the condition that the said explosion would have taken place either surrounding the matter-particles from all around or at least within the new borne matter particles. *The possibility of compression of the matter particles by a big bang doesn't seem realistic because the matter particles didn't exist when the big bang did take place; matter particles are believed to have produced about 100 seconds thereafter. Anyhow, in case low mass black-holes did really produce by the Big-Bang, then the gravitational force of such a black hole won't be able to overcome the mutual repulsion of the subatomic particles because the gravitational force of such a low-mass black hole, would have been very weak.* For example, if a rubber ball is first squeezed to the minimum possible volume, and thereafter, the applied pressure is removed, then the ball would surely bounce back to its original shape and size, likewise, every object could not possibly be turned into a black hole by simply squeezing it to zero size. In case the mass of a body is inadequate to maintain its own compactness against the mutual repulsion of its particles, then soon after the compressive force is removed, the said object might bounce back to its original size and shape. Black holes could possibly be formed when the compressive force, required to squeeze them to the last limit, remains persistently active. *Relativity* also envisages that the gravitational fields of the "black holes," should be so strong that they could create infinite curvature in space-time. *Accordingly, a very large quantity of matter would be required to produce such a strong gravitational force; this doesn't seem possible with the low-mass objects.* Thus, the possibility of the formation of the **low mass black holes** seems very remote, rather **unrealistic.**

In case, the primordial black holes were really created by the "big bang," then possibly, they would not have been able to sustain their sizes and shapes, because particles of such objects would repel each other, and therefore, they will bounce back to their original sizes and shapes. Alternatively, like other black holes, they would have continually sucked material from their surroundings and gradually grow bigger. The period of 13 billion years is more than sufficient for such objects to accomplish either of the aforesaid tasks. Accordingly, they won't be able to continue as primordial black holes for such a long period of time.

The Hypothetical Death Of the Black-Holes

Based on *the Uncertainty principle, the* late scientist **Stephen Hawking**, one of the greatest physicists of this era, did propose that black holes shall also emit some radiation from the regions lying just outside of their *event horizons*. In brief, the *Uncertainty Principle* envisages that there must always be some quantum fluctuations in the (empty) space, even outside of the black holes, too. These fluctuations could be thought of as *the creation and destruction of the pairs of particles and their antiparticles; one of these particles would have positive energy and the other one would have negative energy.* In

Quantum Mechanics, it is further believed that any particle, lying close to a massive body, has less energy than those that are lying far away from that body. *(Please see chapter-3 under "The Positive and Negative Energy").* This assumption is based on the concept, that *some work against gravity would have been done on the particles to move them to their existing locations away from any black hole.* Based on the aforementioned concept, *it is believed that when a black-hole draws any particle closer to it, then the energy of that particle would gradually decrease and might become even negative. Accordingly, out of all the particles lying in the near vicinity of a black hole, the particle having negative energy would fall into the black hole, whereas, its partner, because of its positive energy, might escape the gravitational force and move away from the black hole.* Therefore, on the one hand, it would appear to an observer from a distance that such particles are being emitted from the black holes, and on the other hand, the inflow of the negative energy into the black hole, would gradually reduce the mass of such a black hole. As a result, the black hole in a few billion years would disappear in a tremendous final burst of emission that would be equivalent to millions of hydrogen bombs exploding together.

Hawking further predicted that the particles having positive energy would continually escape from just outside of the event horizon of any black hole. In other words, a shower of particles, having positive energy, i.e., radiation of such particles, would appear to be emitted from just outside of the black holes. This prediction is supported by the fact that: *When focusing a telescope on the prospective location of a black hole, a very faint glow might appear to come from such a place; this glow suggests that Hawking's proposition is correct. In order to honor this great physicist, this radiation has been named "Hawking Radiation."*

My take on the theoretical-Decay of Black-Holes

*The hypothesis of the decay of black holes is based on the idea that the particles located closer to a black hole have lesser energies and also that **the energy of black holes is of a negative nature.** Did the black-holes possess negative energy, then the time in their vicinity should tick at a much faster rate; on the other hand, Relativity envisages that the time does stop near them; both of these ideologies contradict each other on this point.*

Anyhow, contrary to the belief that gravity is negative energy, this energy never performs the negative work; the attractive force of gravity never repeals any object. Nothing can move away from a black hole until a much stronger force compels it to do so. In fact, particles created after the hypothetical "big-bang" were afterwards dragged closer to different gravity centres; they were never dragged away from them. Similarly, black holes draw all sorts of objects

closer to them. Therefore, work done in this process shall be stored in such objects. Accordingly, more energy shall be stored in those particles that were dragged closer to the massive bodies; they must gain energy instead of losing it. Further, no work was ever performed on the particles to drag them away from the black holes. Therefore, the particles located away from black holes should not possess more energy. Furthermore, if any particle possesses negative energy, then the mass of such particle should also become negative because mass is supposed to be the measure of the energy of a particle or an object. However, negative mass is not at all possible. In case particles having negative energy really enter any black hole, then, on the one hand, its mass will continually decrease; on the other hand, its negative energy, i.e., its gravitational energy, shall persist in increasing. Contrarily, when mass decreases, gravity also decreases. However, in reality, both the mass as well as the gravitational force of different black holes always persist in increasing. This fact shows that the idea of ingress of negative energy into the black hole is wrong, being against the reality.

Further, as brought out earlier in chapters-3, 5 and 6, the slowing-down of the atomic clocks near the massive bodies is indisputable proof that massive bodies possess positive energy; in turn, this fact proves that gravitation is positive energy. Had gravity been negative energy, then the region surrounding any massive object should become a region of weak energy, and in that case, the clocks located in that region shall tick at a higher rate; however, this never happens.

In case the pairs of particles having positive and negative energies, are really created near the black holes, then the particles having negative energies should perform negative work and move in the opposite direction to the applied force. Thereby, they should be driven away. On the other hand, the particles, which possess positive energies, would perform positive work and thereby move in the direction of the applied force; accordingly, they would enter the black holes. No particle defying gravity can escape any black hole until and unless it is driven away by a force that is stronger than the gravity of the black hole; however, no such force acts on these particles. *Had the particles having positive energy been capable of moving away from the gravity centre of any massive object, then the formation of stars would have become impossible; all the gas clouds, instead of collapsing* within themselves, *would have scattered away.*

We know that the point-sized cores of the black holes continually accumulate energy in the form of matter. This fact indicates that the temperature inside these cores should have increased infinitely due to the continual compacting of the matter particles. However, since their gravitational force never allows the photons, i.e., the energy-carrier particles, to escape, they shouldn't emit any radiation from within. *Moreover, since the outward pressure generated within the black holes, i.e., the heat energy generated within, would always depend upon it's gravitational force that causes them to do so, such pressure would never be able to overcome the crunch of gravitational force. In view of the above fact, the possibility of the death of the primordial black holes by way of evaporation due to immense heat doesn't seem feasible, at least to me, because gravitation won't let anything escape the black hole.*

Magnetic Fields of the Black Holes

Almost all scientists believe that the Earth's magnetic field is created by the flow of molten magma that flows below the earth's crust, alternatively, due to the flow of the charged particles in the form of the solar wind around the earth. However, this belief is unable to explain the formation of the magnetic fields of the compact celestial bodies like the neutron stars, pulsars, magnetrons, black holes, etc., because no magma flows beneath the surface of these extremely compact bodies. Therefore, a question arises *"How the magnetic field is created in compact bodies like the black-holes?"* The belief that the *matter-particles like the atoms, electrons, protons, and quarks, etc., are crushed and squeezed within the black holes to zero-sized cores attracts a question, "Does the gravitational collapse or compactness of the matter have any relation with the magnetism?"* Whatsoever be the cause of this magnetism, it seems to me that infinite energy, pressure, and temperature that exist within the interiors of the black holes, probably *produces magnetism of a different kind, which is perhaps, not known to us. At such a high energy and temperature, probably all the three nuclear forces, that is, the strong, weak, and electromagnetic forces, unite with each other. This might be the reason that black holes have such a strong magnetic force; however, the fourth natural force, i.e., gravitational force, maintains its separate identity from the rest of the natural forces.* A very long reach of the magnetic force indicates that the force carrier particles of the magnetism, i.e., the virtual photons, do not acquire any extra mass from the stronger energy fields of the black holes, i.e., they do not gain any extra mass. Although black holes are supposed to carve infinite curvature in the space-time all-around their own self, magnetic force, or more accurately, the *force-carrier virtual photons of the magnetism,* surely escape from both the poles of even the most massive black holes, against their immense gravitational force. Since no medium of space-time is supposed to exist between the black holes and their event horizons, a question arises, *"Do the virtual photons of the magnetism don't need the medium of space-time to propagate?* If *magnetism is capable of doing so, why can't all other forms of energy do the same?*

The black holes regularly swallow as much gas as is available in their surroundings; however, the quantity of the infalling matter may vary from time to time. Present-day scientists believe that immense heat is generated in the in-falling gas due to the mutual friction of the infalling particles. Somehow, what I feel, *when the gas molecules are drawn closer to any black hole, they enter an energy field whose intensity goes on increasing. **Accordingly, aside from kinetic energy, the heat energy of the infalling gasses continually increases due to persistent compression and reduction in their volume.*** Whatsoever may be the reason for the heating of the infalling gases, when the temperature of gas attains the value of about 3,000°C and beyond, its molecules get ionized. This ionized gas, rushing toward the black hole, gradually acquires a speed of several thousand kilometres per second. Accordingly, the electrons in the gas-atoms, while orbiting the nucleus of such atoms moving with very high speed, once move in the same direction in which the atomic nucleus is moving, whereas, during the next half-cycle, they move in the opposite direction; **these electrons might maintain the same radial distance from the atomic-nucleus that is hurtling toward the Blackhole.** Accordingly, as shown in Fig.9-A, the speed of any electron while travelling in the direction opposite of the nucleus would decrease due to the drag exerted by the nucleus, i.e., its speed might retard. Even then, such an electron would complete this half-cycle in a shorter time period because it would have to travel through a shorter distance. In the next half-cycle, when the electron and the nucleus both move in the same direction, the nucleus would drag the electron and accelerate its speed. However, it would take a longer time to complete this half-cycle because it would have to cover a longer distance. The electrons, therefore, would not orbit the nucleus in the way in which they orbit any stationary nucleus. Instead, they would probably move in orbits similar to those as shown in the figures: 9A, 9B, and 9C. **The orbital path of any such electron, depicted in the said figures, is imaginative only. I am not sure about what exact paths such electrons might adopt;** however, the distances travelled by them during each half-cycle, would surely undergo periodic changes. Such changes might result in a periodic change in the rate of accelerations as well as in the rate of interaction between the electrical fields of the nucleus and that of the magnetic field of the electrons. Probably, this might be the reason why the ionized gases, falling into the black holes, emit waves of different frequencies rather than photons of varying energies.

Probable path of the electron when the nucleus is moving at 1/2 the speed of the electron

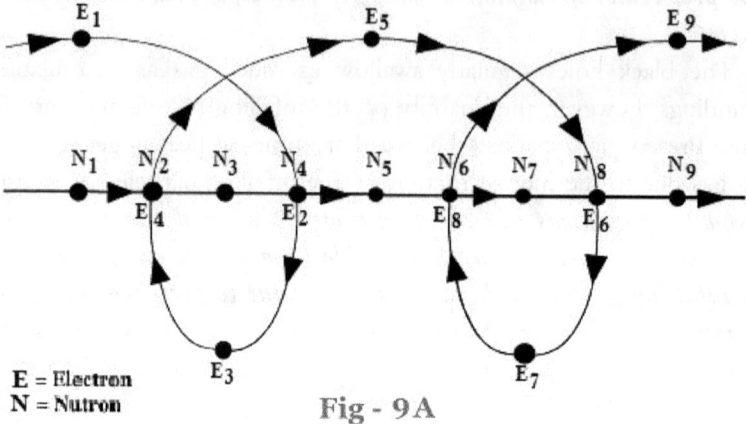

E = Electron
N = Nutron

Fig - 9 A

Nucleus moving with same speed as that of the electron

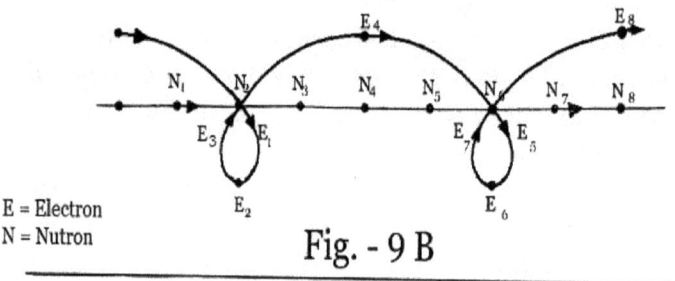

E = Electron
N = Nutron

Fig. - 9 B

Nucleus moving at double the speed of that of the electron

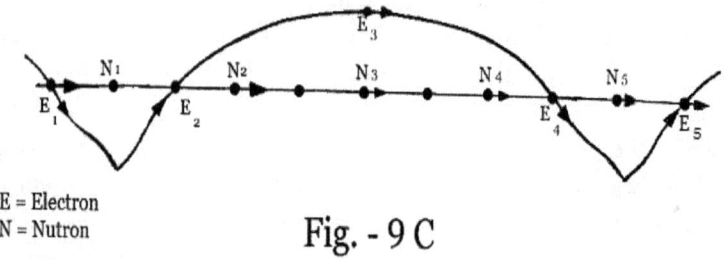

E = Electron
N = Nutron

Fig. - 9 C

When the infalling gas-atoms would move further ahead and reach closer to the black hole, say very near to the event horizon, the gravitational force acting on them would grow so strong that the electrons rushing toward it (the black hole) might not be able to return into their orbital paths; they might break-off from the nucleus and rush toward the black holes alongside the atomic nucleus as independent units, the bare nucleus would also continue their voyage as separate units. The faint glow seen outside the black holes is probably caused by the separation of the electron from the atoms. When the in-rushing particles would reach very close to the event horizon, the particle having negligible mass, such as *electrons* and *photons, etc.,* would also be arrested by the gravitational force of the black hole; due to that, these low mass particles, along with the *bare nuclei,* would also spiral into its core.

After the detachment of the electrons from the atomic nuclei, the spinning axes of both the bare nuclei and the electrons moving independently from each other would be so adjusted by turning the spinning axis of either of these particles upside down, by the magnetic field of the black hole, in such a way that the magnetic field created by the electrons as well as by the bare nuclei, rushing separately toward the core through different spiral paths, would also add up to the pole strength of the black holes. This would mean that *the more would be the quantity of gas that spirals into a black hole, the stronger its (black hole's) magnetic poles would become.* In case the quantity of the infalling matter varies, then the pole strength of the black holes should also fluctuate proportionately. **Based on this fact, it is predicted in the next chapter, "Quasars," that "Any sudden increase in the strength of the magnetic field of any black hole would result in the emission of bipolar jets from its magnetic poles of that very blackhole."**

The Interior of the blackholes- a speculation

Depending upon the strength of the gravitational force of any black hole, its event horizon is formed at a certain distance from its very-very compact core, which, too, may have some size though very small in comparison to its mass. The size of the event horizon, depending on the mass of the black holes, would range from a few thousand kilometres to a few hundred thousand kilometres or even bigger. In case the matter particles reaching the event horizon acquire the speed of light, then immediately after crossing the event horizon, the speeds of such particles while rushing toward the core might surpass the limit of the speed of light. Accordingly, by the time they fall into the point-sized core, their speeds would increase manifold. At this speed, a very strong centrifugal force would be generated in the infalling particles that would keep them whirling for much longer time periods around the core before they fall into it. As a consequence of the whirling of trillions-of-trillion charged particles around the core at such a high speed, the strength of its existing magnetic field would be further augmented. In such a case, a whirl, or a cyclone of trillions of Nano-sized magnetic particles, i.e., electrons and protons, etc., would be created deep within the interiors of the event horizon. A very strong magnetic field of the black hole would also pass through the whirl of these magnetic particles. This magnetic field, analogous to the experiment of magnetic trapping of light*, would trap any kind of energy waves or their force-carrying particles within the said cyclone of these whirling Nano-sized magnetic particles.

Since the black holes, every now and then, swallow fresh matter from their surroundings, a series of new events keep happening within their interiors. As a result, the time would continually keep ticking within their cores, too; this would mean that time does not freeze or stop within black holes or at the edges of their event horizons, as is believed in **Relativity.**

A remote possibility

Sometime during Dec. 2011, when I started to write this chapter of the original Hindi book, I saw the news on the Internet that a giant molecular cloud of gas and dust was approaching the galactic centre of the Milky Way. Accordingly, the black-hole situated in the galactic center might very shortly start to suck matter from this cloud. In case this news is correct, then, in the near future, we might get a chance to verify the above prediction of mine, i.e., *how the infalling gases would affect the pole strength of the black-holes. Now, since* our solar system is located at a distance of about 27,000 light-years from the galactic centre of our galaxy, any radiation emitted from this centre would reach us after a delay of 27,000 years. In case the said gas-cloud had reached the galactic centre 27,000 years in the past, then we would very soon be able to practically verify this speculation. On the other hand, in case this cloud would reach the galactic centre in our present time, then the rays emitted from the infalling gas would reach us after about 27,000 years in the future, not earlier than that. Any effect due to the change in the pole strength of the central black hole of the Milky-way would also be felt after this time interval. Further, in case this cloud contains a very large quantity of gas, then the same might turn into a *"Quasar"* as well. (For more information on Quasars, please see the next chapter).

---

* Please refer to chapter-3, "Some Least-known Properties of the Light"

12: Quasars

"*Quasar*" is a very hot and very bright disk of plasma or ionized gas that rotates around the central black hole of some of the galaxies. This disk emanates very powerful bipolar jets and radio signals from its centre, along both the ends of its rotational axis. Quasars are the most luminous, powerful and energetic astronomical objects, rather regions in the universe, whose luminosity, i.e., its energy output, maybe up to thousands of times of that of an entire galaxy comprising billions of stars. They are even more powerful than hundreds or even thousands of hyper-novae exploding simultaneously. The enormous amount of energy that they generate is not a short-lived one like an explosion. The energy produced by Quasars is generated continuously at such a high rate for billions of years, subject to the availability of an adequate quantity of gas. The radiation emitted by any quasar includes different kinds of radiation ranging from X-rays to far-infrared rays; they also emit ultraviolet, visible light, radio waves and even gamma rays, too.

Astronomers, for the first time, discovered a radio source at the beginning of the decade of the 1950s while they were looking for the extra-terrestrial civilizations that are supposed to live on the exoplanets. Thereafter, by 1960, hundreds of such radio sources were discovered. In this effort, during the years 1962-63, a source of radio waves was observed that was, in addition to radio signals, emitting very faint red light, too. At first, it was thought that this very faint object must be a star located closer to the earth. However, its light spectrum was far different from the other stars. This object was, therefore, called "Quasi-Stellar Radio Source." Such objects, as of today, are still known by the abbreviated form of this name, that is, "Quasar." Very soon, it was established that the hydrogen line in the light coming from this source was redshifted to a great extent and also that it was billions of times more energetic as compared to that of any star. Scientists calculated that this object was located at a distance of about 15 billion light-years, and it was moving away from the earth at an amazing speed of 47,000 Km per second. By 1980, several quasars were detected, which are moving away at speeds that are even much greater than 100 thousand km/sec. This speed is only relative to the earth; their actual speed may be determined by adding the Earth's absolute speed. Up till now, about 1 million quasars have been detected; however, out of all of these quasars, as many as 90% of quasars do not emit radio waves.

Though the nearest quasar is located in the galaxy named Markarian- 231, which is located at a distance of only 600 million light-years away from us, most of the quasars are seen at a distance as great as about 12 to 15 billion light-years. Based on this fact, it is believed that Quasars would have been created when the universe was very young, about only 1 billion years old or so. It is also believed that the quasars are created in the galaxies where *loose gas and dust* are available in abundance and where new stars are being created at much faster rates; such galaxies are called *active galaxies*. The activity of the galaxies increases manifolds when two galaxies collide and merge with each other.

Scientists have found that quasars are always formed in the centres of very hot disks of gas; all of such disks respectively rotate in the centre of one or another galaxy, around their central black holes. These disks are called accretion disks, which are always associated with the powerful bipolar jets, such jets are comprised of subatomic particles and energetic radiations. These jets move with almost the speed of light; they normally extend up to far-off distances, measuring several hundred-thousand of light-years or even beyond. Scientists have found that all the quasars are located in the centres of galaxies, but it is not necessary that all the galaxies should necessarily host quasars. It is believed that quasars are powered by supermassive black holes that rotate at very high speeds. *Any supermassive black hole, irrespective of its mass, is capable of creating a quasar, provided that large quantities of loose gas and dust are available in its vicinity; the gases infalling into the black hole would soon form a whirl, which, in turn, would be converted into a quasar.*

The luminosity of quasars never remains constant; it normally keeps varying and fluctuating. This fluctuation might be caused by the fluctuations in the quantity of the gases that fall into the black holes. On the basis of these fluctuations, scientists have concluded that in spite of the enormous amount of energy that quasars emit, their sizes are no bigger than any solar system, that is, about 1 to 2 light-years. Scientists have discovered that in spite of being hundreds of thousands of times smaller than their host galaxies, quasars can generate several hundred times more luminosity and energy as compared to that an entire galaxy is able to produce. This fact suggests that the density of energy in a quasar should be enormous. In spite of generating energy at such a high rate, quasars can persist in producing such an enormous amount of energy for very long periods of millions of years; however, in case the quantity of infalling gas falls below a certain limit, then the quasars would cease to generate energy. Similarly, when the central black hole of a galaxy is again infused with a fresh supply of matter, the quasar would be reignited.

The source of energy generated in the quasars

Quasars rotate with unimaginable speeds, which is clear from the red-shift observed in the quasar hosted by the galaxy named M-84; half of the portion of the said quasar exhibits redshift, whereas the other half of the same quasar exhibits the blue-shift. This fact suggests that this quasar is not only rotating, but its speed of rotation is also even greater than the speed at which this galaxy is moving away from us; otherwise, blue-shift, in one of its halves, could never be seen. Since the universe is expanding at a very high speed of almost that of the light, it could be imagined that at what speed its accretion disk is rotating.

Scientists have envisaged that such an enormous amount of energy at such a high rate is generated by mutual friction of gas ions and dust particles, etc. *However, it seems to me that this explanation is not satisfactory because* **I feel that the possibility of friction between different gas ions whirling in the same direction and at the same speeds is very remote**. I further feel that energy at such a high rate is produced because the **gases within the accretion disks rush inward at very high speeds, due to which great congestion is created all around the black holes. Moreover, the volume of the infalling gases is rapidly and continually compressed to a great extent; churning and friction between the various particles might also result due to congestion. This might result in the generation of very high pressure and temperature in these gases whirling within the accretion disks.** Whatsoever be the cause of the generation of energy in the quasars, any supermassive black hole, in order to produce energy at such a high rate, should convert matter equivalent to ten to hundreds of Stars every year or 600 to 1000 Earths every minute. This process of energy generation is thought to be about 15 times more efficient than the process of nuclear fusion that powers the stars.

Classification of the quasars

Different quasars, depending upon their energy and luminosity, are classified into different categories. However, scientists believe that all the quasars are of one and the same kind. All of them are like a flat rotating disk that emanates bipolar jets of energy as well as matter particles from their axis of rotation. However, different Quasars, depending on the angles at which we look at them, might appear to be of different kinds. Radio waves travel only in the direction of the jets, which can be seen only when their jets are directed straight towards the earth. This is the reason that only 10% of quasars appear to emit radio waves. *The quasars, which we look at in the direction of their jets, appear to be very energetic and luminous, whereas the quasars that we look at in the directions of the edges of their disks appear to be very faint and weak because a bigger part of their luminosity and energy is obstructed by the stars shrouding them. The quasars, which are located at an angle in between the above two positions, appear to be more luminous and energetic in comparison to the later ones; however, all of them are, in fact, of one and the same kind.*

The secret behind the formation of the bipolar jets

Every quasar invariably emanates bipolar jets along its rotational axis. These jets travel almost at the speed of light. Apart from the jets, sometimes the infalling gasses, while rushing towards the centre, are seen to escape in large quantities from the poles of the quasars. The speed of gas escaping from the centres of different quasars that escapes against the gravitational force of the black holes is estimated to be just about 30,000 km per second as against the near light speed of the jets. Such a vast difference in the speeds of these two kinds of emissions, escaping from almost the same place, suggests that both of these emissions must be caused by different reasons. However, no proper or satisfactory explanation is available for both of these emissions.

Scientists have predicted that when any black hole devours gas in excess of its capacity, then it vomits this excess gas through its poles. Somehow, I feel that this explanation is unsatisfactory; ***"How any nonliving object like a black hole would find out that it has swallowed gas in excess of its capacity, and how much gas should it vomit?"*** In fact, any black hole can suck only that much gas that its gravitational force allows it to suck; ***no excess gas, defying the laws, can fall all by itself in any black hole. Neither any black hole can exert more force than its capacity, nor in the absence of any other force can the gas falling into any black hole would speed up, all by itself; therefore, no excess quantity of the gas can enter the black hole. Moreover, neither gravitation can throw out anything against its own force, nor the gas flowing in any particular direction can change its direction without being acted upon by any external force and flow out against the gravitational attraction.*** Few other questions also arise on this belief: 1) ***"Why does any black hole vomit the excess gas only in the direction of its magnetic poles? Why not in any other direction?"*** And 2) ***"What is that force that acts in that particular direction so that the infalling gas is ejected only in this particular direction?"*** *These questions shall essentially be solved before we can understand how and why these bipolar jets are formed.*

Earlier, in Chapter-2, which describes some of the fundamental laws of science, we have learned that strong magnets are capable of levitating carbonaceous objects, even living creatures too, against the force of gravitation. *This fact indicates that a very strong magnetic force can act as an anti-gravitational force for diamagnetic materials.* Yet another possibility was discussed in the previous chapter (Chapter- 11) that *the strength of the magnetic poles of the black holes could increase due to an increase in the quantity and speed of the ionized gases falling into them,* **the more would be the number of gas-ions falling into any black-hole, the stronger its magnetic poles would grow.** In case the above **speculation of mine** is correct, then *the strength of the magnetic poles of the black holes would fluctuate due to any variation in the quantity of these infalling ionized gases.* Accordingly, **whenever large quantities of gas would fall in any black hole, then its pole strength might grow equal to the gravitational pull, or at times the strength of its magnetic force might surpass even the strength of the gravitational force.** *As a result, during such a circumstance, very small regions would be created at the poles of the black hole, where the strength of the magnetic force, being comparatively stronger, would negate the gravitational force.* **Consequently, gravitation might become ineffective in these point-sized zones, as if gravity-less windows have been opened at both of its poles. These point-sized regions would even repel the charged matter particles against gravity. In case the creation of such windows is possible, then the extreme pressure generated within the interiors of the black holes, or more accurately, the force of the gravity itself, would eject the crushed matter particles through these windows in the form of powerful jets.** The energy trapped within the black holes would also escape through these windows in the form of different kinds of subatomic particles having very high energy (frequency). The particles ejected as bipolar jets would mostly contain protons, neutrons, atomic nuclei, etc., **which would be similar to the composition of the cosmic rays. This fact suggests that the Cosmic Rays, coming from different unknown sources, from almost all the directions, might be coming from the magnetic poles of different Black-Holes that are situated in the centres of different quasars. Further, it can also be assumed that the event horizon, because of the windows created at the magnetic poles of their central black-holes, might curve inwards at both the poles like a "toroid," or an apple, instead of being perfectly spherical; the event horizon might also have small windows, one each at each of the poles, through which these bipolar jets are ejected.**

The jets created just before any hyper-novae explosions, as well as during different stages of star-formation, like the protostars and T. Tauri stars, etc., might be created due to a similar phenomenon, during the heavy inrush of the ionized gases, the magnetic poles of such bodies might become stronger than the gravitational pull. In the event that the quantity of the infalling gasses would deplete below a certain limit, the jets would cease to ooze out.

The Secret of Ejection of the gas From the Quasars

In any galaxy, the disk of stars is not as thin as a two-dimensional fabric; it might be as thick as 1000 light-years or even more. Similarly, the accretion disks of different quasars might have the same thickness. On the other hand, the black holes located in their centres are said to be only atom-sized points. *Because of this vast difference in the sizes of the accretion disks and that of the black holes, accordingly, as depicted in Fig-10 given below, the gas ions that start their journeys from the top and bottom edges of the accretion disk would rush toward the black hole at a very acute angle. Accordingly, the gas ions, while on their way, would surely encounter the repulsive force exerted by the magnetic poles of the black-holes. Any sudden gust of the infalling gas would enhance the pole strength of the black hole to such an extent that the gas infalling from the corners of the accretion disks would either bend a little, toward the centre and would be sucked within the black hole, or as shown in the said fig.-10, it would take a U-turn and would be repelled backwards by the magnetic force.* The gas ions, so repelled, would be thrown out in the space, probably, at the same speed at which they started their inward journeys toward the central black-hole. Alternatively, their speed might be a little lesser than the speed at which they started their journey. The particles, so dispelled because of their spiral motions, would look like an inverted rotating cone of plasma or rotating flame.

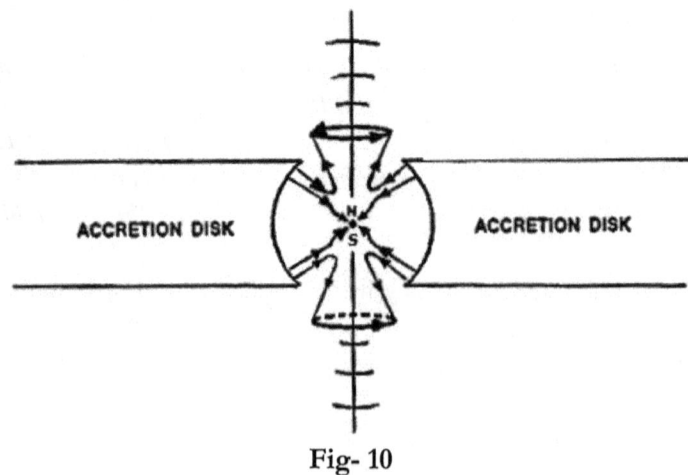

ACCRETION DISK   ACCRETION DISK

292   **Fig- 10**

The matter particles that fall into different black holes would, in due course of time, become part of their point-sized cores; however, whenever possible, these particles would be ejected through the poles in the form of bipolar jets, almost at the speed of light. On the other hand, the ionized gas, which was prevented from entering the black holes and was repelled back into space, would travel on its outward journey at a much slower speed as compared to that of the jets. According to the observations made by the astronomers, such rotating streams of plasma travel at about 10% the speed of the jets, that is, at the speed of about 30 thousand kilometres per second.

13: The Dark Matter

In the year 1929, astronomer **Edwin Hubble** discovered that the universe is continually expanding. At that point in time, scientists believed that the mutual attraction acting between all the stars would one day halt its expansion; thereafter, it would start contracting and eventually end up into a small point with a *"Big crunch".* Later on, when scientists tried to calculate the mass of all the *visible stars,* they found that the estimated combined mass of all the galaxies taken together must be about $1/100^{th}$ times lesser than what is required to halt this expansion. Since then, scientists have believed that the universe must contain a large amount of invisible matter, which is responsible for holding all the stars and all the galaxies together. However, despite their best efforts, they failed to detect the presence of any such matter anywhere in the whole of the universe.

At that point in time, Swiss scientist **Fritz Zwicky**, based on the **Relativity Theory,** was trying to establish a relation between the gravitational force and the process of the formation of galaxies. He, on the basis of the **estimated mass of different galaxies,** as well as **their rotational speeds**, calculated mathematically that the total mass of different galaxies is about 400 times more than the total mass of the **luminous matter** contained within them. It is also understood that sometime during the period 1930 to 34, he made a computerized model of the galaxy, in which he placed some stars and proportionate gravitational force (deep indentation formed in space-time due to the mass of such stars). However, instead of forming a galaxy, the stars in that model straggled away. Subsequently, he included some extra gravitational force in that model; this resulted in the formation of a galaxy. It is believed that on the basis of this model, he concluded that all the galaxies must contain many times as much mass as is the total mass of **luminous stars** so that the total mass of the galaxy may produce an adequate gravitational field (A deep-well of adequate depth) that is essentially required to bind the visible stars in the shape of a galaxy. **Fritz** also tossed the name *"Dark Matter"* for this hypothetical invisible matter.

The aforesaid discovery, made by **Zwicky,** led to the conclusion that this invisible dark matter is the building block of the universe; did the dark matter not exist, all the stars would have scattered here and there in a disorderly manner. Accordingly, it was concluded that in the early universe, the accumulation of the stars over or around the dark matter would have resulted in the formation of different galaxies and clusters of stars. Viewing the fact that several galaxies accumulate around a bigger galaxy to form local clusters of galaxies, it was further inferred that more quantity of dark matter must be present between these galaxies. Similarly, more dark matter must be present between different super-clusters of galaxies. *Since Relativity does not consider gravitation as an independent force, the concept of Dark-matter does not give any clarification on 1) How the stars started to form different whirling structures, that is, what force caused them to rotate in a particular direction and in a particular manner, so as to form a spiral galaxy? 2) How did different stars, in the absence of any force, move and accumulate over the dark matter?*

At the time when the hypothetical dark matter was conceptualized, that is, in the decade of the 1930s, nobody considered the masses of different hard to see objects such as the white, brown, and black dwarfs, neutron stars and quark stars, etc. Although the names such as supernova and neutron stars were suggested by *Zwicky* himself, nobody was serious about them; most of the scientists considered them only hypothetical, not a reality; at that time, the idea of the *Blackhole* was considered merely hypothetical, *nobody was serious about their existence or masses. At that point in time, non-consideration of the mass of such objects that were invisible, as well as, non-consideration of the influence of gravitation caused by such objects, led to the incorrect idea of the total mass or the rotational speed of any galaxy. Accordingly, the inference made at that point in time went totally wrong. When the exact data on the total number of such invisible objects, as well as their combined mass, etc., was not available at that point in time, how could an accurate estimate of the mass of the dark matter required for the formation of the galaxies, could have been worked out?* Even as of today, we don't have any databank on the exact number of such invisible objects, their masses and the effect of gravitation produced by them.

These days, it has been accepted unanimously that in the respective centre of each and every galaxy, a supermassive black hole definitely exists; all the stars in different spiral galaxies rotate around this centre. The mass of such a supermassive black hole alone should exceed billions of stars taken together. Moreover, besides visible stars, any galaxy may contain more than ten billion *white dwarfs,* several billion *neutron stars,* and hundreds of millions of *small and medium-sized* **black holes,** the mass of any such medium-sized black hole might range between tens to thousands of times greater than our Sun. *Accordingly, the actual mass of any galaxy might be several hundred times more than that of the combined mass of all the visible (luminous) stars contained in that galaxy; however, the correct estimation of the total mass of any galaxy is not at all possible. This fact clearly* *indicates that the inference deduced on the basis of the total mass of luminous stars alone was grossly misleading, being based on the incomplete truth.*

Our galaxy contains about 200 to 400 billion luminous stars, as well as numerous other non-visible objects; therefore, analogous to the solar system, it is necessary that the bulk of the mass of any rotating galaxy shall be concentrated at its centre, not anywhere else. As pointed out in this book, at different places, rotational speed in any galaxy has nothing to do with the curvature of space; it depends on the force with which the central black hole attracts the stars surrounding it. This means that a stable and steady whirling motion, in such a big collection of stars that also includes numerous luminous and nonluminous objects, could be generated only on the condition that the mass of the galactic centre of any galaxy must be many times greater than the combined mass of the entire collection of stars and other invisible objects that orbit it. On the other hand, scientists have estimated that the mass of the supermassive black hole located at the galactic centre of our galaxy, the Milky Way, is only four million times greater than that of the sun, which is an average-sized star. Accordingly, the estimated mass of the galactic centre of our galaxy doesn't seem to be adequate to produce a whirling motion in the entire galaxy that contains about 200 to 400 billion luminous stars; the total number of the invisible objects and their combined mass might be several times more than that of the luminous stars. Based on the fact that celestial bodies like planets and their moons, etc., always orbit bigger bodies having many times greater mass than that of their combined mass, it is felt necessary that the hypothetical dark matter shall be located at the centre of any galaxy. This means that in case the dark matter is really holding all the stars together, then the same must be located in the centres of different galaxies. Accordingly, the centres of different spiral galaxies are the best places where dark matter is likely to be detected. To me, another possibility that seems most likely is: *"While estimating the mass of the supermassive black hole, some mistake* might have crept into our calculations."* **In case the mass of a supermassive black hole is greater than the collective mass of all the visible and invisible stars contained in a galaxy, then the dark matter would not at all be required to form any spiral galaxy; the supermassive central black hole of any galaxy would become self-sufficient to form a rotating structure of the stars.**

Ignoring the above-mentioned possibility, scientists are making all-out efforts to find out the hypothetical **Dark Matter** at different places other than the

---

* Please also see Chapter – 10, "The puzzle of Gravitation" & Chapter-11, "The Black holes".

galactic centre; however, it has been eluding them for all these years. We always depend on some kind of energy wave or radiation to see any object or obtain any information about it, whereas dark matter is totally invisible for all kinds of energy waves/radiations. Neither any type of energy wave is reflected by dark matter nor dark matter absorbs any type of wave. Moreover, dark matter itself doesn't emit any type of energy wave; it is, therefore, not possible to directly detect its presence. However, scientists have detected some places in interstellar space where light rays are seen to deflect as if they are bending under the effect of gravity lensing. In spite of the best efforts, no object could be detected in such locations. Scientists believe that such an effect is caused by the curvature created in space due to the mass of the so-called invisible dark matter. Estimation of the mass of such blocks of invisible dark matter is done on the basis of the angle through which the light rays are found to deflect. The astronomers, on the basis of the angle of observed deflection in the light rays, have prepared a few maps of different possible locations of the dark matter and the prospective quantities of the dark matter that might be present in these locations; maps showing locations of prospective locations of different blocks of dark matter, are available on the Internet. To support this prediction of scientists, large accumulations of stars are normally found associated with such locations; however, at some of these locations, negligible or very few stars are observed; the presence of such a small number of stars in these locations has no explanation. On the other hand, a large gathering of stars is sometimes observed in some other places, where the effect of gravity lensing couldn't be observed. This fact indicates that the presence of dark matter is not necessary for any accumulation of stars, particularly when the effect of the gravitational lensing can also be produced by any collection of black holes or even by a single black hole. In this context, please also see the earlier chapters 6 and 11 of this book, wherein it is reiterated that the very existence of space-time is doubtful because no physical proof of its existence does exist.

As brought out earlier, scientists have developed some maps showing the expected locations of different blocks of dark matter. In case these maps are correct, then at least a few stars must be found to orbit bigger blocks of dark matter. Similarly, smaller blocks of dark matter should also be found to orbit the bigger blocks of dark matter. In case any smaller block of dark matter, which orbits a bigger block of dark matter, is detected in any of such locations, then *and only then* the existence of dark matter may be proved. In case the dark matter is really the building block of the galaxies, then the proof of its presence must necessarily be found in the centre of any galaxy. ***The galactic centre of our own galaxy is*** *the nearest and most potent place where the dark matter, if at all it does exist, should be located. On the other hand, if scientists fail to locate the so-called dark matter in the galactic centre of our own galaxy, then its existence would become doubtful. Non-detection of the dark matter in the galactic centre would mean that the ignorance about the existence of supermassive black holes and other hard to see objects existing within different galaxies, which prevailed during the decade of the 1930s, resulted in conceiving the concept rather a wrong concept of the dark matter because the scientists of that era did certainly commit a grave mistake by ignoring the mass of various invisible celestial bodies, though they do really exist. The effect of the gravitational force produced by such bodies was also ignored in that era.*

In case the undetectable space-time, as well as dark matter both, are really non-existent, then **Einstein's** concept of gravitation, as well as his formulae based on the warping of space-time, to calculate the effect of gravity, both would go wrong. In that case, it would not be proper to say that mathematics and computer can never go wrong because mathematics based on a wrong formula and a computer program developed on the basis of such a wrong formula or any other wrong input, both are bound to give wrong results; if wrong data are fed into a computer, then it will certainly give the wrong result. It seems to me that the development of the wrong concept of **Dark matter** is a burning example of the aforesaid possibility.

14: Galaxies- The Formation Thereof And Their Classification

The existing theories predict that initially, the first-generation stars were created approximately 500 million years after the "big bang," in a random manner, one after another, as single units; galaxies were created much later. At that point in time, the newborn stars were randomly scattered all around in a disorderly manner. Galaxies would have been formed within the next few hundred million years due to the gradual accumulation of these stars on/around the dark matter. Scientists further believe that the black holes were created about 1 to 2 billion years after the formation of the galaxies. *However, this concept of the creation of the galaxies does not give any explanation of how these newly created galaxies started to rotate around their centres? This question is important because no direct proof of the existence of dark matter is available* (please refer to chapter-13). *In the absence of any force, neither the newborn stars could have moved from their respective locations nor the rotational motion of stars located within different galaxies could have been generated. The gravitational pull of the supermassive Blackholes is a must for this purpose.* (Please see chapter-9 and Chapter-10)

In the beginning, the universe must have been much more compact and smaller as compared to its present-day size; therefore, at that point in time, the newly created gas would have been much more compact and dense than its present-day density. Scientists believe that the more is the density of a molecular cloud, the slower is the rate of the star formation. *However, I feel that contrary to this belief, stars in the early universe, would have formed in the dense gas clouds at a much higher rate. Giant-sized stars would probably have formed in the early universe due to the availability of dense gases in large quantities; these giant stars, due to their immense masses, would have quickly run out of fuel and become either neutron stars or even black holes directly. The fact that heavier elements are seen in very distant galaxies that would have formed in the early universe supports this presumption because supernovae or hypernovae explosions would have created these heavy elements at that point in time.*

In the early universe, the newborn protostars, after acquiring adequate masses, would have gradually started to attract other protostars located in their close neighborhood (See Chapter-9, under sub-heading "Alternative speculation on the formation of the stars," and also Chapter-10, "The puzzle of Gravity").

The bigger the protostars would grow, the force of mutual attraction acting between these protostars would grow stronger. As a consequence, analogous to the formation of the whirlpool in the washbasins, the most massive protostar would have attracted the neighboring smaller protostars one after another. As a result, the smaller protostars/gas-clumps, along with their protoplanetary disks, would have started to orbit the bigger ones. These gas-clumps, while revolving around the central protostar, continued to collapse inward, due to which the star-embryos were created in the cores of their parent gas-clumps while the said gas-clumps were orbiting the respective bigger stars (central stars). As time continued to pass on, such whirling protostars, due to the gravitational collapse of gas, would have grown into different stars having masses between one to several solar masses. In due course of time, each of these protostars would have transformed into a regular star; each of them would have developed its own planetary system too.

Probably, the small primitive galaxies, rather the mini or dwarf galaxies, were created in the above manner. In the due course of time, the supergiant central star, after consuming its entire fuel, would have directly converted into a medium-sized black hole. At that point in time, numerous mini-galaxies would have been created in different places within a big nebula; all of such mini-galaxies, in the beginning, would have been scattered all around in the infant universe in a disorderly manner.

These mini or primitive galaxies would have continued to grow in size due to the formation of new stars. Their central black holes, too, would have continued to grow more massive by continuously devouring matter from their close surroundings or even a whole star occasionally. Thus, the gravitational force of the central black holes of the newly formed mini galaxies would have gradually gained more strength. Accordingly, their gravitational field would have gradually reached out to different central giant stars or black holes of the other nearby mini galaxies. As a result, the smaller mini galaxies would have started to orbit that particular mini galaxy, which possessed the most massive black hole among all of them. In this way, the local groups or clusters of mini galaxies might also have been formed in the early stage of the formation of the universe. Similarly, different bigger clusters of such mini-galaxies, scattered here and there in the early universe, would have compelled the smaller clusters of mini galaxies to orbit the nearby bigger mini-cluster of galaxies; thereby, bigger clusters of mini galaxies would also have formed. As time passed on, such mini or dwarf galaxies, due to occasional collisions and mergers with other galaxies, would have grown into still bigger galaxies; their central black holes, due to the repeated mergers, would also have gradually grown into supermassive black holes. Such merger of galaxies, as well as that of their central black holes, is one of the normal activities of the universe; it has happened several times in the past and will be repeated several times in the future, too. Almost all the bigger galaxies, including our own galaxy, have grown so big over the past several billions of years due to the repeated collisions and subsequent mergers of different smaller galaxies.

Dark Matter in the Galactic Halo

The expansion of any galaxy or its gamut, its realm, is not limited only to the outer edges of the disk of its luminous stars; a faint glow normally surrounds the entire galaxy, especially the spiral galaxies from all around spherically; this glow extends to a far bigger region, much bigger than the disk of stars. The said glowing region is an integral and extended component of any galaxy, which extends far beyond the main collection of the visible or luminous stars; its radius might be 3 to 6 times bigger than that of the galaxy itself. This sphere of diffused light or the said glowing region is thought to be made of hot-ionized gases. These glowing regions surrounding different galaxies are known as the "Halos" of the galaxies. Such halos can be seen very distinctly in spiral galaxies; however, other sorts of galaxies, too, are surrounded by the halos. It was earlier believed that these halos contain an almost negligible quantity of gas; due to that, star formation is not possible in this region. Contrary to the aforesaid belief, which has now become outdated, the present-day astronomers believe that this region contains very dense gas. They also believe that this accumulation of high-density gas deters the process of the star-formation. *However, it appears to me that gas distribution just after the "big bang" would have been much denser than the gas contained in these halos; even then, the rate of star formation would have been very high in the early universe, such a possibility contradicts the present-day line of thinking.*

Scientists believe that galaxies were formed due to the accumulation of stars over larger blocks of dark matter. *Relativity* envisages that the rotational motion in any galaxy is generated when the stars contained therein move in straight lines along the curved space-time that is warped by the enormous mass of the dark matter over which the stars did accumulate. *Einstein* believed that the rotational velocities of different stars should gradually decrease in the manner in which the speed of different planets of the solar system decreases. Contrarily, the actual measurements taken by present-era astronomers show that the rotational velocities of the stars do not decrease as was envisaged by *Einstein*. *In fact, their rotational speeds, instead of decreasing, flatten out when their distances from the galactic centre increase; rather, the same increases gradually but very slowly.* This fact shows that the actual rotational speeds of the stars don't match with the assumptions of *Relativity*. This deviation is a matter of great concern for the present-day astronomers; *they, based on this fact, believe that such a behavior of the rotational speeds of the stars is resulted due to the additional gravitational force, which acts from somewhere else.* Accordingly, scientists believe that some additional source of gravitational force must exist in the "halos" of these galaxies. However, in spite of the best efforts, scientists have so far not found any additional source of gravitational force that exists in the galactic-halos of even our own galaxy, the Milky-Way. Scientists, due to the non-detection of any visible source of gravity in any of the galaxies, have concluded that dark matter in large quantities must exist in the halos, which should be distributed equally on both sides of the disks of stars. According to this belief, the disk of the luminous stars must be sandwiched between large quantities of dark matter. It is also believed that the highest density of this shroud of dark matter must exist in the galactic center; this density of dark matter shall decrease gradually with an increase in the distance from the center. It is now believed that about 95% of the galaxies are composed of dark matter. *However, in case the presence of dark matter in the halos of the galaxies, in such a large quantity, could not be found, then it would mean that* the prevailing concept of gravity, as envisaged in the *"Theory of Relativity,"* is not correct, the same shall be thoroughly reviewed and modified, if felt necessary.

Although *Relativity* does not consider *Gravitation* as a force, even then, it certainly attracts every object toward any accumulation of large masses of matter. In case the hypothetical dark matter is really present in the halos of the galaxies, and the same is equally distributed on both sides of the disks of stars, then both of these layers of dark matter would either merge into each other due to their mutual attraction or alternatively, they will try to pull the disk of stars toward each other, that is, in the opposite directions. Such a strong force, acting on the disk of stars from both the sides, should tear this disk apart instead of producing a whirling motion of the stars contained in this disk, as did happen with comet Shoemaker Levy-9, which was broken into several pieces by the gravitational force of the planet Jupiter. *And, if at all, these stars start to rotate, even then, as envisaged in Relativity, their rotational speeds shell go-on reducing slowly that would not match with the actual pattern of their rotation. This possibility is a big question mark on the presence of dark matter on both sides of the disk of stars. Moreover, if the disk of stars is really embedded by two separate layers of the dark matter from both the sides, then it would not be possible for the two-dimensional fabric of space-time to warp on both sides around the disk of stars; it is not possible to warp a single sheet of fabric simultaneously, in two opposite directions.*

The similarity in the rotational patterns of any spiral galaxy and that of the whirlpool formed in the washbasin can easily be observed. As we move away from the center of both of these rotating structures, their rotational speeds gradually increase instead of decreasing. Accordingly, as we move closer to the center, the curvature of the paths of the water molecules and that of different orbiting stars, as the case may be, increases gradually due to which the value of the centrifugal force generated in the water molecules as well as in the stars, would go on increasing gradually. This is the reason why these structures maintain definite and stable shapes while rotating. *The pattern of rotation of both the structures suggests that, analogous to the water whirl, the stars in the spiral galaxies rotate due to the gravitational pull acting toward the galactic center, not due to the curvature of space-time;* the balance between this pull and the centrifugal force (produced in different stars) puts them in the stable orbits.

The Classification Of The Galaxies

All the galaxies are not of the same kind; they differ from each other in shapes and sizes. Some of the galaxies are so small that they are known as dwarf galaxies; on the other hand, some of them are much bigger than our own galaxy, the Milky Way. The biggest galaxy seen so far is almost 60,000 times bigger than our own galaxy. All of these galaxies are not spiral in shapes like the Milky Way; the galaxies are, therefore, classified depending on their shapes, not by their sizes. Apart from the spiral galaxies, the galaxies that appear to be perfectly spherical like a football, or elliptical like a rugby ball, are called "Elliptical Galaxies." Likewise, the galaxies whose shapes resemble the convex lenses are called "Lenticular Galaxies," whereas those galaxies, which don't have any particular or regular shapes, are called "Irregular Galaxies;" these out-of-shape galaxies are probably formed due to the collision of two galaxies. Alternatively, a galaxy might be deformed by the gravitational force of any other nearby galaxy.

Astronomers believe that those galaxies that were formed in the early universe would have been elliptical in shapes. Later on, they might have started to rotate within the gravitational fields of dark matter; as a result, they flattened out gradually. In the due course of time, such galaxies, in the next step of their development, might have become lenticular in shapes. Later on, they would have become spiral in shapes due to the continuation of the process of flattening. It is also believed that elliptical galaxies were formed from such molecular clouds, which have been rotating at much slower rates, whereas the spiral galaxies would have been formed in those clouds that were rotating at much faster rates. However, no clarification on the matter is available in these theories regarding *"How and why this whirling motion did produce in different molecular clouds."* Another question that remains unanswered is, *"Why were different clouds rotating at different speeds?" Probably, the rotation of gas clouds was taken for granted.* **Nobody probably tried seriously to find out 1) How the clouds started to rotate and 2) Why different clouds started to rotate at different speeds?** *Since no explanation of the rotational speeds of different clouds is available, this hypothesis of the formation of different types of galaxies doesn't seem to be fully convincing, at least to me.* **Therefore, I have presented below my own speculation in brief:**

Scientists have predicted that numerous protostars and their respective rotating gas disks would have been created in the early universe as independent units. However, as brought out in Chapter-9, under the heading *"An Alternative Speculation on Formation of the stars,"* all these protostars were created in a similar manner. Therefore, their magnetic fields would have similar polarity; as a result, they might repel each other, which is clear from the fact: **"Whenever any moon of the planet "Saturn" happens to pass through its rings, then the ice lumps & particles rotating within these rings, automatically move away from such a moon, instead of colliding with it, as if they are being repelled. The reason for the same might be the similar magnetic polarities of such moons and that of the icy objects, which constitute these rings."** On the other hand, the more mass these protostars would have accreted, the more strength their respective gravitational and magnetic forces would have acquired. As speculated in the aforesaid Chapter-9, the most massive protostar, located within any local group of the newborn stars, would have compelled all the smaller stars to spiral inward, i.e., toward it. However, **the pressure of the star-wind produced by all these stars, as well as by the central protostar; moreover, the magnetic repulsion of all the stars having similar polarity, might have opposed their inward movement. Therefore, analogous to the formation of the water-whirl, these protostars would have adopted spiral paths; the angular moment of these stars, too, might have caused them to adopt curved paths.** *Apart from the above, the magnetic field of the said most massive protostar would have repelled the infalling protostars from both sides and thereby compelled all of them to occupy the plane of its magnetic equator. In case this* **speculation of mine regarding the formation of the galaxies is correct,** *then the mini galaxies, so created, should have adopted* **spiral shapes** *right from the beginning. The fact that the stars in the Elliptical Galaxies are much older than that of the Spiral Galaxies indicates that the Spiral Galaxies were created earlier than the Elliptical Galaxies.*

It was also speculated in chapter-11, Black Holes, that the strength of the magnetic force of any black hole depends on the quantity of the infalling matter. Therefore, in case any black hole is starved of the gas or the quantity of matter falling into its core decreases, then its magnetic strength would also decrease. Accordingly, I feel that when a central black hole of any galaxy consumes almost all of the loose gas from its vicinity, then its pole strength would decrease gradually. As a result, the magnetic force, which repels all the stars from both sides and thereby compels them to form a flat disk, would gradually enervate; as a consequence, the mutual repulsion acting between all the stars, which are rotating and revolving within any such galaxy, might cause them to slowly and gradually move away from each other. As a result, such a galaxy might start to bulge. There may be yet another reason, in case the smaller companion of a binary pair of the black-holes existing in the center of a galaxy is thrown away by the primary pair of black holes, or snatched away by any other nearby galaxy, then the former galaxy would gradually bulge in the absence of adequate magnetic force; the shape of that galaxy would start to fade away, and its stars would lose distinctive spiral structure. The spiral galaxies were probably, transformed in this way into lenticular galaxies. Such galaxies, after a long period of time, would bulge further and, with time, would convert first into an elongated sphere and thereafter into a spheroid. Probably, the elliptical galaxies might have been created in this manner.

15: The Dark Energy Or The Death Warrant For The Universe

In 1929, Hubble discovered that all the stars are moving away from each other; the further away the stars are located, the faster they are moving away; it was inferred from this fact that a force is behind this acceleration seen in the rate of expansion of the universe. Scientists, despite persistent efforts, could not find any reason for this increase in the rate of the said expansion or any source of energy that is responsible for the same. Therefore, scientists have predicted that an unknown force is causing this gradual and continual increase in the rate of this expansion. The said unknown force or energy supposed to be behind this expansion is known as the *"Dark Energy."* However, scientists were confident that one day, the gravitational force would halt this expansion. Later, in 1998, scientists, while observing a very remote supernova, noticed that in the remote past, the universe was expanding at a much slower speed, i.e., slower than its existing speed; however, no information is available on at what rate other stars located at the near vicinity to the said supernova, were moving away during the same point in time? Anyhow, they inferred from the aforesaid observation that instead of slowing down, the rate of expansion of the universe is continually increasing. Therefore, the combined gravitational force of all the stars is unable to halt the expansion of the universe. Further, looking at the speed of the expansion and the total mass of the universe, it is explicit that dark energy must be much stronger than the gravitational force. Scientists have estimated that only 4 to 5% of the total energy of the universe is in the form of normal matter that is contained in all the visible stars, yet another 22 to 25% of the energy is in the form of Dark Matter, and the balance 70 to 74% energy must be in the form of the Dark Energy. In spite of the fact that the said dark energy is in abundance, it is so elusive that so far, we could not find even a trace of it.

I read somewhere in the past that the scientists of the present era have predicted that the repulsive force acting between the normal matter and antimatter may be the cause of this expansion. Galaxies and stars in this universe are not distributed uniformly in all three directions; some vast areas in the known universe are almost devoid of the stars. Some scientists believe that the galaxies made of antimatter might exist in these voids. They believe that the repulsive force acting between the matter and the antimatter is acting like an anti-gravitational force. At times, it is heard that antimatter in very small quantity exists even in the vicinity of our Earth; however, this could not be authentically proved. Therefore, the presence of the antimatter in large enough quantities, adequate to form numerous galaxies, seems only hypothetical and a postulate without proof. Scientists alone can tell what actually the fact is.

Probably, the entire energy of the "big bang" was released instantaneously; scientists have predicted that the cooling-down of this energy resulted in the creation of matter particles, which is supposed to have started after 1½ minutes of the big bang and completed within the next 1½ minutes or so. This means that all the matter particles were not produced in the same instant of time; in case the period of one second could be divided into several billion fractions, then in each of such fractions, the matter particles would have been created gradually in different bunches. Accordingly, the gravitational force would also have gained its strength gradually, in billions of small steps. These particles, created in each of such steps, would have immediately started to move outward at a speed that was supposed to be greater than the speed of light. Accordingly, gravitational force, in proportion to the number of particles created during each of such steps, would have started to act between them with gradually increasing strength. As discussed in chapter-9 under the section "Big bang theory," gravitational force so produced was unable to halt this expansion due to the very high energy of such matter particles and probably, because of their greater than light speeds. In spite of this limitation, the gravitational force might have succeeded in the said slowing down the rate of expansion, probably, to some extent only. Perhaps, due to this reason, or some other unknown reason, the rate of expansion of the universe at that point in time, might have been slower than its present rate of expansion. On the other hand, the energy produced by the "big bang" was only momentary, not continuous. Therefore, the said energy should not have produced any acceleration in the rate of expansion; this is bewildering because, in the absence of any external force, such an increase in the rate of expansion doesn't seem possible. Had no external force been active, gravitation was sure to retard this expansion and thereafter, within the due course of time, squeeze the entire universe back into a small point. This means that some strong force must have been continuously driving the matter particles away; this hypothetical force is known as ***Dark Energy***. However, there is a problem with this idea that ***"It is not possible to produce such a powerful force all by itself."*** *This fact raises the question,* ***"When, where from, and how this so-called "dark energy" was produced?"***

Anyhow, in case the dark energy is behind this expansion of the universe, then such a force must be very powerful because it is hurling all the celestial objects at a speed that is very close to the speed of light. On the other hand, the structure of different galaxies and that of the planetary systems of different stars, including our own solar system, etc., does not seem to be distorted, elongated, or affected by this force in any way. *This fact indicates that the force behind the expansion of the universe, though very powerful, is not a violent or cataclysmic force that would have been produced by a sudden explosion. Had this force been explosive in nature, then the structure of all the galaxies should have shattered, elongated, or destroyed long back ago. Similarly, all the stars would have scattered in a disorderly manner, like pieces of straw. In total contrast to the above possibility, this force is probably pushing all the galaxies and all the stars, etc., very gently, with a very soft hand, to ensure that their structure is not at all disturbed or damaged. This fact indicates that the so-called Dark-energy was probably, not produced by the Big Bang. **Anyhow, this force, immediately after its inception, has been acting so delicately on different celestial objects, including the Earth, that no signs of its effect can be felt, neither on the frail enclosure of the atmospheric air of our Earth nor on any other celestial object; even the pressure of this force that is pushing the Earth so hard; can't be felt by us.** This fact puts a question mark on **the very idea of the expanding universe and also on the existence of the dark energy.***

Further, it is believed that the universe, in the present era, is expanding at an unbelievable speed of the speed of light. However, the mass of any celestial object or that of mankind living on the Earth is not at all affected by this speed, which should have become infinite at this speed; this fact brings the speed of expansion under a question mark. In spite of such a high rate of expansion, all the routine activities of the universe are going on as usual, without any hindrance. The molecular clouds, instead of expanding or shattering away under the influence of dark energy, are still collapsing under gravity to produce new stars. Similarly, the stars, after running out of their fuel, do collapse to form dense objects like neutron stars and black holes, etc. Dark energy being many times more powerful than the gravitational force, shall be able to forcibly increase the sizes of all the gravitational structures, such as different galaxies, the solar system, etc., with the same speed at which the universe is expanding; however, the size of our solar system doesn't appear to expand or distort, even a bit. Likewise, the shapes and sizes of all other similar structures, such as the local clusters of galaxies, etc., appear to be unaffected by this expansion. It seems from this fact that the so-called dark energy is either not able to increase the sizes of any rotating structures that are bound gravitationally or shatter such structures against their binding force. Only the super-clusters of the galaxies, because of very weak gravitational bonds among them, are, perhaps, drifting away from each other. This fact signifies that within the close ranges, dark energy is not effective against strong gravitational bonds. Similarly, different nuclear forces, too, which act within the range of the atomic radii, are probably, not affected by dark energy. *Could it be implied from this fact that "The nuclear force and the gravitational force, both are stronger than the so-called dark energy within short distances? Alternatively, does dark energy act only on bigger structures? How come the inanimate dark energy could select its targets, and also that at what time shall it attack any particular target; it must act equally on all the objects, all the time."* These questions give rise to another question *"does any such energy exist at all?"* In order to find out the answer to these questions, let us analyze the following facts:

The strength of the gravitational force that was produced after the big bang should have been proportional to the total quantity of the matter that was produced after the so-called big-bang. The amount of matter that was produced at that time is still available in the form of trillions of trillion stars and innumerable molecular clouds, etc. However, out of all these astronomical objects, all the stars are collectively burning a very big amount of their fuel, which might amount to trillions of trillions of megatons every second; our own Sun is burning several hundred million tons of hydrogen every second. In the whole of the universe, the matter in several trillions of megatons is destroyed every second; the matter, so destroyed, is converted into mass-less energy. Apart from the above, different stars lose a large quantity of matter every second in the form of the star wind. We are conversant with the fact that the bigger the star is, the higher is the rate at which it burns its fuel; such stars consume more than two-thirds of their stock of matter in less than one billion years. Besides this, *millions of quasars, too,* are converting the matter into energy at a much higher rate. It is very hard to imagine what would be the total quantity of the matter that has been lost during the last 13 to 14 billion years? What one could imagine is, *"the quantity of matter, which we started with, might have depleted to a great extent."* A question arises in this scenario *"Does the quantity of matter and the value of the gravitational force both still the same that we started with?"* Although the amount of the matter so last is still available in the form of energy, no form of energy, except gravitational force, attracts and holds the celestial objects together. *This means that gravity, i.e., the force that prevents the expansion of the universe, is getting weaker day-by-day, because of two reasons: 1) Due to the gradual depletion of the mass of the matter that we started with, & 2) Due to the increasing distances between all the celestial objects.* How this process depletes the strength of the gravitational force is explained with the help of an example given hereunder:

Considering a scenario in which the distance between two celestial objects gets doubled in a certain time period, and during the same time interval, both of them consume fuel equivalent to half of their initial masses. Accordingly, as per Newton's formula, the strength of the gravitational force acting between them would work out to (m'/2). (m''/2) ÷ (2d)^2 = 1/16 (m'. m'')/d^2. In other words, the strength of the force acting between these two objects would reduce by 16 times its original value. Alternatively, if the value of gravitational force varies in the inverse proportion of the cube of the distance between different objects, that is, in proportion to 1/d^3, then the value of the force acting between the said celestial bodies in the time period under consideration, would be depleted by 32 times its original value. It could be seen from the above example that, on the one hand, the binding force is depleting day by day, and on the other hand, the star-wind is pushing away all the stars/galaxies from each other right from the inception of the universe.

Had the entire energy of the "big bang" been converted into the matter particles, then in the absence of any force, the mass of the entire matter that was produced within the first 1$^{1/2}$ minutes should have eventually contracted back into a very small point. However, nothing like this did happen. This fact suggests that only a small portion of the energy released from the big bang might have converted into the matter. *This, in turn, means that the gravitational force, which was produced at that point in time, was not adequate to counter the rest of the major portion of the total energy released during the big bang.* As per Newton's second law of motion, the acceleration of an object depends directly upon the force acting on it, as well as, inversely, upon the mass of that object. This could be represented by the equation f = P/m or **P=mf**, where "P" means the value of the driving force, "m" is the mass of the moving object, and "f" means acceleration produced in that object. Thus, it could be seen from the aforementioned equation that, on the one hand, the force of gravity, which opposes the rate of expansion, is gradually depleting due to the reduction in the mass of all the celestial bodies, and on the other hand, the force responsible for expansion, is probably constant in the absence of any resistance. *This might be the reason why the force causing the universe to expand is getting an upper hand day by day.* Now, since the mass of all the stars, as well as their density per unit area, both have consistently depleted, the rate of acceleration of the universe is bound to increase. Since the distance between different super-clusters of galaxies is far too large to bind them together, the gravitational force acting on them does not seem adequate to prevent them from drifting away freely. Even any weak force, if acting on them, may probably, cause them to drift apart.

The hypothetical End of the Universe

It is not possible to ascertain when the so-called dark energy became effective and how? However, in case the so-called dark energy really exists, then it can be seen that this force didn't interfere with the activities related to the routine works of Nature like the creation and/or destruction of the universe. The formation of different galaxies, the creation of the stars, and their deaths, etc., the formation of the neutron stars, black holes, etc., have continued in the expanding universe in a normal way. In the meantime, this Dark energy simply kept pushing different gravitational structures very gently; all the works of nature related to both creation and destruction have continued side by side. *However, how long this game of expansion, creation, and destruction, etc., would continue?*

Scientists believe that one day, this apparently quite looking dark energy would result in the destruction of the entire universe; even the universe is not exempted from the rule that "Everything that was born shall die one day or the other." Based on the rate of expansion, which is probably, very close to the speed of light, scientists have envisaged that all the galaxies would, first, move apart from each other to such great distances that they couldn't be seen anymore. Next, all the stars would break the gravitational bonds and wander here and there in an unorderly manner; the structure of the universe would be totally disrupted. After the stars use up all of their fuel, the darkness would loom everywhere because all the sources of light would cease to shine. And thereafter, all the material objects, including all the galaxies, stars, and eventually, all the forms of matter, including the atoms, would disintegrate into unbound particles and radiation; eventually, they would shoot apart from each other. The universe would gradually fade away and become so sparse and thinned out that it would completely defuse into an endless void, its existence would be completely terminated; nothing, except a vast emptiness or void, like the one that existed before the "big bang" would be left behind.

My Own Viewpoint on the Dark Energy

It is believed that the universe, since its inception, has been expanding at the speed of light. Further, based on the red-shift seen in the light of furthest galaxies, it is believed that 1) "The furthest galaxies are moving away from us, at almost the speed of light." & 2) "The further the stars are located, the faster are they moving away." In case all of the aforesaid hypotheses are correct, then, over a span of 13.5 billion years, the present-day speed of the expanding universe must have increased manifold than its initial speed. However, neither we shall be able to see the light emitted by the stars moving at faster than light speed, nor any massive object is supposed to move at such a high speed. It could be inferred from this fact that the far-off objects are not moving at light speed. Otherwise, we shall not be able to see them or observe the redshift in the far-remote galaxies. Further, as brought out earlier, in chapter-8, since the Earth is not located at the center of the universe, the amount of redshift seen in the light of the far-remote galaxies would not give a correct idea of their speeds and distances because such galaxies would appear to move in different directions at different speeds.

Further, scientists have observed that the Redshift is seen only in one of the halves of the quasar located in the center of the galaxy M-84, whereas the Blueshift is observed in its other half. Based on this fact, it can be concluded that this quasar is spinning at a speed that is more than the speed at which the universe is supposed to expand, i.e., more than the speed of the light. Accordingly, the spinning speed of the said quasar, also known as NGC 4374, seems to violate the law of science that no massive object can move faster than light speed. This fact, in turn, indicates that **the amount of the Redshift seen in the light coming from the distant galaxies does not give the true picture of the speed of the expanding universe; the distant galaxies are not moving away at the speed of the light.**

Apart from the above, the scientists, during the year 1998, while observing a far-remote supernova, discovered that billions of years in the past, the universe was expanding at a much lower rate; however, I don't have any information about the speed of the nearby stars that were located at the same point in time. I feel that the said supernova must have emitted far-more energetic light, which would have a much higher color temperature as compared to the nearby stars; therefore, over the same span of time, its light would have appeared more energetic, i.e., it would have exhibited lesser red-shift than what was seen in the light of the other stars. **This possibility indicates that the observed redshift in the light of the distant stars/supernovae, etc., might not give a correct estimate of the speed at which the Universe is believed to expand.**

In case the prediction of *Relativity* that *"The curvature of space-time compels the planets to follow its curvature by pushing them back and thereby compels them to orbit the Sun"* is correct, then the space-time, while expanding, shall also be able to compel the Solar-System, to expand continually, or at least, elongate it in the direction of the expansion. I have read somewhere that the Milky-way Galaxy is expanding almost at the speed of sound. Accordingly, the stars located in the neighborhood of the Sun shall be seen to move away from each other with the speed of at least half to one-fourth speed of sound. At this rate of expansion, the size of our solar system shall increase by about 10% every 400 years or so. Moreover, the distance between the Sun and Earth and also between the SUN and its nearby stars, etc., shall be seen to increase; however, during the last 400 years or so, i.e., after the invention of the telescope, none of the astronomers has reported such a phenomenon. It has, therefore, become necessary that *apart from the amount of the redshift seen in the light of the distant stars, we must ascertain: "Whether or not the size of the solar system and the distances between nearby stars are physically increasing in reality?"*

In view of the different anomalies discussed above, I feel that in order to ascertain the existence of **Dark-Energy**, we must necessarily verify the amount of **the Gravitational Redshift** in the light of the distant stars. **Special Relativity** predicts that when any light wave moves away from any strong gravitational field, it would gradually lose its energy, which in turn, would result in the gravitational redshift in the said wave. This gravitational Redshift would be in addition to the Redshift/Blueshift seen in any luminous celestial object due to its velocity. In case this postulate of **Relativity** is correct, then it would become difficult to assess the actual speed of the source of such a light wave. One more possibility that I visualize: *"The photons, after coming out from their source, may gradually lose their energy over a long span of time, due to which, the frequency of light, as well as its color-temperature, both would continually go on reducing; alternatively, their energy may continually attenuate due to the resistance of the interstellar medium; the further the photon would travel, the more would be the amount of the energy it will lose." In case this prediction is correct, then the further away would be the source of light, the more redshift would be observed in the light coming from it. Accordingly, in spite of the fact that the far-remote objects are not moving away with the speed of the light, it might appear to us that they are; the further away would be the light-source it would appear to move away with proportionately greater speed. This would mean that even if the universe is not expanding, it would ostensibly seem to expand. This might be the reason that the so-called Dark-Energy, though, is supposed to cause all the stars to expand at greater than the light-speed, the same, in reality, doesn't produce any destructive effect on any of the stars/planets, even on the atmospheric air-blanket of the Earth.*

In case the above-mentioned proposition of mine is correct, then the concept of the expanding universe, as well as the Big-Bang theory, both would need a thorough review.

There is no direct way to verify the fact that whether or not the energy of the light (Photon) really attenuates while traveling through very long distances. However, in case the distant stars are really moving away with the speed of light, then the luminosity of light coming from them can never remain constant; it shall gradually but surely diminish. The redshift and the dullness of light coming from such a fast-moving source shall continually increase at a very slow rate. In case any record of the luminosity of the distant stars after the invention of the telescope has been maintained, then this fact can be verified; however, in case a constant watch is kept on the luminosity of the distant stars for the next few hundred or few thousand years, it might be ascertained that whether such objects continue to glow with the same intensity, or their luminosity goes on diminishing at a constant rate. In case our scientists consider this point worthy enough for verification, and serious efforts are made all over the world to ascertain this fact, only then this vital fact would be established that *whether or not the universe is continually expanding, and in the case of the universe is really expanding, then at what rate?*

In case my presumption that the "The longer the photons travel, the more they lose energy" is correct, then despite the increasing redshift in the light coming from the distant stars, such stars might be stationary, or at least, they might not be moving away at a very high speed. This would further mean that no energy (Dark Energy) is pushing them away from each other. This fact, if found correct, would mean that the universe is probably, not expanding; and if at all, it is expanding, then not with the speed of light. It would mean that the redshift observed in the light, which is coming from the furthest galaxies, is due to the gradual enervation of the energy of the photons over a long span of time, i.e., the redshift in the light that is coming from distant stars is not produced because such stars are moving away; they might be stationary.

In case my above presumption is correct, then the *frequency or the energy* of the light coming from the far-off stars, due to the elapsing of a very long time span of say 15 to 16 billion years or even more, would drop down to such a great extent that we may not be able even to take notice of such a weak light or its source. In such an eventuality, we will never be able to find out the actual expansion of the universe or its correct age.

In the above scenario, I feel that until all of the above-mentioned possibilities are not examined very thoroughly, it might not be possible to ascertain the existence of either that of the dark energy or the strength of the thrust it produces. Similarly, it can't be concluded that the universe is really expanding, and if it is really expanding, then at what speed?

Coming back to the hypothetical end of the universe: If the *dark energy does really exist, then the same, not being a destructive force, might not disintegrate the matter and its atoms.* Accordingly, when all the stars would consume their entire fuel, then no source of light and heat would be left anywhere; only black holes of varying masses and sizes, ranging from supermassive ones to 3-4 solar masses, as well as the non-luminous dense cores of numerous dead-stars of different sizes, would be left all over. However, even after such an end of the stars, the gravitational force of their remnant cores would continue to act with the same vigor. In this scenario, the black holes and other dense objects, in accordance with the respective masses of such objects and the mutual distances between them, would organize themselves into different families of the black holes in which the less massive ones would orbit the more massive ones. Next, these families of the black holes would start to orbit different supermassive black holes*. Subsequently, different black holes, depending upon the strength and range of their gravity fields, would gradually devour the smaller ones from their vicinities; thereby, they would gradually grow bigger; their mass and consequently their reach of the gravitational force would continue to grow stronger. Therefore, they would devour the smaller ones from even greater distances. In case the so-called **Dark Energy** does not break the gravitational bonds and hurl off all the black holes to far-off distances from each other, then different black holes would continue to merge and grow bigger and yet bigger.

In the year 2010, scientists observed that a large number of galaxies were moving towards an empty stretch of space; this is an astonishing and unexplained phenomenon. It appears to me that a very powerful force of gravity of a super-giant black-hole might be drawing all these galaxies towards the said stretch of the empty space. I understand that several very large voids are observed in the deep space; *these voids were probably created by giant-sized black holes, which might have swallowed all the stars of their host galaxies.* **In case this apprehension of mine is correct, then this would give rise to a possibility that nobody has probably imagined so far, that as the last stage, all the supermassive black holes would merge into each other; they would, therefore, form one single ultra-supermassive black hole, so massive that it would comprise all the matter of the universe. However, what would be the consequences of such an accumulation of the entire matter at a zero-sized point? Would this be the end of the universe? Or alternatively, due to the creation of immense heat, extreme pressure, and extremely powerful centrifugal force generated in such an ultra-massive blackhole, the said black hole might explode because the centrifugal force increases in the proportion of the square of the rotational speed; therefore, after a certain limit, it won't be able to handle the combined effect of the extreme heat, pressure and centrifugal force so produced? This might result in an explosion similar to the "big bang." In case such an explosion would really happen, then this might result in the creation of a new generation of the universe. This is**

---

* During August 2016, I saw the news that a new galaxy has been discovered wherein, bright stars are almost negligible in number. Scientists believe that this galaxy is made of dark matter. However, it seems to me that either star-formation in this galaxy has just started, or alternatively, most of the stars contained in this galaxy, have been converted into black-holes, or other hard-to-see objects. .

probably the game of nature; the cycle of the creation and destruction might be eternal, and such a cycle might continue forever.

It seems to me that there may be one more possibility of the re-birth of the universe: *In case accumulation of the entire matter in a single Ultra-massive Black Hole is not possible, that is, different ultra-massive black holes explode much before the accumulation of the entire matter within a single black hole,* then such ultra-massive black holes would explode at different times to give birth to different universes at different places. This would mean that numerous universes, i.e., the **Multiverse,** would be produced, and each of such multiverse would be separated from each other by un-imaginable distances.

According to a piece of late news that was published in the Astrological Journal, on 28th Feb' 2020, a huge cosmic explosion, the biggest known explosion after the hypothetical Big-Bang, did occur in the center of the **Ophiuchus Galaxy Cluster,** which is located about 390 million light-years away from the Earth, this galaxy cluster is also known as MS 0735+74. The said explosion was so powerful that it punched a cavity as big as 15 times bigger than the *Milky Way Galaxy* in the cluster plasma (hot gases) surrounding the central black hole of the said galaxy cluster. This explosion was so energetic that people were skeptical about this explosion. Probably this explosion did occur in the central black hole of the said cluster. Alternatively, it was caused due to the merger of two binary black holes. Whatsoever be the reason for the said explosion, this explosion from such a large distance, i.e., from a distance of 390 million light-years, seems very slow, like a slow-moving movie, even though the gas released from the explosion, might be moving at a very high speed, in case, the distance moved by these gases, during a certain time-period could be measured, then the speed of the aforesaid explosion could also be measured.

Based on the above phenomenon, it could be said that even after the end of the universe, if at all it so happens, the vast void in which it was born would not be destroyed; that is, space would continue to exist. If so, then it would not be the end of Space; neither space was born with the universe and nor would it end with the universe. And if the universe would be reborn in the future, then the "time" would also not end with the death of the present generation of the universe. Space and time both are non-materialistic entities, both of which are immortal, totally free from the cycles of births and deaths.

16: The Greatest Miracle "The Origin and Evolution of the Life"

Perhaps, out of all the unanswered eternal questions, the answer to the question *"What life is, and how did it begin"* is the most sought-for. Deep knowledge of biology, microbiology, genetics, etc., seems necessary to find out the answer to this question. Unfortunately, I don't have any knowledge of any of these subjects. However, on the basis of the very limited information that I could gather from different sources, I have, in spite of my limited knowledge, tried below to *give my own version of the answer to this question;* this proposition of mine is based *partially on the well-known facts and partially on pure speculations.*

All over the world, a common belief is in vogue that life and death, both, are controlled by a superpower known as *God.* According to this belief, nature, or the universe, doesn't play any role in these matters. Of course, this might be an indirect work of God; however, the chemicals, which are the building blocks of life, were created at the time of the death of different stars, which is a natural phenomenon. Nature, thus, plays a direct and very important role in *the creation of life*. Further, it is also a common belief that life is based on the *spirits* or *souls,* which neither take birth nor they ever die. *In case the aforesaid belief is true, then the total number of the living beings in the whole of the world shall always be limited to the original population of the primitive living beings, or more accurately, equal to the total number of their souls.* On the other hand, the total number of human beings alone, barring other living beings and microbes, has multiplied by billions of times when compared to the original population that existed at the time of the beginning of the human species. *In case the total number of microorganisms and all other kinds of living beings is also considered, then it could be seen very clearly that the total number of souls that existed at the time of the creation of life has grown from zero to infinite in number. This fact raises the question, "Do the souls also multiply?"*

Further to the above discussion, the *microbes* generally reproduce by a process that is known as cell division; in this process, the parent cell first expands and then divides into two new living units of the same kind. This means that multiple living beings, or more accurately, multiple souls are generated from a single soul. Besides microbes, a few creatures such as the Amoebae, Hydras, Planarians, Starfish, and even the common Earthworms (to some extent), etc., do not die even if their bodies are cut into several pieces; instead, each of such pieces, grows into an independent body of the same kind of such a creature. Everybody might be aware of the fact that several living copies of the same plant can be created by cloning; each of such copies does live independently. Each and every one of these examples in which a single living body is multiplied into several living individuals of the same kind casts a question, *"How a single soul could give life to all such living copies?" This question leads to the possibility that the spirits might also multiply or take births with the newborn offspring*; in case they really do so, then this possibility suggests that souls might also die after outliving their lives. Under this scenario, we cannot say that *spirits* are definitely free from aging or the compulsion to die; probably, they are neither immortal nor divine. *The fact that "The souls might also take births" suggests a possibility that life might have originated as a normal phenomenon of Nature without being dependent on the souls.*

Probably, the living bodies are merely electrochemical machines or rather electrochemical systems comprising the building blocks like *carbon, nitrogen, oxygen, hydrogen, silicon, iron, potassium, phosphorus, etc., and their different compounds.* These machines, or systems, are actually a coalition of individual living cells; all such machines carry out their activities with the help of electrical signals. The living bodies differ from the man-made machines in the sense that the living bodies are capable of generating their own electrical signals to fulfill their specific needs; these signals are generated by these machines themselves, in their respective brains. Another point that is important is *"Immediately after the death of any creature, all the elements that constitute its living body don't undergo any change; instead, their brains cease to generate electrical signals."* This clearly means that death is merely a failure of the brain; *more precisely, the failure of the brain to produce the required electric signals results in death. Even the death of the brain does not cause the entire body to die at once; the cells of the fruits and vegetables don't die for several days after plucking them up from the tree. Thus, it could be concluded: "Death is, in fact, a slow and gradual process; different living cells of any dead body (which might be several billion in number) die very slowly and independently; such cells die at different times due to lack of food and oxygen, etc." In fact, a single body dies manifold deaths. Moreover, even the old and worn-out cells of different living beings die regularly after completing their individual lives, which are periodically replaced by the newly created cells. Does this mean that every individual cell of a body is a separate and independent soul? Alternatively, can innumerable souls live simultaneously at the same time within a single body?* This fact makes the very existence of the souls questionable, rather unnecessary.

However, this fact fails to clarify "What is the difference between the living and the dead bodies?" Moreover, the question *"How the life, in the absence of a soul, is infused in the collection of the inanimate chemicals"* makes the matter more complicated. These questions can confuse even the wise of the wisest man. This question points to the fact that *in case the souls do exist, then they must be far different from our existing conception.* However, I personally believe that souls do exist, though it is very hard to prove their existence logically.

Some of the apparent common features in different kinds of living beings, like the similarity in different body parts, suggest that there might be a creator who designed all of them almost on the same lines. In case it is so, then at least the following three possibilities might exist.

1. The creator, depending on his wishes, could have made the environment favorable and produced his best creation in the very first instant; nothing could compel him to go slow.

2. The designer was bound to abide by his own laws, that is, the laws of Nature; he didn't want to take any freedom.

3. Life might have originated as a natural phenomenon; Nature has immense powers in the form of the electric charge contained in countless particles, which permeates all over the universe. Accordingly, Nature, depending upon different environmental conditions, designed different life forms that would suit different environmental conditions.

×   ×   ×

It is evident from the oldest fossils found so far that life didn't originate all of a sudden like a stupefying miracle; instead, its origin was a very slow and gradual process. This fact goes against the first alternate discussed above; the same is, therefore, ruled out. We know that everything in the universe, including different common features in the living beings, follows a definite system. Nobody knows exactly what the mystery behind this orderliness is. However, up till now, we have learned that everything in the world is governed by the four fundamental forces of Nature. This fact raises an important question, the answer to which seems to be beyond the capacity of humankind; this question is- *"Where from such a huge quantity of matter and energy did come into existence, and how the properties of matter particles and all the natural forces, were decided?"*

The oldest fossils reveal that life did evolve after about 1 to 1½ billion years after the creation of the earth. At that point in time, most hostile environmental conditions did prevail on the earth; *however, that particular environment suited-most for the origin of primitive life.* Later on, those creatures that couldn't adapt to the subsequent changes in the environment couldn't survive; they probably were abolished due to this reason only. *This fact indicates that it was the gradually changing conditions that decided: "Which life-form would originate and flourish at what particular point in time and under what particular conditions."*

It is difficult to understand how life would have originated from the lifeless chemicals until and unless we find out under what unfriendly and most difficult conditions can the different life forms survive. Apart from the primitive ages, certain life forms, in the present era, too, flourish even under the conditions which are thought to be much adverse, rather fatal, for the other life forms that we normally see all around us.

It is a common belief that oxygen and mild temperature, etc., are a must for the thriving of life, and also that life is based on the energy we receive from the sun; that is, it is impossible for life to survive in the absence of the oxygen, water, and temperature within a favorable range, etc. Contrary to this belief, some microbes found on the earth can easily flourish in the total absence of oxygen and the most stringent environment that is considered unfavorable, even fatal for life. All the life forms that are generally known to the common man are sure to die instantaneously if they are put under such a harsh environment. Long past, about 20 years in the past, I read that a colony of some unknown microbes was found at the South Pole, beneath a layer of snow that was half-a-mile thick, where neither the sunlight had ever reached, nor the fresh air was available, moreover, the temperature, at that place, was as low as minus 60-70° C, or even less. Besides extreme cold conditions, the microbes also thrive in the total absence of oxygen within the hot rocks found in the depths of about 900 feet below the earth's crust, where the temperature as high as twice that of the boiling water exists. In the year 1966, a new type of microbes was found in the water of the Grand Prismatic Spring in the Yellowstone National Park, North America; the water of the said spring is extremely acidic and almost boiling hot. Besides Yellowstone Park, some species of micro-organisms have been found to thrive in even more extreme conditions of geothermal areas of the "Mono Lake" and Mount Lassen (both in California). The water of Mono Lake is hyper alkaline and extremely rich in salts and other minerals like Calcium, sulfur, Arsenic, etc., whereas water in Mount Lassen is boiling hot and highly acidic. Microbes also thrive in hyper-alkaline waters of the Soda Lakes of Kenya and California. Similarly, microbes also flourish in some lakes of very hot and hypersaline or alkaline water found in the volcanic areas of Iceland, New Zealand, Italy, the Kamchatka Peninsula of Russia, and also in South America. The highly acidic water of the River Rio Tinto (red river) of Spain provides a home to another kind of life. All such microbes, which thrive in the most extreme conditions, are known as the *"Extremophiles."* Apart from the extremophiles, a caterpillar-like microscopic creature named *"Tardigrade,"* which is colloquially known as *water-bear,* too, are also capable of surviving under the most adverse conditions like extremely high or low temperatures and very high rather fatal level of radiation; they can live without food and water for several years and even survive the absolute vacuum of the interstellar space, too. Apart from the *tardigrades,* live *Marine-Plankton,* which is a kind of algae, was also found in 2014 on the outer surface of a Russian spaceship.

One of the most primitive forms of microbes, known as *"Archaea"* or *"Archaea Bacteria,"* is found in the Atacama Desert of Northern Chile, which is the driest place on the earth. This microbe ekes out its living by chemo-synthesis of poisonous volcanic fumes that contain Di-methyl-sulfide and Carbon-mono-oxide, etc. Besides Archaea, other forms of life are found near the *Hydrothermal-vents, which are* formed on the seabed, at a depth of about 2-3 kilometers below the sea level, where temperatures as high as 400 to 700°F, or even more are produced due to the volcanic activities; apart from the high temperature no sunlight ever reaches these places. Creatures like Pompeii worm, and other microbes that live in these places, thrive on different poisonous chemicals like hydrogen sulfide, methane, hydrogen, and carbon-dioxide, etc., which are emitted through the active hydrothermal vents. Microbes have been found even at a depth of 11 kilometers too, where the pressure capable of crushing even an ordinary submarine is produced; only specially designed submarines can survive such high pressure. *NASA* recently discovered a new kind of microbe that ekes out its living by eating poisonous substances like Arsenic. Yet another microbe that lives in very deep gold and platinum mines of South Africa thrives on the energy that is released by the radioactive decay of the old rocks.

Vast diversities in the conditions under which extremophiles are found and also the type of food on which they thrive suggest that "In case, any chemical-food is available even under extremely adverse environmental condition, which can support living hood of any particular kind of micro-organism, then that particular kind of microbe might originate under that particular environment, such microbes would flourish by eating that particular food, may it be poisonous for other kinds of microbes. In case that chemical food becomes scarce, or any change in the environment ceases to support the metabolic process of that particular food, then that microbe would either become extinct or go to sleep; energy is essentially required to sustain all kinds of life and their activities. However, microorganisms are far-more-stubborn and unyielding than what we can imagine. Sometime back, I read a news article that a bacterium was found in Alaska, which was completely frozen for the last 32,000 years; however, in spite of being completely covered by ice for such a long period of time, it was alive and became active after the ice was melted. Microbes can survive, or even flourish, in impossible-looking circumstances. Some of the microbes can survive without food and complete absence of water for a very long period of time; they even tolerate massive doses of radiation up to 3000 times as high to kill a human being; they can even repair major damages to their DNA. Based on this fact, some scientists believe that life might have arrived on the earth through meteors, but no proof to support this belief is available. *If life would have originated elsewhere, then there is no reason why it shouldn't have started independently on the earth, too, under similar environmental conditions.*

Several scientists, since 1920 or even earlier, i.e., much before the discovery of the extremophiles, have been systematically studying the enigma of the origin and evolution of life. *In the year 1924, Oparin suggested that Oxygen-free or reducing atmosphere might have supported the biochemical origin of life; later, Haldane, too, proposed a similar idea independently in 1929. Later on, in 1952-53, Stanly Miller and Harold Urey, based on the aforesaid idea, simulated the conditions of the Early-Earth, to verify the possibilities of the chemical origin of life.* In this effort, they heated a mixture of *Methane, Ammonia, Carbon-di-oxide, and water* to boiling temperature, and simultaneously, they continually passed electrical spark through this mixture for a few days. After a short period of about 2 weeks, 20 to 25 types of amino acids, which are found in various proteins, were produced. However, nucleic acids could not be produced in this experiment; probably a much longer time period is required for this purpose. It is believed that in the beginning, i.e., before the origin of the life, complex molecules similar to the Ribo Nucleic Acids (RNA) and Deoxyribo Nucleic Acids (DNA) would have formed in Nature due to a series of different chemical reactions that would have continued for millions of years; the basic building blocks of life, such as simple sugars or the most basic hydrocarbons (monosaccharides) would have formed by the combination of carbon and water, more precisely, by carbon, hydrogen, and oxygen. During almost the same period, different *amino acids*, i.e., the building blocks of proteins, *glycerol* (the building block of fats), and different *nucleotides* (constituents of nucleic acids), might have formed separately.

Almost everybody might have heard the terms like "DNA," "RNA," "genes," and "chromosomes," etc. Both RNA, i.e., Ribo Nucleic Acid, and DNA, that is, "Deoxy-Ribo-Nucleic Acid," are different kinds of complex molecules of nucleic acids, in which different kinds of nucleotides are attached to each other in different sequences. Both RNA and DNA are the basic and most primitive units of life; DNA, which is found within the nucleus of different cells, is capable of self-replicating, whereas RNA (Ribonucleic acid) is not; RNA can reproduce only in one condition, that is when it (RNA) enters within a living cell; in this process, the RNA breaks the DNA of such a cell and kills it. This means that *from the very beginning of life, this inability of the R.N.A. to reproduce by itself resulted in setting up a cruel system in which the parasites or the mighty ones were given the right to snatch away the life or prey on the weaker ones.* *Thenceforth, the cruelest principle of the life,* **"Might is Right,"** *came into force. Accordingly, Nature gave a free hand to all such creatures that were deficient in any field of life, to snatch away the rights of others, even their lives, too.*

DNAs are the carriers of all the information required to decide how the next generation would look like and support their living. The DNA of the simple creatures may carry very few instructions, whereas the DNA of the more developed beings may carry thousands of instructions. The key to the information-transfer lies in the molecular structure of the DNA; this structure is a special arrangement, or a pattern, made of four different kinds of the nucleotides, each of which consists of a kind of *sugar* (hydrocarbon), which is known as either **Ribose** or **Deoxyribose**; these sugars are bound on the one side to a **phosphate group,** and on the other side, to a **nitrogenous base**. These bases, which are known as nucleotides, are of 5 types, namely: "Adenine" (A), "Cytosine" (C), "Guanine" (G), "Thymine" (T), and "Uracil" (U). The DNA is made of Deoxy-Ribose sugar and four nucleotides, A, G, C, and T, in different combinations. The RNA molecule differs from the DNA molecule in the sense that in the RNA molecule, the Ribose molecule replaces the Deoxy-Ribose molecule, and the uracil molecule (U) replaces the thymine (T) molecule.

In both RNA & DNA, 3'carbon atoms of one molecule of sugar, and 5' carbon atoms of another sugar molecule (Ribose in RNA & Deoxyribose in DNA), are linked together to form *Phospho-di-ester* bonds (the *esters* are those chemicals in which at least one –OH (**hydroxyl**) group is replaced by an –O–alkyl (alkoxy) group). A *Phospho-di-ester bond* occurs when exactly two of the hydroxyl groups in phosphoric acid react with the hydroxyl groups on other molecules to form two ester bonds. The *Phospho-di-ester bond* is central to all life forms; they are the backbone of the strands of different combinations of nucleic acids. In these linkages, strong covalent bonds (molecular bonds) are formed between the phosphate group and two numbers of *5-carbon ring carbohydrates (pentoses)* over two ester bonds. The genetic information in DNA is stored in the form of different sequences or combinations of different nucleotides, i.e., the way covalent bonds are developed between the phosphate group and two hydrocarbons comprising 5 carbon rings each. Enumerable combinations can be formed from the aforesaid 4 nucleotides (A, C, G & T) to form different strands within any DNA; these strands contain the codes for producing a specific protein. Such strands are called *genes, and a set of genes is called a chromosome*. Chromosome, found in the nuclei of both animal and plant cells, is made up of DNA tightly coiled many times around proteins called histones that support its structure. During the process of cell division, these chromosomes ensure that the said nucleotides "A," "C," "G," and "T" always bind with each other *in a particular sequence so that the DNA is accurately replicated. It is this sequence that creates its own copy. The chromosomes are responsible for containing the instructions that make the offspring unique, i.e., different from each other, while still carrying traits from the parent.* Any alteration in any such sequence may result in some change in the next generation. Such changes are known as *"mutation."*

From 1990 onward, scientists have been trying to synthesize complex molecules like Genes, RNA and DNA, etc., by joining their different constituents in different sequences. From time to time, the news is received from here and there that scientists have achieved some success in this field. During 2008 and 2010, I read that some scientists have achieved success in creating some structures that are similar to the RNA of a virus; they have also created synthetic DNA or proto-cell-like structures, which though are not living creatures, still they are capable of making their own replica. *Looking at the efforts being made along these lines, specialists in this field are confident that synthetic life will probably be created in the next 10 years or so, i.e., the puzzle of life will be solved very soon. Looking into these possibilities, it could be concluded that when even mankind is on the verge of creating artificial life, then this work would have been easy for Mother Nature too. However, Nature would have had to wait for the right time to create life under the gradually changing environmental conditions.*

Scientists observed in the year 2011 that even the drops of the oil (nitrobenzene) could perform many acts like living cells; these droplets can move through the water at their will, can sense their fuel, and absorb it from the surroundings. This fact indicates that the reactions between different chemicals that continued over very long time periods might have created primitive life.

In order to understand "How the spark of life was ignited," a broad outline of the environment that existed on the earth at the time of its birth and the changes that took place subsequently is given below:

It is believed that around 5 to 6 billion years in the past, the sun was created out of the gas cover and the debris ejected by a dying star that was probably a second-generation star. Thereafter, the Earth and the other planets of the solar system also did come into existence approximately 4.7 billion years in the past or a little earlier. Initially, all the inner planets in the solar system were made of molten rocks and metals. Scientists have envisaged that the elements needed for the creation of life were in absolute scarcity on these newborn planets. However, different chemicals and water, in the form of stardust, and ice, etc., were available in abundance at the furthest edges of the solar system, in the form of comets, asteroids, innumerable meteors, etc. It's understood that conditions prevailing at that point in time were very violent and hostile. *The total number of planets formed at that point in time would have been far more in number than what we see today; the total count of the newborn planets would have been around a couple of hundreds or even more.* Due to this reason, the distances between these celestial bodies would have been much smaller than what exists between the present-day planets. In that era, the orbits of different planets might have intercepted each other. Such complicated orbits might have resulted in occasional collisions between most of the then planets. A great upheaval was going on everywhere; planets, after such collisions, either merged into each other or shattered into pieces. At that time, the powerful gravitational force of the giant gas planets, like Jupiter and Saturn, etc., would have caused a very heavy shower of comets and asteroids on the earth and other planets. *Perhaps, this was the way in which the raw materials of life and water, etc., reached the earth and other planets.*

An alternative possibility to the above speculation also exists; since both the raw materials, which are required for life and all the planets, asteroids, comets, etc., were created out of the gas ejected from the different supernovae explosions, all the raw materials such as the water and different chemicals, etc., should have been present on all the planets right from their creation. However, since favorable conditions in that era didn't exist anywhere, life couldn't immediately originate on the earth or any other planet. *Much later, when the conditions would have become favorable to at least some primitive forms of life, the seeds of life would have been produced after a very long chain of different chemical reactions.*

Scientists have now come to the conclusion that water would have existed on the earth from the very beginning.

As time passed on, those planets, which survived the said upheaval, would have adopted stable orbits; other environmental conditions also started to change slowly. The Earth, in due course of time, cooled down a little, due to which a thin crust would have developed over the molten lava. During that era, the inner heat of the earth would have created thousands of active volcanoes almost everywhere. These volcanoes would have continually emitted large quantities of water vapor. In addition to the volcano eruptions, a continual shower of numerous comets might also have brought the bulk quantity of water to primitive Earth. Besides water vapor, volcanoes would also have continually emitted columns of volcanic ash, different poisonous gases, fumes, etc., which might have formed the first atmosphere of the Earth. The atmosphere in that era was much different from what it is today; it was made up of poisonous volcanic gases like Sulfur-oxide, Hydrogen-cyanide, Hydrogen-sulfide, Carbon-mono-oxide, Methane, Carbon-di-oxide, Nitrogen, and its different oxides, etc. and of course, the water vapor. At that point in time, the *Ionosphere* that protects Earth from excessive and harmful radiation had not yet formed. Therefore, powerful radiations would have been reaching the Earth unobstructed. This radiation might have facilitated the formation of some chemicals that couldn't have formed in a radiation-free environment; this could be understood by the process known as *photosynthesis,* which takes place only in the presence of light, which, too, is a sort of mild radiation. Powerful radiation might have disintegrated and oxidized Methane gas present in the upper layers of the atmosphere to produce a large quantity of water. Besides radiation, repeated discharge of lightning was also responsible for the production of different types of new chemicals such as different carbonates, hydro-oxides, oxides, nitrates, etc. At that point in time, different chemicals, such as Arsenic, Phosphorous, etc., and different compounds of these elements, were in abundance, except for free Oxygen, which was almost absent because of its very high chemical affinity toward other substances; at high temperatures, it readily oxidizes most of them. *The creation of life in such unfavorable conditions would have been a very complicated process; even nature, too, might have to wait for the conditions to become favorable.*

In due course of time, condensation of water vapor in the upper layers of the atmosphere would have caused repeated rains; however, Earth's surface was so hot that drops of rainwater would have evaporated even before reaching the earth's crest. However, the continual repetition of this process must have helped to take away some of the atmospheric heat gradually but regularly. Within the next 200 to 300 million years or so, when the temperature of the Earth's surface dropped down to about 100°C, a sea of boiling water would have formed, which covered almost the entire Earth's surface because continents might not have formed by that time. Volcanic gases and fumes emitted by different active underwater volcanoes and thermal vents would have dissolved in the water of that boiling sea. As a result, the entire seawater would have been converted into a soup or broth of the primitive chemicals that were necessarily required for the chemical origin of life. On the other hand, ice caps might have formed over both of the geographic poles of the earth. At that point in time, different types of acids were also produced by dissolving different oxides of phosphorus, sulfur, carbon, nitrogen, etc. in water, and on the other hand, elements like Sodium, Potassium, Calcium, Lithium, etc., would have been producing different alkalis by reacting with the water. This, in turn, would have produced some new salts and other chemicals. *Russian scientist, Alexander Oparin, envisaged in 1924 that atmospheric oxygen prevents the synthesis of certain organic compounds that are the necessary building blocks of life. This fact implies that the oxygen-free or reducing atmosphere formed in that era would have helped the synthesis of the building blocks necessary for primitive life, especially when the chemical activities of all the substances did increase a lot due to high temperature and the excessive-radiation.*

×   ×   ×

As brought out earlier, around 4.4 to 4.5 billion years in the past, a sea of boiling hot primitive soup was formed on the Earth. In the due course of time, this sea of primitive soup might have become ice cold at both the poles. The temperature difference so created would also have helped the formation of some different kinds of chemicals at different places. Boiling water is essentially required for the synthesis of Cytosine (a constituent of DNA) and Uracil (a constituent of RNA); accordingly, both these nucleotides would have formed in the hot regions of that sea. On the other hand, the freezing temperature, which existed at the poles, would have facilitated the synthesis of Adenine and Guanine. This means that different constituents of different kinds of nucleic acids would have been formed at far-off distances from each other. However, the natural sea currents formed in the oceans, probably, due to the temperature difference, would have washed them off from their places of origin, due to which they could have come in contact with each other. Accordingly, about 4.1 to 4.2 billion years ago in the past or so, the aforesaid phenomenon, as well as a long chain of different chemical reactions, would have resulted in the formation of pre-cell-like structures of the non-living chemicals in the said sea of the primitive soup.

The DNA was not produced all of a sudden or by any miracle; instead, a long time period of several million years was required in this process: different *phosphates, nitrogenous compounds, and different types of amino acids, etc.,* would have formed as a first step. The lack of oxygen would have facilitated the formation of different kinds of hydrocarbons and sugar molecules. In the beginning, a chain of different chemical reactions would have resulted in the formation of numerous simple molecules, which would have been based on *carbon, hydrogen, nitrogen, etc.* Later on, different types of base chemicals, similar to the amino acids and nucleotides, would also have formed. Numerous different permutations of chemical reactions would have resulted in the formation of different combinations of these building blocks. In the beginning, some non-living chemicals that were the precursor to the living cells, such as different combinations of *amino acids* and *hydrocarbons, etc.,* would have probably, formed around the underwater active hydrothermal vents. Subsequently, as a result of different long chains of chemical actions, different kinds of *nucleotides* that formed earlier would have combined together to form different sequences or chain-like structures. In due course of time, molecules similar to *nucleic acids* but simpler in constructions would also have formed. Later on, such chains would have combined together in numerous different permissible ways. As time passed on, the development of more complex molecules would have continued. Gradually some lifeless, pre-cell-like molecules and various structures resembling the RNA would have been created, which would have been much simpler in construction in comparison to the RNA and DNA molecules. How life got infused in those lifeless chemicals is not known; this fact appears to be beyond the perception of humankind. *However, I feel that the solution to the mystery of the origin of life is probably, hidden in the intrinsic properties of the matter particles; this solution is discussed below:*

We have learned that matter, whether living or inanimate, consists of negatively charged electrons that orbit the positively charged nucleus; that is, *the flow of electrical energy is intrinsic to all kinds of matter, even including the living beings, too. Electrons within different atoms or molecules always move in a specific and prefixed order; that is, every substance, in accordance with nature's law, comprises a specific circuit of electricity that is typical to only that particular substance.* When any chemical reaction takes place, then electrons of the final product share the orbits of more than one atomic nucleus; thereby, electrons adopt different but definite prefixed and more complex paths. Regardless of the quantities of different chemicals put together, molecules of different chemicals would always combine in a prefixed order that would be decided by the properties and molecular structure of those chemicals, as well as by other conditions also, such as the pressure, temperature, the intensity of the radiation, etc., to which these chemicals are subjected to. *During such reactions, different paths of electrons in different molecules of the end-product would be automatically reorganized in a prefixed order following the laws of nature.* Each of the molecules of these chemicals so formed would also have a fixed order of movement of electrons specific to that particular chemical only. *It is believed that when the electrons orbiting any nucleus undergo any change of direction, then they emit different types of waves or radiations.* *Therefore, specific paths of electrons in different chemicals might play an important role; every substance, probably, for this reason, has its own unique spectrum.*

*We may get another clue, to the riddle of life, from some single-celled microbes: S*cientists have observed that whenever a unicellular organism, known as *Amoeba,* senses food, it reaches out to that food, engulfs it through its cell walls, and eject out the waste. *From this fact, it could be inferred that the body of all the single-celled microbes might be capable of performing all the functions of different limbs that are necessary for moving, eating (as the mouth), sensory organs like eyes, or some other sensory organ to sense the availability of food, as well as its direction and distance, etc., and of course, the entire body of such microbes might act as the brain, too.* All other kinds of unicellular creatures, too, might possess similar abilities. Primitive microbes might also possess the senses to find out food as well as favorable conditions for their living. This fact gives rise to the question, *"How do they do this?"* In order to understand and solve the enigma of *how life would have infused in the lifeless chemicals,* we need to understand how the microbes can sense the exact direction and distance, etc., in which they have to move to find their food, more precisely, *how do their brains function?* A *speculative solution* to this question is given below.

In the preceding paragraphs, we have already seen that the *electrons in different molecules orbit their nuclei in different but definite ways. In case of any change in the energy levels of these electrons, each one of such electrons might emit weak radiation or a definite electromagnetic signal.* Any DNA molecule comprises numerous genes; all of such genes contain different but definite sequences of nucleotides. Probably, similar to the above property of the subatomic particles, each gene produces a definite electromagnetic signal. *It is a well-known fact that all the electromagnetic fields and radiations of different frequencies, etc., have a definite effect on the orbital paths of the electrons. Since the intensities of these fields and radiations, etc., vary with distances, any change in the intensities of such energy fields and/or the distance of the source of any radiation, etc., should affect the paths of orbiting electrons differently, rather than to the different degrees.* It might, therefore, be possible that the electromagnetic signals emitted by different substances or by the sources of different radiations, which come from any distant object, might make some changes in the paths of the electrons orbiting within different constituents of the genetic sequences of different microorganisms. These signals might alter the paths of the electrons orbiting within the genetic sequences of different microbes or probably divide the orbitals of some electrons into hyperfine structures (Kindly see Chapter-6, under "Ticking speed of the time"). Each signal, following the laws of nature, might produce a definite reaction. Microorganisms, because of the above reason, might cognize these signals and react accordingly. They might have learned to use these changes in the orbital paths of the electrons to differentiate between different substances and make an accurate estimation of the directions as well as the distances of these substances. Human-skin and eyes are also capable of sensing such alterations in the orbital paths of the electrons; thereby, humankinds decode the information carried by the infrared radiations and light waves, etc. Our sensory organs are connected to the brain through a chain of interconnected nervous tissues or neurons. When we touch, hear or see any object, then an electrical impulse, or signal, is generated in the first nerve-tissue. This signal produces some chemicals called neurotransmitters between two interconnected tissues, which in turn generate a similar electrical signal within the next tissue. In this way, such signals are transmitted to our brains through the chain of nervous tissues. *Single cells or microbes, too, might be using a similar electrochemical process to transmit or decode various information; they may even communicate with each other by electrochemical language.* This fact indicates that *a particular electromagnetic signal is capable of assembling particular chemicals found within the cytoplasm (cytoplasm is the fluid that exists inside the living cells) and thereby converts such chemicals into neurotransmitters.*

Now, based on the above inference, it may be imagined that under the gradually changing conditions, which prevailed at about 4 to 4.2 billion years in the past, different chemicals might have combined together due to continued chemical reactions, and thereby different sequences of nucleotides, or inanimate protocells, probably having *Phospho-di-ester bonds*, might have formed. Since the electrons always orbit the nucleus of different atoms at some definite energy levels, the electric fields created by different constituents of such a protocell would fluctuate with a definite frequency. Based on the capability of the electric signals that create Neurochemicals within the neurons of the living beings, it could be further imagined that different constituents of the said protocell that was created in the primitive soup would have produced a specific electric field. This electric field might have interacted with the electric fields of different chemicals dissolved in that hot primitive soup and thereby broken down some complex molecules into two simpler molecules. The energy waves released from this process might have synchronized with the electrical field produced by the said protocell. As a result, the resonance (increase in the amplitude of these waves due to frequency matching) created among these two frequencies might have intensified the signal produced by that protocell. This intensified field might have attracted identical constituents of the said protocell, dissolved in that hot primitive soup, and assembled them in that particular sequence to create the exact but inanimate replicas of the parent molecule. What actually happened thereafter is beyond imagination; *probably at a particular environmental condition, electrons of different constituents of the said inanimate protocell might jump to such energy levels that the waves emitted by them would have bestowed a capability to such a protocell to gain continual energy by repeatedly breaking-down the chemicals available in their close vicinity into simple constituents; the energy (radiation) so released might have enabled the said protocell to repeat this process again and again. This capability might have provided continual energy to that particular chain of chemicals, i.e., the said protocell, to continue this process till available raw material didn't exhaust. Probably, life might have been infused into those inanimate replicas by a similar process.* As a result, primitive forms of life might have been created under some favorable environment as a natural phenomenon. *In case the above presumption of mine is correct,* then different chemicals created during the deaths of different stars are definitely capable of creating life in due course of time, under a particular environmental condition. Such a possibility indicates that life can definitely be created artificially by arranging different chemicals in a correct sequence *under the appropriate environmental conditions; appropriate conditions that support the origin of life, and a particular sequence of nucleotides, both are perhaps, equally important for the creation of life, either naturally or artificially within a laboratory.*

Based on the above discussion, it seems that the origin of life was purely a Natural phenomenon; God had nothing to do with this process. Some of the scientists also agree with this idea. One of the greatest scientists of this era, Late Mr. Stefan Hawking, was probably an atheist; he perhaps believed that with the proper knowledge of Physics, Chemistry, and some other branches of science, each and everything in the world could be understood. Hawking had his own reason for this belief because, based on the expanding universe, he himself devised the **Big-Bang Theory** that describes the process of creation of the Universe (please see chapter-8). Similarly, the Nebular Theory discussed in chapter-9 gives the details of the Life-cycle of Stars. Almost all of the scientists, based on the above-mentioned theories, believe that the creation of the entire universe and all the stars along with their planets, etc., was a natural phenomenon, and God didn't have a hand in this phenomenon.

Anyhow, maybe some of the scientists (not all) don't believe in God; perhaps they don't have any logical proof based on science to prove that God definitely does exist. However, believers of God have no reason to feel gloomy because science has no solid proof of the non-existence of God either. As per the Big-Bang Theory, the universe began about 13.8 billion years in the past due to a massive explosion of energy. However, science doesn't have any explanation that in case the universe really began in an aforesaid manner, then wherefrom that initial energy got produced and why? What was the root cause of that explosion? And, on the other hand, in case the universe didn't start from the so-called Big Bang, then how, wherefrom, and when the entire matter, which constitutes the universe, did get produced? And how were its properties which resulted in the creation of the universe decided? Finding out answers to the above questions seem beyond the capability of science.

In fact, the Big Bang Theory was based on the *continuously Expanding Universe;* in turn, the idea of expanding-universe itself is based on the *Red-Shift* observed in the light coming from the furthest stars. Regarding the said redshift, I have proposed an alternative possibility in the chapter-15, Dark-Energy, of this book; the said proposition is: "The universe is not expanding; instead, the red-shift observed in the light of distant stars is probably, produced due to the gradual attenuation of the energy of the light-photons, i.e., the gradual cooling-down of the photons over a long time-span of billions of years, which these photons take to reach us from the far-off stars." Now, in case the aforesaid speculation of mine about the alternate cause of the said red-shift, is correct, then the concept of the expanding-universe will go wrong, and in turn, the concept of the Big-Bang that is based on the expanding universe will also go wrong. In that case, the prediction of the scientists that the creation of the universe was a natural phenomenon would come under doubt. In the absence of any irrefutable theory about the creation of the universe, a situation of uncertainty will be created; a possibility that "The universe and the life, both, were created by God" will rise to about 50 percent. However, to prove whether God exists or not; seems beyond the capability of humankind or science. The concept of creation will become a matter of faith; anybody will be free to choose his or her option. However, in case God does exist, then he might be far different from the belief of different sects of religious people.

×   ×   ×

In whatsoever way life did originate, the fossils of the earliest Archaea and Cyanobacteria (blue-green algae) found so far are only about 3.5 to 3.8 billion years old. However, life on the earth might have evolved a little earlier than that, that is, within the first ½ billion years or so after the creation of the earth. *The primitive environment that seems very harsh to us would have been the most favorable one for different types of living beings that evolved under that particular environment.* The life-forms, which originated at the seabed, i.e., Archaea, were adapted to generate energy for their living and multiplying by breaking down the molecules of poisonous substances like Hydrogen-sulfide, which were in abundance in that environment. During the same era, another kind of life, Cyanobacteria, would have originated on the sea's surface, where the sunshine and carbon-di-oxide would have supported their lives by photosynthesis, which is a process that breaks down the molecules of Carbon-di-oxide into Carbon and Oxygen. *Likewise, depending upon acidity, alkalinity, and different temperatures, etc. that prevail in different places, different kinds of life forms, which might have been well adapted to the environmental conditions prevailing at that particular place, would have originated in that place as if they were specially created for that particular environment and to eat that particular food. In other words, if a particular kind of microscopic life-form can support living in any specific environmental condition by metabolizing any particular substance that was available at that place, then a microbe of that very kind would originate and evolve in that particular environment. Both a particular environment and a specific kind of food are necessary for the origin and evolution of any particular kind of life form.*

Initially, under that primitive environment, the food was probably, available in abundance in the close vicinity to the colonies of the newly originated life forms, which did originate to thrive on that very food. And as such, mobility would not have been necessary for them. *However, favorable conditions can't prevail forever at any particular place;* closing down of a thermo-vent would have created a scarcity of food. Since Nature has bestowed microbes with the capability to mutate their DNA, such a calamity might have forced those life-forms to either change their ways of living by either eating the dead or living ones of their own kind or develop the ability to move toward the place where that specific food was available. *The law of Nature, "The mighty ones have the right to rob others," did come into force during such a calamity. The life-forms that evolved during the eventuality of acute scarcity of food were forced to adapt to the new circumstances and start to eke out their living by snatching away the food from others or even devouring them. Even as of today, all the life forms (the pure vegetarians too) except a few most-primitive ones, such as the Archean-bacteria, Cyanobacteria, Algae, Plants, etc., cannot generate energy directly from the non-living chemicals such as Carbon, Hydrogen, Nitrogen, etc., they have to drive energy for their living by eating the enzymes obtained from other living beings or their dead bodies. Those who failed to change their food habits or could not*

develop the ability to move would have perished or become inactive. Under any calamity, microorganisms are capable of changing themselves or mutating their genetic codes according to the needs; this ability enables them to cope up with almost any situation. This is evident from the fact that different kinds of bacteria and fungi, etc., develop resistance or immunity toward different antibiotics within a very short period of time.

Almost for the next one billion years after the origin of life, the environment continued to change, and the temperature continued to fall gradually. During this period, different microbes continued to produce "Oxygen" as a biogenic waste, which was probably harmful to the microbes which did evolve in the Oxygen-free environment. As a result, some genes of some of the microbes would have become inactive, whereas other genes mutated their genetic codes to adapt to the changing conditions. By the end of the first billion years or so, different life forms would have adapted themselves to survive the new conditions. By that point in time, the excess of oxygen probably brought such changes in some of the microbes that their offspring, after cells-division, couldn't move away from each other; instead, they got stuck up together. Probably multicellular creatures would have emerged only due to similar changes. *Apart from such mutations, different kinds of microbes might have developed the skill to live together to avoid extinction under the new conditions; such relations between two or more microorganisms are known as* **symbiotic relations**, *i.e., the art of living together.* In the present era, almost every living-being depends on one or the other sort of such relations. In one kind of symbiotic relation, both the partners depend on each other for their survival; they gather food independently from different sources and share the energy obtained by metabolizing different foods in different ways. In another kind of relationship, only one partner is benefited, and the other partner is neither benefited nor harmed. In yet another kind of relationship, one member benefits, whereas the other one is harmed. In due course of time, different kinds of symbiotic relationships would have paved the way for the development of complex and improved life forms.

A good number of examples of such relationships can be seen all over in nature. Lichens, which are composite organisms in which algae or cyanobacteria or both, are embedded between two layers of filaments or fungus. In such a relationship, algae or cyanobacteria produce food by photosynthesis; in turn, fungus protects them from environmental hardship and also provides moisture, salts, and other nutrients to them.

One of the best examples out of the lot is that of hundreds or even thousands of species of the gut -bacteria, fungi, viruses, and other microbes, living within the intestines as well as all over the bodies of different animals - *including ourselves*. These gut microbes may outnumber the body cells of the host by 10 to 1. Though the digestive system of the hosts can function independently, it cannot break down the sugar that is present in various fruits, grains, vegetables, etc. In such a relationship, these gut bacteria get safe home to live in; in addition, they also get a regular supply of free food as a bonus. In return, they break down the food of the host into useful nutritious components and simple glucose, etc., which provides energy to the host to live. In addition, the Gut bacteria also protect the host from harmful microbes; they also train his immune system to fight the harmful microbes. However, the unbalance in the number of different types of these gut bacteria in the host's body may cause insulin resistance, obesity, asthma, etc., and even cancer. The absence of any proper type of gut bacteria may adversely affect our learning process, motor control, etc., and even our genes too; such an unbalance in the number of gut bacteria may eventually pave the way for the mutation in our genes.

It is a general belief that the brain of any creature is the organ in charge; however, its functioning also depends on our gut flora. The brains of all the hosts, whether humans or other animals, are directly connected to their stomach and intestines; the brain sends chemical signals to our guests, and the gut bacteria also communicate information to our brains, of course, in the chemical-language. This is the reason that mental stress affects our digestive system, or the food we eat has a direct influence on our mood and even the way of our thinking too.

The story doesn't end here; the host animal eats only that kind of food that these guests can digest. Different body parts of the host are also shaped according to the food these guests need to eat. Carnivorous animals have different types of teeth and claws, whereas cattle are very much different from them in body construction and also in their nature; all this is the magic of the electrochemical communication between the brains of the host, and their guests, the gut-bacteria.

Although different types of the gut-bacteria live within the intestines of different species of animals, different individuals of even the same species don't have all the gut bacteria of the same type and in the same number. Probably this is the reason that all the individuals of the same species are different from each other in their nature and behavior.

It is manifest from the above description that gut bacteria are the natural partners of different creatures, including ourselves; they are essential for our living. Contradicting this fact, I have seen an advertisement on the internet that invites people to get remedial action for getting rid of millions of harmful gut-bacteria living within their intestines. I understand that, along the same lines, some researchers/

doctors are planning to develop different types of medicines that would target only the suspected culprit. *I, somehow, feel that such a line of treatment might further increase the imbalance in the number of different types of the gut bacteria; this may even result in the serious side effects because an imbalance in the number of different types of these gust bacteria may cause insulin resistance, obesity, asthma, rheumatoid arthritis, depression, ulcers, autism*, etc., and even cancer, too, in the host. This is evident from a very simple fact that antibiotics upset our digestive system. In view of this possibility, I feel that instead of killing the suspected gut-bacteria, efforts shall be made to restore their balance in the body of the host, by way of the gut-bacteria transplant or by the food supplements that are rich in the type of the gut-bacteria that are lacking in number.*

×   ×   ×

---

* A mental condition existing from childhood, characterized by great difficulty in communicating and forming relationships with others and also using the language and expressing the abstract concepts.

Are We Alone In The Universe?

Generally, we people believe that life, in the whole of the universe, exists only on the Earth because the conditions favorable to life don't exist anywhere else. However, the three most likely places, apart from some other places, exist in our own Solar System, where life can flourish in one or the other form; one of such places is "Europa," the moon of the planet Jupiter and the remaining two places are "Titan" and "Enceladus," respectively which are the moons of the planet Saturn. The water is in abundance on Europa and Enceladus, and active volcanoes are also present on both of these moons; on the other hand, some different forms of life can flourish on Titan, here too, volcanic activities are going on; however, life if originates here, then the same might be based on the liquid methane in place of the water because the water in liquid form cannot exist in the severe cold of this moon. Scientists are, therefore, hopeful that apart from the Earth, life in the form of microorganisms might possibly flourish on these moons, too. Of late, the scientists have also found several proofs that about 4 billion years in the past, water in the liquid form was available in plenty on the planet Mars. At that point in time, this planet also did have an atmosphere of dense air as well as a magnetosphere too. Accordingly, at that point in time, this planet was capable of supporting and protecting different life forms. Later on, due to some unknown reason, Mars lost its magnetic field; subsequently, the atmospheric air was blown away, probably, by the solar winds. About 10,000 years in the past* liquid water, too, completely vanished from this planet. Both the Earth and Mars did come into existence almost at the same point in time. However, Mars, being much smaller in size, must have cooled down much earlier. Since similar environmental conditions might have prevailed at the beginning on both the Earth and Mars, there are strong possibilities that life might have evolved on Mars earlier than it did on the earth. This prediction is supported by a meteor known as "Martian Meteorite ALH 84001," which was found in December 1984 at the Earth's South Pole. Supposedly, this meteor did dislodge from Mars about 4.5 billion years in the past; later, about 13000 years ago, this meteorite fell on the earth. Researchers, in the above-mentioned meteor, have found some organic carbon compounds and also biogenic crystals of magnetite that many microbes and animals produce to detect Earth's magnetic polarity. The carbonaceous matter found in the said meteor has some resemblance to a microbe

---

* Almost at the same time period, or a little earlier, advanced civilization suddenly evolved on the Earth.

known as GFAJ-1 that was, during the beginning of 2009, found in Lake Mono in the USA. Thus, there is a very remote possibility that the compounds found in this Meteorite might be the fossil remains of some microorganisms. Moreover, it is understood that NASA has, of late, found evidence of organic matter on Mars. *I also understand that within another Martian rock, a fossilized microorganism, which was undergoing the process of cell division, has been found.* If this information is correct, then this might be undisputed proof that life did originate sometime in the past on the planet Mars, too. In case some life forms were, in fact, originated on Mars about 4.5 billion years in the past, then there is a remote possibility that their fossil-remains might be found on Mars in a future date. Even as of today, methane gas has been seen coming out of some of the areas of this planet. The presence of methane suggests that either some volcanic activities or alternatively, some biogenic activities are still going on under the surface of the planet Mars.

Since the extremophiles can originate and thrive even in the most adverse conditions, it is most likely that at least some of them could still flourish on the planet Mars, under the conditions prevailing at the present time, or at least on the moons of Jupiter or Saturn, etc. Recently a new kind of microbe has been discovered in the upper layers of the Earth's atmosphere, where the air is very thin and extremely cold; this variety of extremophiles is capable of surviving and flourishing in the atmosphere of planet Mars. A new species of the bacterium has also been found to flourish inside the caves of Iceland, where the ice in the absence of the sunrays has not melted since the last several thousands of years. The species of the similar bacterium can flourish in very cold places where though sunlight is scarce, water is available in the form of the ice; places like different comets, "Europa" and "Ganymede" - the moons of Jupiter, our own moon, dwarf planet "Ceres" and some planets such as Mars, Uranus, and Neptune, etc., are some of such places within our own solar system where this kind of microbe can easily flourish. Scientists believe that there are fair chances that one more planet (hypothetical ninth planet) might also exist, far beyond the dwarf planet Pluto; this planet is supposed to be a rocky planet, almost 10 times as massive as the Earth*. In case such a planet does really exist, then this planet and/or its moons might support some forms of life because such a massive planet might be a potential source of heat energy. Moreover, water, probably in the form of ice, might also exist on such a planet and its moons. Similarly, the microbes found in different lakes of hyper-acidic or hyper-alkaline water, and those that thrive around the hydro-

---

* Please see Addendum- 1.

vents found at the bottom of the sea, might also survive on the moons of Jupiter or Saturn, where several active underwater volcanos/hydro-vents might exist. One of the most primitive forms of life, an "Archaea," found about 2-3 meters below the surface of Atacama Desert, could probably also survive on Mars at about 2-3 meters below its surface.

During the period of May 2012 to December 2015, the Mars Express Orbiter, a European spacecraft, has detected a possible underground water reservoir on the south pole of planet Mars, at a depth of about one mile (1.6 Km.) beneath the solid ice. Although the presence of underground liquid water on Mars could not yet be confirmed, its possible presence gives rise to a hope that microbes may originate and flourish at such a water reservoir, too, because one species of microorganism has been found to live in a similar underground water reservoir located at the south pole of our own Earth. Of late, scientists have discovered the presence of **Phosphine** gas in the upper layers of the atmospheric air of the planet Venus, which is normally produced by the living beings *(extremophiles)* during some biological process; this fact indicates that some sort of life-forms might exist even on the planet Venus. Moreover, some time back, NASA scientists are said to have discovered an entirely new microbe, which is known as GFAJ-1; the said microbe is believed to thrive on Arsenic. This creature is entirely different from the other varieties of microorganisms found on the earth because Phosphorus in its DNA has been replaced by Arsenic. Other scientists have though failed to discover any such microbe anywhere; however, if such a creature does really exist, then this is a very important discovery. If, life could be made by the combination of entirely different chemicals, then such an entirely different life-form might thrive in an unimaginable rather alien environment and in the most unfriendly conditions that might exist on different exoplanets.

Astronomers believe that at least 17% of the total stars in our own galaxy, the Milky Way, possess their own planetary systems. In case life is thriving only on 1% of these planets, even then, the life similar to that is known to us, or in some other form, might be thriving on several hundred million planets in our own galaxy alone. In case of life can prosper in our solar system, then there are all the chances that it could also originate and evolve in other galaxies too, which are several hundred-billions in number, because all the galaxies are made of similar chemicals and by a similar process. This means that there exists a very strong possibility that we are not alone in this vast universe. No wonder civilizations more advanced than us might be living on some of the exoplanets of even our own Milky Way. The count of such planets, in the whole of the universe, might reach up to a few billion or even far more than that.

Epilogue

17: Closing remarks

In short, long back in the prehistoric ages, the curiosity of mankind had put him on a long journey of discoveries and inventions; Mother Nature had, perhaps, made him for this purpose only. Some curious and crazy people, defying some undeclared prohibitions that would have been ruling in those days, started their endeavor to know the unknown, to unveil the mystery behind various natural phenomena. Since then, the paths of *religion* and *science* have separated out from each other. Accordingly, the **science** set out on its onward march on an endless path staggeringly with the help of the crutches of the rudimentary wisdom, immature speculations, and raw predictions that the primitive man could have made during that point in time.

Mother Nature had though provided food to all the living beings; it conferred them only that much wisdom as was necessary for maintaining their living and existence. Mankind, too, was not different from other wild animals; they were not wise enough to unveil any mystery in their very first attempt. The journey of science was, therefore, not an easy and smooth-going one; in fact, it was a blind journey. The early men, in the absence of proper means, proper guidance, or any definite strategy, had to proceed on their path of discovery barely with the help of very crude speculations and immature imaginations. In that era, the power of imagination of our ancestors was, unfortunately, very blunt; they didn't have any heritage of the pre-earned knowledge of the laws of nature or any idea that could help them to solve complicated matters. The unplanned and aimless efforts of the early men led them to innumerable misconceptions. However, he somehow muddled through. Though he, from the very beginning, started his journey in search of the *"truth,"* but because of his impromptu attempts, he kept divagating and aberrating from one direction to the other; the *"true knowledge"* remained as elusive to him as it was before.

A long time thereafter, some renowned philosophers and scientists, such as Aristotle, Galileo, Newton, etc., gave new directions to the gradually growing knowledge of science. The scholars and scientists of every era made their best efforts to explain the **unknown.** However, theories made by them soon had to be either revised or abandoned. Even the modern theories, which have replaced the old ones, are unable to give clarifications to some of the mysteries. Accordingly, the search for a more comprehensive theory is still going on. The science had to keep swerving from one direction to the other before reaching its present stature. However, no one knows where this blind journey will lead us to?

It is true that during the last couple of hundreds of years, science has surged ahead, galloping at a very fast pace. However, in view of different doubts raised in this book, we can't rule out the likelihood that we, at some unknown point in time, might have gone off-track. Anyhow, it is not easy to realize this fact or ascertain whether this doubt of mine is correct or not; because it is not easy to ascertain whether the knowledge that we have earned during all these years is comprehensive and absolutely correct in all respects. This doubt of mine is based on the question, *"Have we, prior to the formulation of different theories for explaining the unknown, really acquired the absolute knowledge? Whether these theories are really capable of unveiling the absolute truth or understanding the truth comprehensively?"* What I want to stress upon is, *"Whether our theories are based on adequate facts? Whether our findings are backed by comprehensive knowledge, or we have yet to discover further facts that still remain undiscovered?"* In case some of the riddles have not yet been solved, then how could it be ascertained *"The new facts if ever revealed in the future, would not give new direction and shape to the existing scientific theories?"* Alternatively, *"Have we, without any doubt, established the correct root cause of the puzzling problems, which we are trying to solve?"* If not, then how can it be ascertained that the theories developed in the recent past are absolutely perfect in all respects, i.e., **no further improvement is needed in the existing theories?**

As was brought out in chapter-1 under the story of *"Two Wise Professors," "in case we are not able to establish the root cause of any puzzling problem before making any effort to solve the same (problems), then it might become impossible to deduce the correct solution."* Similarly, the story of *"The Elephant and the Blind students"* tells that *"The wrong inputs, miss-estimation, incomplete knowledge, etc., might lead to the formation of misconceptions, wrong conclusions, and even the blind faith."* Of course, the above conditions don't necessarily lead to wrong findings always; for example, about 2000 years ago in the past, when mankind didn't know anything either about the internal structure of the atoms or about the different kinds of charged particles that comprise the atom, even then he learned to generate electricity by rubbing two dissimilar materials. Likewise, in the year 1660, in spite of our total ignorance about the electrons and their properties, we succeeded in making the machines for generating static electricity. It could be inferred from the facts brought out above that inventions can be made at times by chance without having absolute and pure knowledge. *However, proper knowledge is absolutely necessary to thoroughly understand the true cause of any deep mystery or to unveil the truth. On the other hand, incomplete knowledge leads to misconceptions and blind faith; such blind faith, whether it relates to the field of religion or science, overwhelms our minds to such an extent that even very clear evidence and proper reasoning, etc., fail to attract our attention. Such circumstances may normally lead to the deduction of wrong and misleading conclusions.*

Unfortunately, in our world, only those persons who blindly follow renowned personalities are generally considered wise, whereas those who have a different opinion are considered unwise or non-believers. In spite of the above fact, most of the changes in the world, as well as in the way we think, etc., were brought out by these *so-called unwise* persons. Ignoring this fact, the common man thinks that he simply has to blindly follow the thoughts of the renowned persons/scholars; he need not take any initiative of his own. Our thoughts are still unknowingly governed by some conceptions that were conceived long past by prehistoric man. I feel that even today, this propensity of humankind prevails in the fields of science as well as philosophy, too; from the long past, scientists of the new generations have been following the theories formed by their predecessors. It means that once we conceive a conception, then we take it for granted that the same is the ultimate truth. We normally avoid deviating from such concepts; probably, our belief and blind faith, etc., prevent us from doing so. This tendency of humankind is, however, detrimental to the development of science and all other fields as well.

As we know very well, the truth is always hidden behind the multilayered veils; this fact indicates that the conclusions we have arrived at during all these years might not necessarily be absolutely correct and ultimate. We, therefore, should never be satisfied with the achievements we have made to date because deeper and unimaginably weird mysteries might still be waiting their turn to be unveiled. This means that this matter might not be as simple as we have assumed; it might be far more complex and complicated than what we could even imagine. Therefore, we shall continue to put on uninterrupted efforts to identify the probable shortcomings that might have stealthily infused into any one of our existing theories so that such shortcomings, if any, could be sorted out and removed at the earliest possible opportunity.

In the above context, it may be seen that most of the prominent scientific theories of the modern era have grown almost 100 to 200 years old. Although subsequent to the discoveries made after the formulation of these theories, our knowledge has grown manifold, even then, none of these theories have been modified even a bit. We, the humankind, probably believe very firmly that these theories are absolutely correct and perfect in all respects. Accordingly, we are carrying on our research work only on the basis of these theories without any doubt. The students of science, from the very beginning, are made to inculcate and assimilate these theories by heart, without any question. Their minds, as well as their line of thinking, are both molded in line with these theories; they are trained not to think on any other line. This situation leads to unshakable rather blind faith in these theories. Our scientists are the hardcore believers of the mainstream scientific theories that are mainly based on the belief that **light is a disturbance or a wave formed in the medium of Aether/Space-time;** *they have, without any doubt, spent their whole lives working with these theories. Their minds are besieged by the ideas prevailing in the present, to such an extent that they are sure to oppose any other idea that contradicts the so-called mainstream science. Perhaps, during the past 150 years or so, nobody has ever raised any doubt that some deficit might exist in these theories, or these theories also need to be reviewed, modified, or even abandoned.*

One thing could be noted very explicitly; our modern theories were devised on the basis of one or another unproven postulate. These theories were developed at different periods of time, on the basis of a particular school of thought or on the result obtained from some particular experiment. However, any fact or finding, which contradicts such results, and/ or the very school of thought on which the theory was based was not considered. Rather the same was ignored willingly (such facts are brought-out in several places, from Chapters- 3 to 15 of this book). Under such circumstances, it is possible that at some unknown point in time, some shortcomings might have crept in into these theories. However, we were not able to take any notice of such an untoward happening. **As brought-out earlier, different modern scientific theories are based on a few unproved predictions; however, when different predictions of any of such theories are deeply analyzed, then it might be found that such predictions contradict either the reality or any one of the other predictions or both. Sometimes, some of such predictions contradict even the basic idea of that very theory (please see Chapter- 6). This fact reveals that there might not be any consistency in different predictions of any of such theories. Deep contemplation indicates that while devising these theories, the revered originators of such theories did not cross-examine these predictions from all possible angles. The scientists of the later generations, too, ignored such anomalies due to their (blind) faith in these theories and, of course, due to their one-sided thinking. The following example will clarify this point -**

The theory of relativity predicts that the *fabric of space-time* warps below the massive celestial objects like the stars; any object, including light, moving through the warped space follows the curvature of space and deviates from its straight path. In case this prediction is correct, then a very large and deep dimple shall be formed below the Sun, and the outer edges of this dimple shall extend beyond the scattered-object disc. On the other hand, the light passing through the said dimple deviates from very close vicinity from the Sun; this fact indicates that the dimple formed below the Sun shall be very small and shallow. This anomaly in the expected size of the said dimple clearly tells that this prediction contradicts reality.

Several such contradictions are highlighted in Chapter-6 of this book. Similarly, Quantum-Mechanics predicts that the matter possesses positive energy, whereas gravity is negative energy. Had this prediction been correct, then the formation of stars would have become impossible; the gas clouds, instead of collapsing, would have staggered away. Apart from this, the stars shall not be able to radiate energy; instead, the energy from all around should rush toward the stars (see Chapter-3). However, this never happens in reality. This fact clearly tells that the Quantum-Mechanics, too, defies Reality.

The great scientist *James Clerk Maxwell* established in the year 1864-65 that *electricity, magnetism, and light,* etc., are the disturbances in the same medium. Accordingly, the conception of the medium *"Aether"* was conceived, which was later on replaced by **Space-time.** *Contrary to the **Maxwell's theory**, I proposed, in chapter-4,* that light probably, *propagates in the form of a shower of fluctuating, rather spinning or rotating particles, namely* **photons;** *I also pointed out in the same chapter that "Had the light been a wave propagating through the very tension fabric of space-time, then it won't be able to travel unrestrictedly for billions of years; its entire energy would have dissipated very soon; consequently, such a wave would have died-off soon (please see Chapters-5, 6, 9, 10 & 11). Although the aforesaid proposition of mine is merely a speculation, the same is capable of explaining the **duality of particles and waves that is found in light and all other kinds of energy radiations, whereas Maxwell's theory can't explain it.** In fact,* at that point in time, when Maxwell made the aforesaid discovery, no knowledge about the photons was available; therefore, wave-nature in light, as discovered by Maxwell, was misperceived as the disturbance in some unknown medium. Thus, *Maxwell's Theory* was, perhaps, formulated on the basis of *incomplete-knowledge or partial-truth, and without establishing the root cause of the wave-nature observed in the light.* This might be the probable reason why we still have no explanation of the fact that *"How and why light and matter-particles possess properties of both particles and waves?" Or how do the three-dimensional waves of light propagate undulating in an "up and down" or in a "to and fro" motion in the form of the particles (photons)? In case the existing theories are unable to clarify the above points, then the wave-nature of light, as well as the necessity of a medium of space-time for its propagation, both, would need a thorough overhaul.* Thereafter, during the period of the next 150 years or so, the knowledge bank of mankind has increased manifold; even then, aforesaid theories have not been modified even a bit. However, *in case the medium of Aether or Space-time doesn't exist, or the light doesn't propagate in waves,* then the concept of *the wave-nature of Light* (on which the foundation of modern science was laid) would have gone wrong since the very beginning; this was probably the very first mistake in the history of modern science. As a result, science was put on the wrong track since then onward. Consequently, the theory of **Relativity** and the concept of **"Gravity"** (this concept is based on the notion of the formation of indentation in space-time) would also have gone wrong; unfortunately, modern science is mainly based on these two theories. Aforesaid mistakes, if committed at that point in time, are going on multiplying thence onward. *Quantum Mechanics and all other theories that were developed subsequently,* on the basis of the wrong inputs of *the wave-nature of light, and the medium of space-time, etc.,* might also have gone wrong. In case it is so, then the *theoretical wave-nature of light might have caused the science to aberrate from its path of finding the truth; it seems to me that the journey of science has not only taken a wrong turn because of different misconceptions formed during that time period, the same is moving in a wrong direction thence onward. This prediction of mine, if correct, attracts a question: "How far will we allow science to go in the wrong direction?"*

In the year 1887, *the duo of scientists Michelson and Morley,* on the basis of the *wave nature of light,* conducted their famous experiment *in which an effort was made to compare the speed of light running in the direction of the Earth's rotation with that of the speed of the light-beam running in the direction perpendicular to it. However, the speeds of different light beams, i.e., the beam running in the direction of Earth's rotation and the beam running in the direction opposite of it, couldn't be measured in the said experiment exactly and directly, i.e., separately and independently of each other. Instead, the average speed of light, which was first, made to move in the direction of Earth's rotation and then in the opposite direction, could be compared with the speed of another beam running in the direction perpendicular to it. In this experiment, the speed of light running in both these directions was found to be almost equal. Based on the aforesaid experiment, a stupefying conclusion was deduced that light maintains its normal speed even in the reference frame of different moving objects, i.e., light doesn't have any relative speed when viewed in the reference frame of the speed of any moving object; in other words, light moves past the moving objects with its normal speed even if measured relative to the speeds of such objects.* In order to justify this false belief, **Lorentz** came up with another **miraculous** solution, '**the distances contract and the time dilate at the speed of light."** This solution was readily accepted by everybody; the same is still in vogue because miracles always overwhelm mankind to such an extent that they lose their power of reasoning or to see the plain and simple truth; they normally ignore the truth without recognizing it.

Although light, which comes from the most distant stars, takes an unimaginably long time period of 13.2 billion years to reach us, ignoring this truth, it is believed in *Relativity* that *time totally stops at the speed of the light*. However, had this presumption been correct, then *either the light coming from the distant stars should not have taken any time to reach us, or more preciously, it shouldn't reach us at all because no event can happen in the frozen time; nothing would be able to move in the still time.* *This fact indicates that the ticking speed of "Time" doesn't stop even for the light; it continues to lapse at its normal rate. Instead of the freezing of time, the information about that very instant at which the light did emit from its source travels with the light.* Accordingly, while light rays move past one end of the sun to its other end, which lies apart at a distance of about 1,500,000 km, the sun, moving with a speed of about 250 km/second, moves through a distance of almost 1250 km. Light, in order to travel through this extra distance, would take an extra time of about 0.00416 seconds. This fact very clearly reveals that neither the distances contract nor the time dilates even at the speed of the light. *The same result should have been inferred from* **Sagnac's experiment (Chapter-6);** *however, this experiment is considered proof of relativity.* Earlier, in chapter-5, *I brought out similar doubt on the conclusion deduced from the aforementioned experiment that was carried out by Michelson and Morley in 1887.* **In the said experiment, the light beam running in the direction of the Earth's motion and in the opposite direction of it was allowed to travel both ways; therefore, the effect of Earth's speed was nullified for that particular beam.** *I feel that the conclusion drawn on the basis of the aforesaid experiment should have been taken only after measuring the speed of light in both the directions separately, independent of the other beam; this was, however, not possible in those days.* As brought out earlier in chapter-4, the Danish astronomer *Olaus (Ole) Roemer* was the only man who, in 1676, measured the unidirectional speed of light; *thereafter, no one has made any effort to measure the speed of the light beam running on the Earth's surface in a single direction alone.*

Based on the assumption that the light waves propagate rippling through the medium of aether, which is considered to move relative to the Earth, in the opposite direction, *Michelson-Morley envisaged* that *"Time gained by light-wave while traveling downwind, would be lesser than that lost traveling upwind,"* however, *this assumption could never be proved physically.* We are now in a better position to verify the aforesaid assumption; in such an attempt, we will obtain only one out of the following three results:

1. Such an experiment would confirm the correctness of Michelson-Morley's assumption.

2. As proposed by Lorentz, in case distances do really contract at the speed

of light, then the light will take exactly the same time to travel equal distances in any direction.

3. In case, as proposed by me in Chapter-4 of this book, light travels in the form of a shower of photons, then while the light will travel a distance of 3 kilometers in the direction of the Earth's motion, the Earth itself moving with the speed of 30 km per second, will move ahead by 30 cm. Accordingly, light traveling at a speed of 300000 km per second will take a time period of tenthousand+1 nanoseconds to cover a distance of 3 kilometers in this direction, whereas, in the opposite direction, based on the actual distance travelled by the photons, the light will take approximately 9,999 (10,000-1) nanoseconds to cover the same distance. Such a result will prove that the speed of the light, too, ostensibly changes in the reference frame of the speed of all the moving objects.

Whatever result we obtain from such an experiment, the same would establish the truth beyond any doubt; however, if no effort is made, then the doubt will remain unsettled.

Although the measurement of the unidirectional speed of light is considered impossible, I have an idea by which this feat might be achieved. Although scientists are in a better position to conduct such an experiment, I do not see any possibility that scientists would, in the near future, take any initiative in this direction because they have full faith, without any doubt, in the conclusion drawn from the Michelson-Morley's experiment. On the other hand, until the expensive equipment required to perform a very simple experiment is sponsored by someone, I won't be able to conduct such an experiment, the same being beyond my financial resources. However, neither will I approach anybody to help me out, nor will anybody offer help all by himself; the possibility of performing this experiment seems very gloomy.

Anyhow, in case the predictions of **Relativity** are true, then the curvature of space-time around the Earth must be exactly the same as that of the Earth's outer surface. Accordingly, the light rays, following such curvature of space-time, shall encircle the Earth and come back to the point of their origin. In such a case, we shall be able to see any event that had happed just behind us after a time lag of about 0.133 seconds; however, nothing like such a phenomenon ever happens. *This fact indicates that the light neither follows the curvature of space-time nor it propagates through it.*

Einstein, in his **Special Theory of Relativity,** asserted that nothing could move faster than light. However, **the Newtonian concept of gravity necessitated the infinite speed** of gravity. In order to remove this anomaly, **Einstein,** *with a special* *purpose to negate* **the Newtonian Concept of gravity**, *proposed in his* **General Theory of Relativity** *that* **gravitation is not a force;** *it is merely a consequence of the warping of space-time.* **Although the very large size of the event-horizon around any black hole is indisputable evidence that space-time doesn't warp (see Chapters 6 & 11), we have, without any hesitation, accepted** that **(1) the said fabric of space-time does really exist. (2) Gravitation is not a force; instead, the effect of gravitation is produced due to the warping of this fabric.** Contrary to the above proposition of **Einstein**, everybody must have *practically experienced* the **Force of Gravity** *as well as its effect.* Anyhow, after the introduction of the *"Verlinde's Theory"* in 2009, the scientists of the later generations believe that **Gravity is an Entropic Force,** not a physical force. According to this theory, massive bodies do not exert any sort of gravitational force on any object; such objects move toward any massive body due to an increase in the level of their entropy (disorder). In case this theory is true, then its reverse shall also be true; objects moving away from any massive body shall not move away under the influence of any force; this happens because their entropy goes on reducing. We know that whenever any object is heated, then its entropy is said to increase; however, such an increase/reduction in entropy doesn't endow such an object with the ability to move. Defying this fact and ignoring the effect of **buoyancy, it is believed** that the increase of entropy of hot-air balloons causes them to soar in the air; moreover, it is also ignored that the hot-air balloons cannot soar up in a vacuum chamber. This fact casts a question *"How can it be ascertained that the objects fall toward Earth due to an increase in their entropy? It is also possible that their entropy increases under the influence of the increasing intensity of Gravitational-force."* Explanation of this point is a must; nothing can move all by itself; there must be a force behind this phenomenon.

There is yet another anomaly in the theory of Relativity: *The actual (measured) rotational speeds of the stars rotating in any spiral galaxy do not match with their speeds as calculated by the theory of Relativity.* This fact indicates that this theory needs a review; however, instead of reviewing this theory, *we are trying to explore the existence of the doubtful rather non-existent dark matter (see chapter-14) in the galactic halos. Similarly, so far, we have not made any effort to review the conventional concept of gravitation.* In case the theory of **Relativity** is correct, and the fabric of space-time does really exist, then any coin onboard a spaceship shall, all by itself, be able to orbit any massive object floating in the space. However, Einstein's concept of gravitation could be verified very easily in a spaceship by rolling a coin along the curved surface of the equipment, as shown in Fig-3B. *In case the said coin continues to move in a straight line instead of moving in a circular motion* *along the curved surface of this equipment, then it would be proved that the concept of Relativity that "Gravitation is not a force; it is merely a consequence of the warping of space" is totally wrong. In case the coin fails to follow the curved surface of the said equipment, then it will put the very existence of the continuum of space-time under a question.*

Different scientists, during different time periods, verified the theory of Relativity by their own methods. The Theory was found correct on all such occasions. All the experiments that are conducted to verify *Relativity* are focused on obtaining a particular result that agrees with the predictions of *Relativity*. However, any other factor that might be responsible for the result so obtained is not at all considered. Some of such experiments conducted to verify relativity, as well as the factors that were not considered, were discussed earlier in chapter- 6. One more attempt was made in 2004 to test two unverified predictions of *General Relativity* when NASA established a satellite named *"Gravity Probe B"* in a polar orbit around the Earth, at an altitude of 642 km. The mission of this satellite was to prove that the fabric of space-time really warps around the Earth and also that the spinning Earth drags and twists this fabric around it. This mission was to be accomplished by measuring any tiny change that may occur in the direction of the spinning axes of four numbers of ultra-precise gyroscopes that were installed in the said satellite; the spinning axes of these gyroscopes were very precisely set to point in the direction of a binary pair of stars named *"IM Pegasi."* After an observation over a time span of 17 months, and very precise measurements, it was found that the spinning axes of the said gyroscopes have deviated slightly from the set direction; this deviation very closely conformed to the prediction of *Relativity*. Based on this experiment, it was concluded 1) The massive objects do create dimples in the sheet of space-time, & 2) The non-static or rotating masses, or the mass-energy currents, produce an effect on space-time due to which an anti-Spinward *torque* is exerted on the moving or spinning mass. This effect is known as the *"Frame Dragging"* or the *"Gravitomagnetism." Accordingly, relativity envisages that in the direction of the earth's spin, the light shall move at a higher speed; however, in such a case, the light shall not be able to spread in the direction opposite of the expansion of the universe.*

Although the conclusion deduced by this experiment has been accepted worldwide, I feel that the rotational axis of the gyroscope could have been deviated due to some other reasons that the scientists didn't consider before arriving at the aforesaid conclusion, the factors that, apart from the non-static space-time, could have produced the desired result, are discussed below:

A rotating medium/energy-field can definitely exert a torque on the objects located within such a medium or a field; however, it is necessary that either such a medium shall have adequate viscosity or the field shall exert a force on that object. Torque cannot be created without the application of force. Although gravitation is not considered a force in **Relativity**, the Moon, due to the interaction with the Earth's gravity, has been in **tidal-lock** with the Earth for several billion years; that is, the moon always has the same face toward the Earth. This does not seem possible that merely a dimple made in space can forcibly hold the Moon in the tidal-lock; only a real force of adequate strength can compel the Moon to do so. Based on the aforementioned possibility, a question arises "When gravitation of the Earth can forcibly hold the moon in the tidal-lock, from a distance of about 380,000 km, then it may also be possible that the gravitational force of the Earth might have dragged the spinning axes of the said gyroscopes located at more than 500 times closer as compared to the moon. In such a case, the near end of the gyroscope would be dragged along with the spinning Earth to a greater degree as compared to its far end. Therefore, it would seem that the spinning axis of the gyroscope is torqued in an antispinward direction. This can be understood from the fact that a moving magnet has the ability to drag or tend to drag magnetic materials from quite a distance in the direction of its motion.

Apart from the above possibility, the following factors shall also be considered:

1. The effect of the frame-dragging shall not be produced by the spinning Earth alone; apart from the said spin, the Earth also orbits the Sun, it also moves along with the milky-way in the same direction as well. The Earth on its axis rotates at a speed of about 1600 km/HR., or 0.48 km/sec. As against this speed of the spin, the Earth also orbits the sun in the same direction, but at a much higher speed of 30 km per second, which, being about 60 times higher than its spinning speed, should produce a far greater deviation in the spinning axis of the said gyroscope, which should have been 60 times higher than that produced by the spinning of the earth. Apart from above, the sun, along with the Earth, orbits the galactic center of the Milky-way at a speed of about 250 km/Sec. Moreover, the Milky-way galaxy orbits the Andromeda galaxy at a speed of about 590 km/sec. This speed is about 1250 times higher than the spinning speed of the Earth. This fact moots the conclusion deduced, **"Does only the slowest movement of the earth produce the effect of the frame-dragging?"** *Had the direction of the spinning axes of the said gyroscopes been really dragged by the Earth's real speed, then* the spinning axes of the said gyroscopes should have deviated by a far greater degree.

2. In case the spinning axes of the gyroscopes can be torqued by the drag

produced by the spinning of space-time, then a similar effect produced by the speed of the Milky-way shall be able to shift the said star "IM Pegasi" gradually from its location. Similarly, the effect of Frame-Dragging produced due to Earth's spin, if combined with the torque produced by the Earth's speed in its orbit, should have torqued the entire body of the said probe along-with the gyroscopes kept within it, instead of the spinning axes of the gyroscopes alone. I have no idea whether calculations for such an effect were made beforehand.

3. The Earth has been continuously spinning for the last 4.5 billion years or so. Accordingly, under the effect of anti-Spinward torque acting on the air blanket shrouding the Earth for such a long time period, the entire blanket of the atmospheric air should continually move with the speed of 1600 kmph, in a circular motion around the Earth, in the direction opposite of its spin. Moreover, the skyscrapers like the Burj Khalifa, Empire State Building, etc., shall bend a little in the direction opposite of the Earth's spin.

In case the above-listed possibilities were not reckoned (calculated) beforehand, then the so-called success of the mission *"Gravity probe B"* shall not be considered the proof of Relativity. In fact, establishing the satellite Gravity probe B in the pole-to-pole orbit itself defies the prediction of *Relativity;* in case the fabric of space-time is really warped below the Earth, then the shape of the indentation formed in the ecliptic plane shall resemble to that shown in the fig-3A/3B. In such an indentation, the satellite can orbit the Earth in only the horizontal plane, that too, along the walls of the indentation; however, it will not be possible for the probe to orbit the Earth in a vertical plane. The pole-to-pole orbit clearly shows that neither the space-time warps below the celestial objects nor the celestial objects follow the curvature of the space.

Apart from *Relativity, the Quantum* Theory is also one of the most important theories on which modern science is pivoted. Another important theory is *String Theory;* efforts have been made to keep both, *Quantum theory* and *String theory* consistent with *Relativity*. However, it seems to me that it is difficult to prove the existence of the fabric of space-time; in case the existence of the hypothetical entity of *space-time* couldn't be proved, then all of these three theories would need a thorough review; rather, they shall be revised at the earliest.

Although both, *Quantum Theory* and *String Theory,* were formulated in conformity to the *Theory of Relativity*, all these three theories differ from each other in many aspects. In the *Theory of Relativity*, interstellar space is considered a fabric that is woven from the warp and weft of space and time; on the other hand, to the best of my knowledge, space, in the *Quantum Theory,* is considered to be made of *grains,* whereas, according to the *String Theory* the same is supposed to be made of *vibrating strings*. Although all the above three concepts are contradictory, all of them are considered correct. All the aforesaid theories differ not only on this point alone, but they also differ on many other points as well. We all know that the truth cannot differ from reality; accordingly, only one out of the above three concepts regarding interstellar space can be correct. Otherwise, all three of them may be wrong. This fact indicates that at least two or all three of these theories might have some loose ends which need to be tied up.

In *Quantum Theory*, the subatomic particles are considered waves. Accordingly, it is believed that in case a single electron is fired from an electron gun, then it would travel simultaneously on 2, 3, or even more paths at the same time instant. *This belief can only be verified by tracking the path of such an electron by applying the attosecond technology; however, till to date, nobody has made any attempt to find out the truth.*

Quantum theory *further envisages that the locations of the subatomic particles do not remain fixed all the time; they may be found in one place at a certain point in time; however, during the very next moment, they may be found at any other place.* However, in case, we keep a constant watch on the corners or edges of any solid object, then it might be seen that excepting the case of abrasion, erosion or corrosion, etc., the particles in the solid objects, at the molecular/atomic level, occupy fixed locations, they never wander here and there randomly. Similarly, the molecules/atoms of the liquid and/or gaseous substance do not change their locations until an unbalance in their energy, i.e., in their temperatures, is created in different places of that object, which, subsequently, results in the creation of unbalance in the pressure in the different places of that substance. In such a case, molecules/atoms of that substance start to flow toward the area of low energy so as to maintain the balance in energy. However, the locations of the subatomic particles never remain fixed or stable because the electrons always keep orbiting the nucleus at a steady speed at a given quantum state. Thus, although the locations of electrons never remain fixed, the same are not uncertain; they always obey the rules. As a result, the atom/molecule may appear to fluctuate like waves due to the rotation of the electrons in their orbit; an angular-momentum might also be created in these particles due to the electrons revolving in orbits. However, in case an unbalance is created in the energy level of electrons, they, following the laws of nature, may break-open the bondages of the nucleus and wander randomly to dissipate the extra energy.

Scientists believe that gravitation is negative energy; however, this concept does not seem to be consistent with the famous formula $E = mc^2$; *according to this formula, energy can become negative only on the condition that the mass of any object shall become negative, which seems absolutely impossible.* **Further, had gravity been negative energy, then the stars would have never formed; the gas clouds, under the influence of the negative energy, should scatter away. Moreover, in such a case, the stars would never emit energy; instead, energy would have rushed toward the stars. However, this is against reality.**

It is also envisaged in the **Quantum Theory** that the pairs of the positive and negative energies are always created and destroyed in the vacuum or interstellar space. *However, no direct proof of such a phenomenon is available; moreover, there is no explanation in this theory of* **1) How the pairs of particles and antiparticles are created out of nothing. 2) How these particles, in the absence of any force, move away from each other, that too, against their mutual attraction.** *Both of these assumptions are against the fundamental laws of science;* **neither can any particle be created all by itself, nor can anything move without being acted upon by force.** *In spite of the above fact, nobody ever raised any question about this concept.* The aforesaid concept of **Quantum Mechanics** is purely based on the **Principle of Uncertainty,** which envisages that nothing, no phenomena, in the world is definitely certain; there must be a definite amount of quantum uncertainty in everything or every phenomenon. In case this principle is correct, then it is not certain that the Sun will always rise from the East or objects will always fall on the Earth; different objects shall sometimes levitate against gravity. It shall also not be certain that every time when a piece of **Potassium Metal** is added to water, then the same chemical reaction will always be repeated; accordingly, instead of Potassium-hydroxide, some other salt of any other element may be produced, or alternatively, no chemical reaction will at times, take place. In case of everything is uncertain, then how does this principle *certainly* apply to everything? The central idea of this principle seems to violate the principle itself.

On the other hand, the fact that has not been given any importance in the **quantum theory** is **"Subatomic particles as well as the celestial bodies, both orbit only the objects having a greater mass than their own."** This fact clearly indicates that the **mass** also plays a direct role in the construction of the atoms too; however, the Standard Model of Particle Physics doesn't give any weightage to this fact.

After May 2015, a series of experiments were carried out at the Large Hadron Collider by smashing the high-energy protons. I understand that the results of these experiments have bewildered scientists all around the world; these experiments have hinted that a new particle may also exist, which immediately after its creation, decays into two high-energy photons of 750 Giga electron-volts each. *The very high* energy of the particle so produced is far more than that the **Standard Model of Particle Physics** *can predict.* The energy of these photons indicates that the suspected new particle is created at the energy of around 1500 GeV or above. Further, scientists have estimated that the mass of the said particle might be about 6 to 12 times that of the Higgs-Boson. In case any such particle does really exist, then, after the big-bang, such a particle, because of its high energy, should have been created much earlier than the Higgs Boson because the so-called Higgs Boson is created at much lesser energy of, say, 126 GeV, as against 1500 GeV at which the particle discovered in the said experiments is produced. This scenario gives rise to an unanswerable question: ***"Wherefrom this newly detected particle did acquire mass even before the creation of the Higgs-Boson?"*** Further, I have also gathered from an unauthentic source that *the decay of the unstable particle* **B-meson** *('B' meson is a pair of one down quark and one bottom antiquark) behaves in a manner different than the Standard Model of Particle Physics can predict.* **In case the aforesaid discoveries are found correct, then the "Standard Model of Particle Physics" would need a complete overhaul.** At present, the scientists are examining the data generated from these experiments; to the best of my knowledge, they have not yet deduced any conclusion.

Humankind has, of course, made numerous discoveries on the basis of our modern scientific theories; even then, we still don't have an answer to a few important questions. *We have discovered that light has dual properties of waves and particles; however, we don't know the reason behind this duality. Similarly, we don't know how the light-waves, while expanding spherically in the three dimensions, would fluctuate?* The **subatomic particles, when observed by the most powerful microscopes, appear as point-like objects; even their photographs can also be taken, they even cast shadows too, even then all the sorts of particles, such as atoms and electrons, etc., are, since 1920 onward, considered the matter waves, or disturbances in some sort of energy field; in the String Theory, they are considered strings that persistently vibrate.** However, our theories are unable to explain that *1)* **In case the particles are disturbances in the energy fields, then wherefrom these energy fields are created? 2) Why do the particles behave like waves? 3) How any particle in the form of a wave can have any mass?** *And 4)* **In case the particles are waves in reality, then why do different particles (waves) have different kinds of spins?** All the above questions indicate that our theories are not perfect; they either need a thorough review and/or further improvement. ***This is very important because the scientists believe that in the case if any system is left to itself, then any disorder (Entropy) existing in that system tends to increase gradually.*** *In view of this fact, it is necessary that every system, or theory, shall be reviewed from time to time and modified if needed.*

In this book, under Chapter- 13, I have pointed out how the concept of **Dark Matter** was conceived. During the decade 1930-40, scientists of that era were unaware of the existence of different hard to see celestial objects such as the White/Dark dwarfs, Neutron stars, small, medium-sized, and bigger black holes, which along with the other visible stars, rotate in the galaxies around the central Supermassive Black-holes. The cumulative effect of the summed-up gravitational force of such objects was also not known at that time. In that era, the concept of **Dark matter** was conceived under the above circumstances. Accordingly, the combined mass of all such objects was mistaken as that of the **Dark Matter**. Similarly, in Chapter- 15 of this book, I expressed a possibility that probably, the light (coming from the far-off celestial objects) loses its energy in the direct proportion of the time taken by the light to travel through this distance. In case the light actually loses its energy, then in the absence of any knowledge about this fact, we have mistaken that the entire Universe is expanding with the speed of light rather than at a speed more than this limit. Almost all scientists all over the world have been searching for the **Dark Matter** and the **Dark Energy** for almost the last 70 to 80 years or so. However, both of these entities have been eluding the scientists so far. A very large sum of billions of dollars, besides a very precious time and energy, has been expensed so far in this effort to no avail. Under this situation, it has become very important to verify both of my aforementioned propositions and ascertain whether the universe is really expanding or is it stationary. In case both of my aforementioned speculations are correct, then modern science would have to be revised drastically.

I have no idea how our scientists plan to resolve various problems that couldn't have been resolved until now. In spite of the fact that the present-day theories have not yet succeeded in giving any dependable clarification on such unresolved problems, the scientists still follow them and are probably, dealing with these unresolved problems because of the same theories, which have been in vogue ever since the long past. Some of the probable shortcomings in these theories have been repeatedly brought out in this book in different places. However, in case any of these theories really do have any shortcomings, even then, nobody would probably trust an unknown layman in this field like me. The fact that *"the points brought forth in this book are totally against the widely-accepted conventions or the existing theories"* gives rise to a possibility that specialist-scientists and the common men, both, without any differentiation, might reject these points in a similar manner, without even trying to understand or scrutinize them; *thus, the strongest point of this book is its weakest point, too. Because nobody would believe in the points raised by an unknown man*, this possibility leads to a very unfortunate situation; nobody would ever review these theories; the doubts raised in this book would remain unresolved. Such doubts could only be removed when some renowned scientists would seriously consider these doubts and try to find out the answer.

Our past history reveals that *any new thought that is inconsistent with the prevailing ideology is normally totally disapproved by the so-called experts on that subject.* In 1912, Alfred Wegener's theory of *"continental drift"* was adjudged ridiculous and was rejected because it was inconsistent with the prevailing school of thought. There are strong chances that this book might also meet a similar fate. However, this possibility cannot stop me from presenting my viewpoint; even if the whole of the world makes a mockery of me, I am ready to pay this price too. Another possibility that I foresee is that this book will be considered so worthless that nobody will bother to take any notice of it; in that case, no one would ever try to re-evaluate the existing theories. *However, suppose the points raised in this book are not examined carefully by the renowned scientists; in that case, there is a strong possibility that the science, for an infinitely long period of time, would continue to move on the same path that has probably diverted into a wrong direction since the long past.*

I feel that the whole of the world has undergone a deep indoctrination into the so-called mainstream scientific theories; everyone respects these theories like "Bible," "Quran," "Gita," or "Guru Granth Sahib," and all the other religious books. Everybody has full faith in these theories; they are confident that these theories are absolutely correct, totally immaculate, and perfect in all respects. Under this scenario, scientists, in order to prove their point, might have deliberately ignored some inconspicuous proofs found from time to time, which might have gone against any one of these theories. As enthusiastic supporters of these theories, they interpret such findings in such a way that the correctness of the existing theories could anyhow be established; the facts going against these theories were deliberately negated, ignored, and suppressed. Till such time, we will continue to give preference to our theories and their revered founders over such unexplored truths; it would not be possible to unveil the deep mysteries that our existing theories have failed, so far, to resolve. Our belief that our modern scientific theories are absolutely correct, as well as our propensity to stick to them, and our habit of hero-worship, etc., are the obstacles in our path to the development of the perfect theories. Even the undetected errors committed by renowned personalities of the yesteryears might be considered great achievements of humankind; *who knows?* If we can get rid of these tendencies, only then we would be able to realize that the existing theories need a thorough review, and if deemed necessary, only then efforts to revise these theories would be made. However, no conclusion shall be deduced by neglecting even an apparently trivial fact; otherwise, we would never be able to unveil the complete truth.

Anyhow, in spite of the very grim chances of meeting the public approval, I have written this book because I simply want to push a general awareness among the people about the probable shortcomings that might exist in our theories; this is my main purpose. Up till the time when any doubt won't bother our heads about the probable undetected flaws in our modern scientific theories, nobody would try to examine these theories with the alternate angle that is discussed in this book. I earnestly desire that in spite of the best efforts put up by the renowned scientists to find the faults in these theories, not even a single fault could be found. However, if any hidden flaw really exists in any of these theories, then the same should be identified and eliminated at the earliest possible opportunity. If necessary, we should modify these theories or discard the old ones and formulate new and improved theories that might explain the puzzles that hitherto remained unresolved. The earliest this could be achieved, the better it will be for mankind and the smooth development of science as well.

Although strong possibilities exist that the points brought forth in this book would be ignored in a likewise manner by both specialists and commoners; I am still hopeful rather I have a firm belief that someday, maybe after 100 years or so, someone would seriously examine these points with an open mind, without any prejudice or bias toward the existing theories; I hope that such a person would critically analyze the logic/reasoning behind each point raised by me. I can't say what would be the outcome of any such scrutiny if, at all, one is conducted.

No one can predict what will happen in the future; however, I would have the satisfaction that in the capacity of an ordinary man having no standing in any of the fields, I have done the best at my level to achieve my goal of spreading awareness toward the probable shortcomings in our existing theories. Anyhow, the chances of my success in achieving my goal depends on the fact that whether the readers, especially the curious younger lot, will take my thoughts seriously or reject these thoughts out of their prejudice toward the existing theories.

Under the above scenario, all of my hopes rest on the possibility that some renowned scientists would critically examine at least some of my arguments, some of which at least may be found practically correct. In chapter-9, under the heading *"The Puzzle of the Whirlpool,"* I envisaged that *probably, the direction of the magnetic field of the earth controls the direction of the rotation of the whirlpool created in the wash-basin.* I feel that out of different presumptions made in this book, the above presumption of mine is the easiest one to verify; other points are not so easy to prove. Anyhow, in the same sub-section of this book, I insisted that *"such a whirlpool of water will not be produced in the micro-gravity conditions,"* however, *in case of water could be sucked through the drain, then a whirlpool might be created within a space station, too.* In case this assertion of mine could be verified, then it could be proved that *"Creation of the galaxies, as well as induction of rotational motion within them, were both caused by the Gravitational-Force that acts toward their Centers, not by any curved indentation created in the space-time."* At present, only the big agencies like NASA and some other similar agencies are capable of verifying the truth of my aforesaid assertion. However, the question is, *"Will they ever bother to verify this point?"*

Anyhow, in chapter-3, "Quantum Mechanism," under the section "The Duality of Particles and Waves," I proposed that the path of a single electron fired from the electron gun should be tracked down with the help of Attosecond technology. In case this could be done, then and only then, the fact "How a single electron can move on more than one path at the same time instant" could be examined; only specialized agencies are capable of conducting such an experiment.

Further, in the chapter-6 of this book, under the section "Analysis of different predictions of *Relativity,*" I made emphasis on the necessity to determine the following important facts, too:

1. As suggested under the sub-heading *"Gravitational Force and the Ticking-speed of Time,"* efforts shall be made to find out "does the energy (frequency) of light really increases when it enters the higher gravitational field of the earth. For this purpose, a beam of light/laser of the known frequency shall be sent to the Earth from a spaceship, and the frequency of this beam shall be measured at its point of origin as well as at the Earth also. Such an experiment might clarify this point. Similarly, such a beam shall also be sent from a space station toward both the poles of the Earth, to find out whether the direction of the magnetic polarity of the Earth has any effect on the frequency of light, i.e., does the magnetic field really increase/decrease the same.

2. By sending a piece of equipment like the one shown in Fig-3B to a space station, we must determine whether celestial bodies really orbit the more massive bodies due to the curvature of the indentation made in space-time. However, only the big and specialized agencies like NASA, having the ability to send spaceships into Earth's orbit, are capable of conducting such experiments.

This book includes many other presumptions, too; *all of them, including the phenomenon of change in the ticking speed of Time (due to any change in the strength of gravity-field and/or the speed), need thorough verification.* Sometime during 2016, I questioned this possibility on *Google Search*; at that instant, Google reproduced an excerpt from the earlier edition of this book, as well as the conventional viewpoint prevailing in the present era. Probably, Google has now removed the reference to my book from their search engine; seemingly, my viewpoint has been considered frivolous or worthless by them. However, the possibility brought out by me shall not be considered worthless unless the same is critically examined. For this purpose, the distance between the nucleus of an atom and the electron orbiting it shall be measured (by using the attosecond technology) at Earth's surface as well as onboard a spaceship orbiting the Earth; in case any change is observed in the radius of the orbit of the electrons moving around the nucleus, at both of these places, then it may be concluded that the scale of the atomic-clocks, for measuring the ticking speed of time, definitely changes in according with the change in the speed of the satellite, and/or the strength of the gravity-field around these clocks. However, the time interval that the electron takes to complete one orbit of the nucleus at each of these places may not give any clue because the scale of measuring *the ticking speed of time* might differ in both of these places.

I feel that out of all of my prepositions, ***determining the unidirectional speed of a light-beam,*** as proposed in chapter-5, is of the utmost importance. Though this experiment is probably considered impossible to perform, I have an idea* by which this experiment may possibly be conducted successfully. *In case the light is found to propagate with different speeds in the direction of the Earth's motion and in the direction opposite of it,* then it could be inferred:

1) Neither the distances contracts nor the time dilates at the speed of light.

2) "Light does not necessarily propagate in waves, i.e., it is not necessarily a disturbance created in any medium; because the similar result would also be obtained even in case light travels in the form of a shower of particles that vibrate at different frequencies." A similar assertion was made by me earlier, in Chapter- 4 of this book.

---

* I have not disclosed this idea in this book, because I hope I will, someday perform this experiment myself, however, if, I couldn't do so, then I will surely disclose the said procedure before my death.
** Please refer to Addendum 3.

I do not expect that any individual or any organization will take interest or put up any effort to conduct any such experiment. On the other hand, it is almost impossible for me to conduct any such experiment because of the paucity of funds and limited resources, especially in the dusk of my life; I feel that not much time is left for me to prove my point.

During the last 4 to 5 years, I did approach a good number of Universities and other institutions all over the world which are capable of helping me to conduct such an experiment. However, none of them has responded or shown any interest in helping me out in this matter. Thus, the situation looks pretty gloomy; even then, I have not lost any hope even a bit. I know for sure that neither my goal is far away nor the same is impossible to achieve; though there are some obstacles between me and my goal; what I need to do is only to overcome these obstacles, that is, to conduct the aforementioned experiment, its result is sure to fetch the success to me.

Famous Indian Urdu poet, Late Mr. Shakeel Badayunee has said that: -

"No matter how far away your goal is, you just need to put up continual efforts; you are sure to achieve the desired goal very soon."

I know for sure that I will have to prove my point alone; why anybody else would bother to take up this apparently unworthy-looking task. If any person has conceived a dream, then he himself, not anybody else, is responsible for fulfilling that dream. Therefore, I have decided not to give up till the end of my life; I will do whatever is possible within my own resources. In case I'm not able to do anything during my remaining lifetime, then this book will cast a question for future generations: -

"Who will find out and remove the probable shortcomings that might exist in our prominent scientific theories discussed in this book? Who will take up this task? Who…?"

Will anybody ever take up this challenge to unveil the truth?

I am still hopeful that sometime in the future, someone who would feel that the logic brought out in this book makes some sense, **or the one who would independently come to a similar conclusion, would fulfill my dreams.**

However, the game is not over yet; though circumstances don't seem favorable, I am still alive and healthy. Up till the time I am alive, I will put up all-out efforts to prove my point. However, under the scenario explained above, I am now closing the book with the following verse that was written by honorable Indian Shayar (Urdu Poet) Late Mr. JAN NISAR AKHTAR-

"पता नहीं कि मेरे बाद उन पे क्या गुजरी?
मैं चन्द ख़्वाब जमाने में छोड़ आया था।"

The above verse means: -

"I am not aware of the fate of the dreams,
That… I left behind in this (cruel) world."

In other words, these lines mean: -

"I am anxious to know the fate of those dreams-
That, prior to my death, I left behind in this world."

— • —

Addendum

Addendum- 1

These days, the information about different planets, Solar-System, Milky-way, etc., is included in the syllabus of even the primary schools; however, these lessons normally cover very limited information; therefore, some additional information, in brief, is given below: -

The Milky-Way - Our Own Galaxy

Astronomers have predicted that the Universe comprises about 200 to 400 billion galaxies. All these galaxies are spread all over within a vast three-dimensional spherical region measuring about 13.8 billion light-years; however, we cannot see all of them with our naked eyes. Our Milky-way galaxy is one of them; its estimated age is about 13.2 billion years. An average Galaxy might contain about billions of stars, out of which the approximate percentage of the Red-Dwarfs may be as high as 80%, and the rest of the stars may be about a little bigger or a little smaller than the Sun. Astronomers believe that galaxies were created in the early universe by the accretion of the stars over Dark-Matter. However, as described in this book, different galaxies were created during the early stage of star-formation due to the force of attraction exerted by the first-generation stars on each other; the stars and the galaxies were created side by side, but not due to the accretion of stars over dark-matter.

Normally, galaxies are not found alone; smaller galaxies start to orbit a comparatively more massive galaxy. Milky-way galaxy and some other smaller galaxies are orbiting a bigger galaxy named Andromeda Galaxy, which is located about 2.5 million light-years away from us and is about 2.5 times more massive than the Milky Way galaxy. The local cluster of galaxies so formed is a member of a super-cluster of galaxies. There are countless super-clusters in the Universe, which are connected with each other by very weak gravitational bonds.

The Milky Way consists of a disk of gas and dust; in turn, this disk consists of about 2 to 4 billion stars, and our Sun is one of them. All these stars orbit an invisible center. This disk appears as a band of stars because its disk-shaped structure is viewed from within; its true shape can be seen only from any other place that lies outside. This disk consists of billions of stars, which (the disk) in turn, consists of different spiral arms that branch-off from a central bar-shaped core. Though it is believed that Milky-Way is made of 2 or a maximum of 4 main arms, there are, in all, about 8 to 9 arms, including different branches, spurs, and extensions. Like all other galaxies, the Milky-Way is also surrounded by a spheroidal region of a faint glow known as the Gas Halo. This component of the Milky-way consists of very thinly diffused gases, which may be about 3 to 6 times bigger in size in comparison to the size of the disk of luminous stars.

The diameter of the disk of luminous stars comprising this galaxy is about 1 to 1.5 light-years, and its thickness is about 1,000 light-years. However, this disk of stars is not perfectly flat like a coin; the stars located in the central part of this disk form a bulge. This bulge is about 10,000 light-years in size. This part of the disk is very densely populated with stars; most of them are much older than those located further away from the center. The stars comprising Milky-way revolve around the galactic center with a speed of around 225 to 250 Km/Sec. Whereas, the stars located further away from the center rotate with a slightly higher speed. However, on moving closer to the center, the curvature of the orbital paths of the stars increases sharply; accordingly, it might be mistaken that such stars are moving at higher speeds. The Milky-way, in spite of spinning at such a high speed, takes about 225 to 250 million years to complete its one spin. Such a long time-period required to complete one rotation gives an idea about the size of the milky-way. However, the milky-way was not so huge from the beginning; it has grown so big, probably due to repeated mergers of different smaller galaxies over the past billions of years. The Milky-way is not the biggest galaxy in this vast Universe; it is merely an average-sized galaxy, the known biggest galaxy is almost 60,000 times bigger than the Milky-way.

The Milky-Way
An Imaginary Depiction

Like all the other galaxies, a supermassive black hole known as *Sagittarius-A* exists in the center (the galactic center) of the *Milky-Way*. The mass of this black hole is about 4 to 4.5 million times more than our Sun. The estimated diameter of the *Event-Horizon* of this black hole is approximately 22.5 million Kms, which is almost ½ the distance between the Sun and the planet Mercury. The light, in order to travel across this event horizon, would take a time period of about 1¼ minutes. The Sun orbits the galactic center with such an angle (about 8°) to the galactic plane (the average plane in which all the stars are supposed to orbit the galactic center) that while completing its one orbit, it fluctuates on both sides of the galactic-plane, by a distance of about 500 light-years each side.

The Location of the Sun in the Milky-Way

An imaginary diagram of the Milky-way, depicting the location of the sun, is given on the previous page; as depicted, the Sun is located in the inner rim of the "Orion" spur that branches off from the Sagittarius arm. Sun is located at a distance of about 27 to 30.000 light-years from the Galactic Center. Since relative locations of different stars in the Milky Way remain almost unchanged with respect to the Sun, humankind, right from the prehistoric ages, imagined some shapes/figures in some of the groups of the stars, such groups of stars are known as the constellations of the stars. While the Sun moves ahead in its orbit around the galactic center, it faces different **constellations of stars** (Zodiac Signs) during different periods of the year. The primitive men with naked eyes could identify only 12 Zodiac signs in the pre-historic ages; however, in the modern era, Astronomers have identified a total of 88 constellations of stars within the Milky-way.

The immediate neighborhood of the Sun

There are almost 2 to 4 billion stars in the Milky-way including all its arms; however, in the near vicinity of the Sun, only very few stars are located within a distance of 10 to 15 light-years from the sun. Since the thickness of the Milky-way is about 1 to 2 thousand light years, all the stars in the Milky-way, are not located in the same plane, i.e., in the galactic plane, some of the stars are located above this plane, whereas others are located below it. There are about 115 stars within the three-dimensional space of around 20 light-years around the Sun, out of which 52 stars are located within a distance of 16 light-years, and only 33 stars are located within a spherical distance of 12.5 light-years. Most of these 33 stars are red dwarfs, whose mass is less than half (½) of the Sun or even lesser than that. Out of all these 33 stars, only 6 stars or the systems of stars are orange or yellow stars, which though are bigger than the red dwarfs but are smaller than the Sun; whereas only 4 star-systems have equal or greater mass than the Sun. These nearby orange/yellow stars are listed below-

Name of the star	Category	Mass of the Star	Distance from Sun
Procyon A (α CMi A)	Yellow Star	1.6 Solar Mass	11.4 Light-year
Procyon B (α CMi B)	Yellow-White Dwarf	0.6 Solar Mass	11.4 Light-year
61 Cygni A	Orange Star	0.7 Solar Mass	11.1 Light-year
61 Cygni B	Orange Star	0.6 Solar Mass	11.1 Light-year
Epsilon Indi	Orange Star	0.76 Solar Mass	11.2 Light-years
Epsilon Eri	Orange Star	0.82 Solar Mass	10.8 Light-year
Sirius A	Yellow Star	2.1 Solar Mass	8.6 Light-year
Sirius Centauri B	White Dwarf	1.1 Solar Mass	8.6 Light-year
Alpha Centauri A	Yellow Star	1.1 Solar Mass	4.4 Light-year
Alpha Centauri B	Orange Star	0.9 Solar Mass	4.4 Light-year
Proxima Centaury	Red Dwarf	0.12 Solar Mass	4.2 Light-year

Out of the above-listed stars, *Epsilon Eri* is the only single or lone star, whereas *Epsilon Indi* is a system of 3 stars; this star-system consists of one orange star and two Brown-Dwarfs; the remaining others are systems of 2 or 3 stars. Out of the rest of the star-systems, Alpha Centauri, which is the nearest neighbor of the Sun, is a system of 3 stars; out of these stars, "Alpha Centaury-A" and "Alpha Centaury-B" consist of a binary pair of stars that orbit each other, the third star of this system, "Proxima Centaury," is a Red-Dwarf that orbits the pair of Alpha-Centaury A & B. Proxima-Centaury is also known as "Alpha-Centaury C." An Earth-sized planet orbits the red-dwarf Proxima Centaury in the habitable zone. This planet is known as "Proxima-b."

Out of all the stars, Sirius, which is the brightest star, rather than a binary star-system, had always been of the utmost importance for all of our ancient civilization. This star-system is a member of the famous Star-Tringle *(Winter Triangle),* which is an imaginary equilateral triangle; 3 bright stars, namely *Sirius, Procyon, and Betelgeuse,* are located at the vertices of this triangle. Out of these 3 stars, *Sirius,* which is about 8.6 light-years away from the Sun, is located at the bottom-most vertices of this triangle; Procyon, which is located at a distance of 12 light-years from the SUN, is located on the upper left vertices. *Betelgeuse,* which is located at a distance of about 650 light-years from the sun, occupies the third vertices of this triangle; **Betelgeuse** is a red-giant star having almost 11 times more mass in comparison to the Sun. Although all these stars seem to form the winter triangle, they are located in different constellations of the stars.

Probably, since 3,000 B.C, a very strange legend about the brightest star *"Sirius,"* has been in vogue among the **Dogan People,** the primitive tribe of Mali, West Africa. According to this legend, *"Nommos,"* the Devine People, who did come from an exoplanet that orbits a very dense and invisible star (a black or white dwarf) of the star-system *"Sirius,"* visited the Earth in the remote past. These visitors gave knowledge to humankind, taught them the construction methods, and developed an advanced civilization on Earth. Similar stories are included in the ancient religious books and/or myths of Babylonia, Mesopotamia (Akkadian), Sumeria (Sumer), Assyria, and ancient Egypt, etc. Similar legends are also in vogue among the tribal people (Kayapo tribe) of Brazil. Scientists, however, do not consider such legends reliable or true because the White-Dwarf, which is known as Sirius-B, was not known to the astronomers before the year 1862 when it was first seen; the said invisible star **Sirius-C,** or its planet **Xylanthia**, from where the Nommo People did supposedly come, could not yet have been found. However, even in the case the stories of Dogan People are really several thousand years old, there is no explanation for the same.

The Sun

Only two elements, Hydrogen and Helium, were created after the Big-Bang. Therefore, early stars would have been created out of these two elements only. Heavy elements such as Silver, Gold, and Platinum, etc., were not created at that point in time. However, the Sun, in addition to Hydrogen and Helium, also contains small quantities of some heavy metals. This fact indicates that the Sun was created much later; it must be a star of the second or even third generation. The early stars must have been very big and massive, whose lifespan might have been very short, ranging from around a few hundred million years to 1 to 2 billion years at the most. Heavier elements were probably, produced by supernovae like explosions in those early stars. Scientists have predicted that a massive star, just before its death, might have ejected its gas envelope into interstellar space. Later, Sun was created about 4.6 billion years ago from the remnant gas and debris of the said supernova.

The Sun is located in the center of the solar system; the average distance between the Sun and the Earth is approximately 150 million kilometers. Sun is a star of average size whose diameter is approximately 1.5 million kilometers, i.e., its diameter is about 110 times bigger than that of the Earth. The Sun completes its one rotation about its equator in about 25.6 Earth-days. Volume-wise the Sun is about 1.3 million times bigger than the Earth, whereas its mass is about 3,30,000 times more than that of the Earth. About 99.88% mass of the entire Solar-System is concentrated in the Sun. The temperature of the Sun's core is about 15.7 million degrees on the Kelvin scale; this temperature is adequate to switch-on the *process of fusion.* The core of the Sun is about 20% of its size, even then it is about 10,000 times bigger than the Earth. The surface temperature of the Sun is about 5,800 degrees, whereas the temperature of its atmosphere is about 20,000 degrees. Every second the Sun converts about 600 million tons of Hydrogen into Helium. About 4 million tons of matter is converted into energy in this process. However, the Sun has accumulated such a vast stock of Hydrogen that it will continue to produce energy at such a high rate for about the next 5 billion years or so.

Although the heat and energy of the Sun are continuously transferred by radiation, however, the intense heat of the core produces internal convective motion and violent turbulence in the ionized gases contained within it, due to which numerous eddies of plasma are formed. Such eddies move outward toward the surface of the Sun. This convective motion results in the formation of different areas on the surface having reduced temperature and reduced brightness. Such areas of reduced brightness are known as *sunspots.* Sunspots are often associated with *Coronal Loops* that are formed on the surface of the sun. Individual and independent magnetic fields due to fast-moving charged particles are produced in these loops. Such spots and loops do not have a very long life, they die off after some time, and the plasma loops sometimes burst, due to which a sudden flash of increased brightness is produced, which is known as *Solar Flare*. These flares are associated with *Mass Plasma Ejection;* huge magnetic energy and a stream of *Charged Particles* like electrons, protons, etc., are also released regularly from the surface of the Sun.

Different Planets and other Members of the Solar System

It is believed that the Sun was created due to the condensation of a stellar gas-cloud; it is also believed that numerous asteroids were also created at the same time due to the accretion of dust particles. Later, planets were created due to repeated collisions and mergers of these asteroids; small terrestrial (rocky) planets did create near the Sun, whereas the gas-giants were created far away from the Sun.

Contrary to this belief, planets in the other star-planetary systems do not follow this rule. This fact tells us that Nature might have its own rules; it might not follow the man-made rules. As brought out in Chapter- 9 of this book, the direction and the speed of rotation of all the stars and their planets, etc., is decided by the rules of nature; barring some exceptions, all the celestial objects rotate in the same direction. Almost all the celestial objects, such as the asteroids moving in the asteroid belt, and icy objects moving within the rings of the planet Saturn, etc., move in the same direction, at almost the same speeds. Therefore, the chances of their collisions are very remote. Based on this fact, an alternate possibility of the formation of planets was also given in the said Chapter- 9, according to which planets were probably, formed due to the accumulation of gas and dust over the star-debris ejected after different supernova explosions.

In whatever way the planets might have formed, there are total 8 known planets in the solar system, excluding Pluto, which has now been demoted as *a Dwarf Planet*. A brief introduction to the different members of the Solar-Family is given below:

Planet Mercury

Planet *Mercury* is the nearest and the first planet of the Sun; after the re-designation of Pluto as a dwarf planet, Mercury is now the smallest planet. Its diameter is approximately 4,800 Km, and its mass is almost 16.5 times lesser than that of the Earth. This planet, in comparison to the Earth, is about 3.8 times smaller in size. *Mercury* is a rocky planet; about 70% of its core contains metallic iron, and the rest is silicate rock. It is believed that heavy metals like Silver, Gold, and Platinum, etc., are in plenty on this planet. The average specific density of this planet is 5.4, which is marginally lower than that of the Earth. The field strength of its magnetic field is only 1.1% of that of the magnetic field strength of the Earth. This planet orbits the sun once in 87.97 earth-days; during the period of its 2 revolutions around the sun, it spins only 3 times; however, in the reference frame of the Sun, it completes two rotations in its one year. This planet doesn't have any atmosphere, and it doesn't have any natural satellite or moon.

Planet Venus

Venus is a territorial planet, and it is the second closest planet to the Sun. This planet, having been almost similar to the Earth in size, density, and strength of the gravitational field, is called the sister planet to the Earth. The diameter of planet Venus is about 12,104 Km, which is almost 95% of that of the Earth, and mass-wise, it is almost 80% of the Earth. Its average distance from the Sun is about 108 million kilometers, i.e., in comparison to the Earth, it is about 70% closer to the Sun. Planet Venus takes about 224.7 Earth-days to orbit the Sun, and it spins around its axis once in 243 Earth-days. Thus, one year of this planet is smaller than its one day. However, the length of its one day seems much longer due to its onward moment on its orbit as well as its *retrograde spin*. Accordingly, the time interval between sunrise to sunset is about 1.92 times greater than what it takes to complete its one spin. This planet, too, like Mercury, doesn't have any natural satellite.

Venus has an extremely dense atmosphere that is mainly composed of carbon dioxide, a small quantity of Nitrogen, Sulphuric Acid, and traces of other gases. The atmospheric pressure of this planet at its surface is about 92 times of what exists at the Earth's surface. Thick clouds of Sulphuric Acid floating above the dense layer of Carbon-di-oxide make it impossible to see its surface. Since the Carbon-di-Oxide gas generates a very strong Greenhouse Effect, the surface temperature of this planet rises as high as 462° Centigrade, which is hotter than even that of the planet Mercury. Much of the surface of Venus appears to have been shaped by volcanic activity; evidence of fresh volcanic eruptions was also seen in the years 2008 and 2009. It is believed that water was available in abundance in the long past on this planet. However, the greenhouse effect resulted in the gradual heating of the planet and subsequent vaporization of the water. The absence of a magnetic field enabled the ultraviolet rays to ionize the atmospheric molecules, which were later, washed away by the solar wind. Since there is no water on this planet, it cannot sustain any form of life under the present conditions.

This planet rotates about its axis in a direction that is opposite to that of all other planets. Uranus is the only other planet that spins differently; its axis of spin lies almost in its orbital plane. Actually, its axis of spin makes an angle of 97.77° with its orbital plane; therefore, both of its poles appear to lie in its orbital plane. Because of its retrograde spin, the Sun on the Planet Venus appears to rise from the west and set in the East. It is believed that earlier, this planet did spin like all other planets; however, in the ancient past, this planet turned upside down due to a collision with a huge celestial object. As a result, the direction of its spin seems to have been reversed.

The Earth: Our home planet

Earth is the 3rd planet from the Sun. It was created about 4.5 billion years in the past. Its average distance from the Sun is 150 million Km; accordingly, the light coming from the Sun takes a time period of about 8.3 minutes to reach the

Earth. The diameter of Earth, at its equator, is about 12,756 Km, its circumference is 40,000 Km, and its total mass is about 4.8×10^{24} Kg. It is a little flattened out at its poles. The Earth, on its orbital path, moves at a speed of about 30 Km per Sec. Moving at this speed, it completes one revolution around the sun in about 365.24 days. It is tilted by an angle of 23. 45° about its rotational axis. It completes one rotation about its axis in 23 hours, 56 minutes; however, since the Sun also keeps moving ahead on its path, its same face again comes in front of the sun after a time period of 24 hours exactly.

At the time of the creation of the Earth, its atmosphere was as harsh and unsuitable to any sort of life as that of the planet Venus. It is believed that during the first few hundred million years, the Earth received a huge quantity of water through a continuous shower of comets. As a result, some quantities of Carbon-di-oxide gas did get dissolved in the water, which in turn resulted in the formation of carbonate-rocks. It is also believed that the magnetic field of the Earth (Magnetosphere) was created about 3.5 billion years in the past, which shielded the Earth from fatal radiation and solar wind. This helped in the creation of conditions that were favorable for the origin of life on the Earth; had the magnetic field not been created, then the evolution of the advanced forms of life on the Earth would not have become possible.

The majority of scientists believe that Earth's magnetic field is created due to the circular motion of the molten iron in the Earth's core. However, molten iron also exists in the core of the planet Venus, but this planet doesn't have any magnetic field, probably, due to its slow speed of rotation. Another question that comes up is, "Though the molten iron was available in Earth's core since the beginning, then why did it take about 1 billion years to develop the magnetic field?" There is, probably, no explanation for the same. As per another opinion, Earth's magnetic field is probably, created due to the flow of charged particles in the ionosphere. However, in spite of the fact that there is no ionosphere on the planet Mercury, even then, it has a magnetic field. This fact indicates that there is probably no hard and fast rule for the creation of a magnetic field.

The Moon

The moon is the only natural satellite of our Earth. Size-wise the moon is about ¼ of the Earth; its mass is about 80 times lesser than that of the Earth. The Moon is about 384,402 Km away from the Earth. Light from the Moon takes a time period of about 1.28 seconds to reach us. Moon's rotation is tidally locked by the Earth's gravity. The moon goes around the Earth once in about every 27.3 days; however, since the Earth, at the same time, keeps moving in the same direction in its orbit around the Sun, it takes slightly longer for the Moon to show the same face to the Earth, which is about 29.5 days. Since the Sun is about 400 times the lunar distance and diameter, the Moon covers the Sun nearly precisely during a total solar eclipse. A solid sphere made of iron exists in the core of the moon, which is covered all around by liquid iron, which in turn is again covered by a mantle and crust. It has a very weak magnetic field and a very thin atmosphere, just like a vacuum. Its surface is covered by numerous craters, some of which are impact craters, and some are of volcanic origin. Water-ice exists in its polar craters.

Scientists have predicted that a little after the creation of the Earth, a Mars-sized celestial body named *Theia* collided with the Earth at such an angle that this impact caused the Earth to tilt on its axis. The same impact also blasted a portion of the Earth's crust into Earth's orbit; later, this material accreted in one place to form the Moon. The Moon's far side has a crust that is about 50 Km thicker than that of its near side. On the basis of this fact, scientists believed that, out of the said impact, two moons might have been created initially, one following the other from behind. Later, the one that was following the other one might have collided with the latter from the behind and fused with the second one to form the existing Moon. Alternatively, the Moon, in the beginning, must have been completely melted; the solidification of this molten matter would have created a uniform crust. However, the Earth's gravitational force might have elongated and thinned out the face of the crust that always faces the Earth.

Mars, the Red Planet

Mars is the fourth planet of the solar system, whose average distance from the Sun is about 230 million kilometers, i.e., in comparison to Earth, this planet is about 1½ times further away from the sun; due to this reason, it receives about 43% energy in comparison to the Earth. The diameter of this planet is about 6,792 Km, which is about 65.3% of that of the Earth. Planet Mars moves in an elliptical orbit around the Sun; Mars, when furthest from the Sun, is about 250 million Km away, whereas, when closest, it is about 210 million Km away from the Sun. It completes one orbit around the Sun in about 686.97 Earth-days; on the other hand, Mars takes a time period of 24 hours 36 minutes to complete one spin around its axis, i.e., one day of Mars is almost equal to that of the Earth. It looks red because Iron-Oxide is in abundance on its surface. It has a very thin atmosphere that consists of mainly Carbon-di-oxide, very little oxygen, and traces of other gases.

Mars has two known natural satellites, which are not so massive to become spherical; they look like big rocks, rather like asteroids. It seems that Mars has captured a few asteroids from elsewhere. The names of these moons are respectively

"Phobos" and *"Deimos."* Phobos, whose diameter is roughly 22 Km, orbits Mars at a very short orbit measuring only 9377 kilometers; it orbits Mars thrice in one day in a retrograde direction, i.e., from West to East. Every year, it is gradually coming closer to Mars, at a rate of 1.8 Km per year; it might crash into Mars in a few million years. The interior composition of Phobos is similar to that of Mars, suggesting that it might have been created from the materials ejected from Mars. The second moon, Deimos, which is 12 km across, orbits Mars from a distance of 20,062 Km, in every 30 hours. Like Earth's moon, both of these moons of Mars are in the tidal lock by the gravity of Mars; they always present the same face toward Mars. Mars may have some other moons smaller than 50 to 100 meters in diameter; apart from them, a dust ring is also predicted to exist between Phobos and Deimos.

Recent pieces of evidence suggest that a few billion years in the past, Mars might have its own magnetic field, liquid water, and an atmosphere similar to that of the primitive Earth. However, due to some unknown reason, its magnetic field died off, due to which its atmosphere was blown away by Solar-wind. This planet, being further away from the Sun, receives much less heat energy and light. Moreover, the very thin atmosphere of this planet is unable to retain much heat; therefore, it is far cooler than the Earth; during the summers, the temperature at its equator doesn't rise beyond 20°C, whereas, during winters, its temperature falls as low as -140°C.

The Asteroid Belt

The distance between the Sun and the inner planets increases gradually in a definite order, but there is comparatively a large gap between the planet Mars and the next planet, Jupiter. This gap is almost 4 times larger than the distance between the Sun and the Earth. On the basis of this fact, some of the scientists believe that there should have been another terrestrial planet between these two planets, whereas the others don't.

The first asteroid situated in this gap, after great difficulty, could be seen on 1st Jan' 1801. Later, after continual efforts, millions of asteroids were discovered in this gap, which surrounds the Sun, in a doughnut-shaped circumstellar disc. This disk of asteroids is known as the *Asteroid Belt,* or more precisely, *"Main Asteroid Belt,"* this name is given to this belt of asteroids to differentiate them from other asteroids that orbit Planet Jupiter, Earth, and other planets.

This belt contains numerous irregularly shaped celestial bodies called the asteroids or the minor planets; by count, they may be several million in number. The sizes of different asteroids moving within this belt, maybe anything between dust particles to as big as 700 to 800 Kms. This belt is spread in a very vast region measuring roughly two ***Astronomical Units**** or even more.

The inner rim of this belt is about 1.8 AU away from the SUN, whereas its outer edge is spread up to about a distance of 4.2 AU from the SUN. This belt is spread over such a vast region that the asteroids are very thinly distributed in this belt, leaving very large gaps between them. For this reason, numerous unmanned spacecraft have passed through this belt without any untoward incident. Different asteroids orbiting within this belt orbit in elliptical orbits with different eccentricities and angles; moreover, they move at different speeds. The asteroids orbiting the sun from comparatively closer distances move with comparatively higher speeds. Probably, due to this reason, occasional collisions between different asteroids do occur because the distances between different asteroids do change continuously. In case they collide with very high speeds and force, they may break into pieces, which may shatter here and there randomly. On the other hand, if the collision is not so hard, then they may fuse together. Sometimes, a small asteroid may start to orbit a bigger and more massive one.

The total population of asteroids comprising this belt may be as high as 0.7 to 1.7 million; in spite of such a large number of asteroids that constitute this belt, the collective mass of all these asteroids is only 4% of that of our moon. Almost half of the entire mass of this belt is concentrated in 4 big asteroids, the names of these four asteroids are given below in the descending order of their sizes:

1. Ceres, 2. Vesta, 3. Pallas, and 4. Hygiea

Out of the above-listed asteroids, Ceres is the biggest one, its diameter is about 950 km, and it has an adequate mass to acquire a spherical shape. Ceres is the only Dwarf Planet that is located in the asteroid belt; it is almost as massive as one-third (1/3) of the entire mass of the entire asteroid belt. The remaining three of the above-listed asteroids are less than 600 Km. in size. Apart from the aforesaid 4 asteroids, there may be about 200 asteroids whose sizes range between 100 Kms. or more.

The astronomers believe that asteroids are the leftover precursor planetesimals, which later accreted at different places to form different planets; it is also believed that the leftover asteroids, after the creation of different planets, are roaming in the asteroid-belt and elsewhere. They further believe that the formation of a new planet from these leftover asteroids is not possible because, on the one hand, the

---

* The distance between the sun and the earth, which is roughly 150 million Km., is known as one Astronomical Unit or simply by AU.

total mass of remnant asteroids is too less to form a planet, and on the other hand, asteroids located in this belt, were too strongly perturbed by Jupiter's gravity to form a planet; they continued to orbit the Sun as before. However, the following facts indicate that there might be some other reason for the non-formation of a new planet:

1. In case all the asteroids were created in the same period, by the same process, then all of them should have been rotating in the same direction, however, contradicting this possibility, different asteroids rotate about their axes in different directions, and at different angles and speeds; they even rotate in opposite directions too.

2. Had all of them been created from the same interstellar cloud of dust and gas, then the chemical composition of all of them should also have been similar. However, geologically, these asteroids are of mainly the following three different kinds: -

a) Carbonaceous, b) made of Silicate Rocks, and c) Metal Rich, made of Nickel and Iron.

The Above-mentioned facts indicate that these asteroids are, probably, not the remnant planetesimals. The reason being that metals like nickel/Iron are found in the cores of the planets; metals are normally not formed by the coalescing or accretion of the dust particles. On the other hand, silicate and carbonaceous materials are found in the mantel and crusts of the different planets. It seems to me that the metal-rich asteroids can only be created due to the shattering of a planet. In case a planet orbiting at this location was destroyed due to any reason, then the planet Jupiter is likely to sweep away the major portion of its fragments; another bulk portion of the fragments might be attracted by the inner planets like Mars and Earth, etc., the Sun also might have ingested numerous fragments. Probably this might be the reason for such a low mass being left behind this belt after the destruction of a probable planet of the long past.

"Ceres" The Dwarf Planet

A brief introduction to Dwarf Planet *Ceres* has already been given under *Asteroid Belt.* However, Ceres is a member of the asteroid belt; it is much different from the asteroids. Its composition is more like the comets. With a diameter of 945 kilometers, it is the largest object in the asteroid belt; it acquired a spherical shape due to its gravity. Ceres is composed of rock and ice and is estimated to comprise approximately one-third of the mass of the entire asteroid belt. Its outer surface is made of a mixture of ice, water, various hydrated minerals, and clay. Its rocky core is covered by an icy mantle, beneath which there may be an ocean of liquid water. During the year 2014, emissions of water vapor were detected from several regions of the surface of the dwarf planet Ceres. In March'2015, a NASA spacecraft named **Dawn** was established in orbit around Ceres; by now, its fuel has been completely exhausted. The outer surface of Ceres is covered with numerous craters; inside such a crater, scientists have seen two bright spots comprising ice, minerals, and salts. The presence of **liquid water, salts, and minerals** suggests that microbial life may originate and flourish independently on this minor planet.

Planet Jupiter

Jupiter is the 5th planet from the Sun and the 1st amongst the outer planets, situated outside the asteroid belt. It is the largest and most massive planet in the solar system; it is about more than 2 times massive in comparison to the combined mass of all the other planets taken together. Although the mass of Jupiter is one-thousandth that of the Sun, it is about 318 times more massive than the Earth; even then, the field strength of its gravity field on its surface is only 2.5 higher than that of the Earth because the distance between its center and the surface, is about 10 times greater than that of the Earth. Its diameter is about 143,000 kilometers, i.e., it is about 11.2 bigger in size as compared to Earth; however, in comparison to Sun, it is 10 times smaller; though, volume-wise, it can hold 1321 Earths. Because of its immense gravity, the planet Jupiter shields the inner planets from possible comet/asteroid impacts.

Jupiter is primarily composed of *Hydrogen,* whereas a quarter of its mass consists of *Helium;* even then, its average specific gravity is 1.3 (1.3 times heavier than the water). It is believed to have a rocky core of heavier elements, but, like the other gas/ice giants, Jupiter also lacks a well-defined solid surface. This planet moves in an elliptical orbit around the Sun; its furthest distance from the Sun is about 5.46AU, whereas, when nearest to the Sun, it remains about 4.95 AU away from the Sun. It takes about 11.9 Earth-years to complete one orbit of the Sun, and in spite of its giant size, it rotates once about its axis in less than 10 hours; it is the fastest rotating planet in the Solar system. Its magnetic force is also the strongest among the other planets; it is about 20,000 times stronger than that of the Earth's magnetic field.

The atmospheric temperature of this planet is as low as −163°C. However, its core is much hotter, as hot as about 24,000°C. Heat in its core is generated due to gravitational collapse; It radiates more heat than what it receives. Apart from *Hydrogen* and *Helium,* small quantities of *Oxygen, Nitrogen, Water vapor, Ammonia*, and *Methane,* etc., are also found in its atmosphere; this planet is covered with clouds of ammonia. Violent storms are very often created in Jupiter's atmosphere; these storms are normally associated with lightning strikes; these electric discharges are up to 100 times more powerful than those that occur on the Earth. A persistent cyclonic (rather anticyclonic) storm, as big as twice the size of the Earth, has been in existence for the last 350 years or so; this storm is known as the *Great Red Spot.* Like all other outer planets, Jupiter also has a faint planetary ring made of dust and Icey lumps.

Out of all the planets in the solar system, planet Jupiter has the greatest number of natural satellites; as many as 79 known moons orbit this planet. Excepting the 4 big (Galilean) Moons, the remaining moons were discovered later during 2002-2003 or earlier. Most of these newly discovered moons are less than 10 kilometers across, mostly ranging between 2 to 4 kilometers in size; some of them have retrograde orbits. This fact indicates that these moons are not the natural satellite of this planet; they were not created along with Jupiter. Instead, they were captured by Jupiter from elsewhere, probably from the asteroid belt.

Apart from the smaller moons, Jupiter has 4 main moons, which Galileo Galilei discovered in the year 1610; on his name, these moons are known as Galilean Moons. Out of these moons, the one named "Io" is the innermost moon of Jupiter; the size of Io is about 3643 kilometers. The moon Io is primarily composed of silicate rocks, which are supposed to surround an iron core. It is in tidal lock with Jupiter; therefore, its same side always faces Jupiter. This moon orbits Jupiter on an elliptical path; accordingly, when it is nearest to Jupiter, a powerful gravitational force (of Jupiter) produces a tidal effect on the surface of the Moon Io. As a result, its solid surface bulges out up to 100m as compared to when it is furthest away from Jupiter. The repeated tidal effect produces immense heat, due to which Io has become the most geologically active celestial object in the Solar System; around 400 active volcanoes might exist on this moon. The second nearest moon is Europa, which is as big as 3,122 Kms dimeter wise. Though the surface of Europa is covered with thick ice, it sometimes ejects jets of liquid water. Probably it is also heated up by the tidal effect. Astronomers believe that a vast sea of liquid water might exist beneath its frozen surface. This possibility makes this moon a potential place that can support life. Ganymede is the third Galilean Moon of Jupiter, with a diameter of 5262 km., it is the biggest moon in the solar system, it is even bigger than the planet Mercury. Callisto is the fourth Galilean moon of Jupiter, its diameter is 4821 Kms.

Planet Saturn

Saturn is the sixth furthest planet of the Sun; it is mainly known for its beautiful Rings. Size-wise, it is the second biggest planet in the Solar system. Similar to Jupiter, this planet is mainly composed of Hydrogen, Helium, water vapor, and some other elements/compounds, etc.; its core is composed of metals like Nickel and Iron and silicate rocks. The diameter of its core is about 25,000 Km., which is almost 2 times bigger than the Earth, whereas, mass-wise, its core is about 9 to 25 times more massive than the Earth. The diameter of this planet is about 1,20,540 Km., which is about 9 times bigger than that of the Earth; accordingly, it can contain about 764 Earths. However, this planet, having been constituted mainly of gases, its mass is only 95 times more than that of the Earth. It is the only planet of the solar system whose average specific gravity is lesser than that of water. Although the temperature of its atmosphere is as low as -183°C, its core is very hot at 1170°C.

The average distance between the Sun and planet Saturn is about 9.6 A.U.; this planet goes around Sun in an elliptical orbit. When on the nearest point of its orbit, the Saturn remains 9.04 AU away from the Sun, whereas its furthest distance from the Sun is 10.12 AU. It takes about 29.5 Earth-years to complete one revolution around the Sun. However, it spins at a very high speed about its axis; it completes one rotation in Earth's 9 hours and 33.5 minutes.

The total number of the known moons of Saturn is 62, out of which 34 moons are lesser than 10 kilometers in size, and 14 moons are below 50 kilometers in size. Titan is the biggest moon of Saturn; it is just next to Jupiter's biggest moon *Ganymede*. Titan is the second biggest moon in the Solar System; the diameter of Titan is about 5,149 Km; this moon, too, is bigger than the planet Mercury. Titan orbits Saturn from a distance of 1,221,870 Kms. The surface topography of Titan is very much similar to our own planet; it has high mountainous terrain similar to that of the Earth and large lakes of liquid methane. Scientists believe that some other kind of life based on liquid methane can evolve on this moon of the planet Saturn.

Rhea is the second-largest moon of Saturn; its diameter is about 1528 kilometers. Rhea may have its own planetary ring system. Three other moons of Saturn that measure over 1000 Km are Lapetus, Dion, and Tethys. Another notable moon is Enceladus, which is though very small in size (only 504 Kms. dimeter-wise); this moon is supposed to be a potential place that is capable of harboring microbial life. Enceladus is very cold and is mostly covered by fresh ice; its surface temperature, even at noon, touches −198 °C. However, jets of water, gas, and dust are sometimes ejected from the region lying near its south pole; these jets indicate that liquid water may exist in huge quantities under its frozen surface.

The planet Saturn is surrounded by a **Ring System** that looks like an annular disk; this ring system is made up of innumerable different rings. This planet was first seen by Galileo in the year 1610. This ring system comprises three main rings, each of which is made up of countless ringlets, each one of them is made up of countless small particles of iced water and a small quantity of dust and rocky minerals. These rings are roughly 20 Km wide; they start at a distance of about 6,630 Km from Saturn and are spread up to a distance of 120,700 km. from it.

Planet Uranus

Uranus is the 7th planet from the Sun. Sir William Herschel discovered this planet in the year 1781. The diameter of the planet Uranus is about 51,120 Km., which is almost 4 times bigger than the Earth; it can hold about 63 Earths. This planet is known as an Ice-Giant. In spite of being so big, its mass is only 14.54 times that of the Earth. Its average distance from the sun is almost 19 times more than that of the Earth; when it is nearest to the Sun, it is about 18.33 AU away from the Sun, whereas when furthest from the Sun, it is 20.11 AU away. It completes one revolution around the Sun in 84.2 earth-years and rotates once around its axis in 17hours, 14.5 minutes. Its spinning axis is tilted at an angle of 97. 77°, therefore, its poles seem to rotate almost in its orbital plane, in a direction opposite to other planets. Uranus has a total of 27 known moons. Like all other outer planets, Uranus also has its own planetary Ring-System.

Uranus consists of three layers: a solid core in the center, an icy mantle in the middle, and an outer gaseous envelope. The core of Uranus is much smaller in comparison to that of other outer planets; it is made of Iron, Nickel, and Silicate Rocks; the mass of the core is only 0.6 of that of the Earth. Its mantel that surrounds the core from all around is made up of various ices, such as water, ammonia, and methane, etc., the bulk of the mass of Uranus, which amounts to about 13.4 Earth-mass, is centered in the mantel, and the rest of the mass of this planet weighing about 0.5 Earth-mass, is centered in the upper atmosphere, which is mainly constituted of Helium and Hydrogen. Being far off from the sun, its atmosphere is much cooler, as cool as -224°C.

Planet Neptune

Neptune is the eighth and furthest known planet in the Solar System; it too is known as an Ice Giant. Although Neptune is slightly smaller than its nearest neighbor Uranus, it (Neptune) is a little more massive than the latter; Neptune is 17 times more massive than the Earth. On the other hand, the diameter of Neptune is about 49,528 kilometers, which is about 3.9 times bigger than that of the Earth. Volume-wise it can hold 57 Earths. This planet revolves around the Sun in an elliptical orbit with an average distance of about 30 times more than that of the Earth. When at the furthest point of its orbit, it is 30.33 AU away from the Sun, and at the nearest point, it is 29.81 AU away. It takes 164.8 years to go once around the Sun; in order to spin once around its axis, it takes a time period of 16 hours and 6.5 minutes. The axial-tilt of Neptune is 28. 32°.

The internal structure of this planet is very much similar to that of Uranus. Its core, which is made of Nickel, Iron, and silicate rocks, weighs about 1.2 times that of the Earth. In spite of the fact that its core is as hot as 5,400K, the same is surrounded by a thick mantle made of Icey cold water and Ammonia; the mass of this mantle is 10 to 15 times greater than that of the Earth. The rest of the mass of this planet is contained in its gaseous envelope. The outer atmosphere of Neptune is mostly made-up of Hydrogen, Helium, and traces of Nitrogen, Ammonia, and Methane, etc. Total 14 known moons orbit planet Neptune. Out of them, Triton, the biggest and the lone spherical moon, rotates in the retrograde direction; this fact suggests that this moon has been captured by Neptune, from elsewhere, probably from the Kuiper belt. This planet, too, has a faint planetary ring system.

Minor Planets – "Centaurs"

Centaurs are small to medium-sized mysterious Solar system bodies, which are though very small on the planetary scale, known as minor planets. Normally, Centaurs orbit the sun between the outer planets Jupiter and Neptune. These bodies are named after the Greek mythological creature "Centaur," having the upper body of a human and the lower body and legs of a horse. Centaurs have mixed properties of the planets and asteroids as well as those of the comets. Although centaurs are known as minor planets and also asteroids, most of them are too small to be observed; even the largest ones are much smaller by planetary standards. All the centaurs do not have similar orbits; some of them revolve around the Sun in orbits that can be relatively stable, like those of the planets, whereas others go around the

Sun in extremely eccentric orbits, like comets. The total number of centaurs may be as high as 44,000 or even more; out of all these centaurs, a few may be as big as 250 Kilometers in size. At least a dozen of centaurs orbit the sun in a retrograde direction. Centaurs orbit the Sun on a path that takes them across Neptune and Jupiter. This path can be perturbed by the gravitational effects of Jupiter and other massive bodies; as a result, they may change their orbital paths and sometimes collide with other celestial objects or even get captured by other planets to become their moons. The orbital paths of some of the centaurs are highly eccentric so that they may orbit the Sun from well beyond Neptune to inside the Mars's orbit or even inside the Earth's orbit. Whenever such centaurs reach closer to the Sun, a comet-like tail may appear behind them. Like outer Gas-Planets, the bigger centaurs like Chariklo and Chiron may also have a planetary ring.

The Kuiper Belt

The average distance between the Sun and the planet Neptune is about 30.33 AU; the region beyond Neptune is known as *Trans Neptunian Region.* In this region, a little beyond Neptune lies a Doughnut-shaped circumstellar disk consisting of millions of Icey and rocky stellar objects that are supposed to be the remnants of the formation of the Sun. This disk is known as the Kuiper belt, which is a great ring of debris that is similar to the asteroid belt, but it mainly consists of objects composed primarily of ice. It is estimated that about 250,000 objects over 100 kilometers in size might exist in this belt. This belt is similar to the asteroid belt but is far larger, about 20 times as wide and 20 to 200 times as massive. The inner edge of this belt extends from a little outside of the orbit of planet Neptune (30 AU) to approximately 50 AU from the Sun. Those Kuiper-belt objects (KBOs) that go around the Sun at the near edge of this belt move at different speeds when compared to those that move at its far edge. Similarly, different KOBs orbit the Sun at different inclinations; some of these objects, while moving on their orbits, makes a very narrow-angle (2.6°) with the plane that contains most of the objects orbiting the sun, whereas others make a much larger angle of about 16 to 17° with the same plane known as *the ecliptic*. Astronomers have predicted that this belt might contain thousands of dwarf planets. However, more information is being collected before such objects can be put under any specific category. At present, there are only three known dwarf-planets in this belt, which are- Pluto, Makemake, and Haumea, out of which Pluto, with a diameter of 2,378 Km, is the biggest and most massive Kuiper-belt object. Dwarf planet Pluto has five (5) moons; its biggest moon is known by the name Chiron.

The Kuiper belt is considered the home of comets, especially of that of the short-period comets that have orbital periods under 200 years or so. It is believed that those comets, which reappear after the time-gaps of much longer than 200 years, might originate from a much distant place, probably from the "Oort Cloud;" such long-period comets have highly eccentric orbits.

The Scattered Disc

The *scattered disc* is a distant circumstellar disk in the solar system that overlaps the Kuiper belt at a distance of 30 AU from the Sun; it is though similar to the Kuiper belt; it extends much further outwards, up to about 150AU or even beyond; the scattered disc objects have much larger orbits having very high eccentricity as well as large orbital inclination. These extreme orbits are thought to be the result of gravitational "scattering" by the Gas/Ice Giants, especially by the planet Neptune. This belt is the home of millions of Icey, Rocky, and other objects made of frozen volatile material. Although it is much bigger than the Kuiper belt, different members of this belt are populated very sparsely, having large gaps in between. The Scattered disk objects go around the Sun with different orbits that are highly inclined to the ecliptic plane, up to an angle of 40° or more; some of them often orbit the Sun at almost perpendicular to the Ecliptic. The *Scattered Disk Objects,* being far off from the Sun, are probably, as cool as 30°K (- 243°C) or even much cooler than that. Because of its unstable nature, astronomers consider the scattered disc to be the place of origin for most periodic comets in the Solar System; it is believed that perturbations from the giant planets send such objects either toward the Sun or away from it.

Dwarf Planet- "ERIS"

Amongst all the known objects of the Scattered Disk, Eris is the biggest and most massive one; it has been classified as a Dwarf Planet. The diameter of Eris is about 2330 Km., which is almost equal to that of the dwarf planet Pluto; it is about 27% more massive than Pluto; this makes Eris the most massive and most distant dwarf planet. Eris goes around the Sun in an elliptical orbit that is inclined to the ecliptic at an angle of about 44°, at this orbit, its shortest distance from the Sun is 38 AU, and longest distance is 97.6 AU. Eris has a moon named *Dysnomia.*

The Heliosphere

Every star, including our own Sun, emits a constant stream of charged particles known as the star-wind (in the case of the Sun, it is called Solar wind). The said stream of charged particles spreads out in the space that surrounds the respective star until it collides with the **Stellar wind.** The said stellar wind is different from the solar wind/star wind in the sense that the charged particles contained in it do not originate from any particular star. Moreover, such particles are much colder and more concentrated. The region of space filled with stellar wind is known as **Interstellar space.** Like the stellar wind, the magnetic field permeating the interstellar space does not originate from the Sun or any other star.

Apart from the charged particles, a rotating magnetic field also extends out from each and every star. The rotating magnetic field created by the Sun is known as the **"Heliospheric Current Sheet,"** which is the largest rotating structure within the solar system.

The solar wind emitted from the Sun flows out spherically, at a speed of about 400 Km per second or more; this flow pushes against the stellar wind and causes it to bulge out like a balloon or a bubble. In turn, the pressure of the stellar wind, which opposes the growth of this bubble, also increases. Finally, a balance is established between the outward and inward pressures. As a result, the speed of the outward flow of the solar wind first slows down, becomes turbulent, then suddenly stops; an increase in the intensity of the Sun's magnetic field is also noticed in this region. The bubble-like region formed in the interstellar space is called the **Heliosphere,** and the region where the flow of solar wind slows down, is known as the Termination shock; this region extends far beyond, even beyond the orbit of the dwarf planet Pluto. The termination shock is supposed to be created at a distance of 75 to 90 AU from the sun. Beyond the region of the terminal shock and before the outer edge of the hemisphere lies the Heliosheath, which is a broad transitional region between the inner heliosphere and the external environment. The outermost edge of the heliosphere, where solar wind finally terminates, is called the Heliopause. A marked increase in the level of cosmic rays and other radiations has been noticed beyond the Heliopause.

The Detached Objects

Far beyond the Heliosphere, a few stellar objects are seen, which orbit the Sun in exceptionally long and elongated orbits; Sedna is the biggest object amongst them. Sedna's orbit is so highly eccentric that it ranges between 76 AU to as far as 940 AU; it completes its one orbit around the Sun in 11,400 years. Apart from Sedna, at least two more objects orbit the Sun from far beyond the orbits of the scattered disk objects. All of these detached objects may be dwarf planets; however, more information is required about these objects.

The Oort Cloud

The *Oort Cloud* is a hypothetical cloud of Icy planetesimals, which is supposed to surround the Sun at different distances ranging from 50,000 to 150,000 AU (0.8 to 2.5 light-years). The total number of objects contained in this cloud may be as high as hundreds of billions. The temperature of these objects, being so far off from the Sun, may fall very close to absolute zero. This cloud is divided into two regions: 1) a disk-shaped inner Oort cloud and 2) a spherical outer Oort cloud. Both regions lie far beyond the *Scattered Disk*; both of them lie in the *Interstellar- Space*. The outer limit of the Oort cloud defines the boundary of the solar system and also that of the reach of its gravitational field. At such a far-off distance, the outer *Oort cloud* is only loosely bound to the Solar System; thus, the gravitational pull of different nearby stars may occasionally dislodge different *Oort Cloud Objects* from their locations and send them toward the inner planets, these objects are known as the long-period comets that orbit the sun in highly eccentric orbits.

The Hypothetical Ninth Planet

Our knowledge about the region of the solar system that lies beyond the Kuiper belt is very limited rather than almost nil. Based on the abnormal orbital paths of the Kuiper belt objects as well as that of the scattered disk objects, some researchers have found mathematical evidence suggesting that there may be a far-bigger undiscovered earth-like planet deep in the solar system, which is a must for the stability of the Solar system. This hypothetical Neptune-sized planet probably orbits our Sun in a highly elongated orbit that should lie far beyond the scattered belt. This hypothetical planet has been named the "Planet Nine," or "Super-Earth." This planet might have a mass about 10 times that of the Earth, and the same orbits the Sun from an average distance of about 600 AU. It may take between 10,000 and 20,000 Earth-years to make one full orbit around the Sun. Some of the astronomers are making day and night efforts to locate this hypothetical planet, however, no success could yet be achieved. Naturally, any object that far away from the Sun will be very faint and hard to detect in an absolutely dark region, however, an extensive search is still going on.

Addendum - 2

The information given in this part of the book has no relation to cosmology or astrophysics. During the early 1990s, I found that the odd numbers bear a definite relation to the value of different exponential powers of the natural numbers. Mathematicians must be aware of this relation; however, during my school days, I didn't read anything about such a relation. Therefore, I want to share my observation with the readers:

Value of the squares of different natural numbers

The relation between odd numbers and the squares of different numbers is given below:

$1^2 = 1$

$2^2 = 4 = 1+3$

$3^2 = 9 = 1+3+5$

$4^2 = 16 = 1+3+5+7$

$5^2 = 25 = 1+3+5+7+9$

$6^2 = 36 = 1+3+5+7+9+11$

$7^2 = 49 = 1+3+5+7+9+11+13$

$8^2 = 64 = 1+3+5+7+9+11+13+15$

The value of the squares of different numbers can be derived in a similar manner.

×   ×   ×

Above relation is explained below graphically-

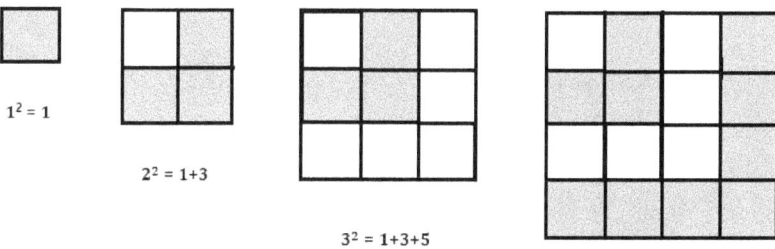

$1^2 = 1$

$2^2 = 1+3$

$3^2 = 1+3+5$

$4^2 = 1+3+5+7$

The Relation between odd numbers and the third power (Cube) of the natural numbers

$1^{3=1}$

$2^3 = 8 = 3+5$

$3^3 = 27 = 7+9+11$

$4^3 = 64 = 13+15+17+19$

$5^3 = 125 = 21+23+25+27+29$

$6^3 = 216 = 31+33+35+37+39+41$

$7^3 = 343 = 43+45+47+49+51+53+55$

$8^3 = 512 = 57+59+61+63+65+67+69+71$

And so on…

Value of the 4th Power of the natural numbers

$1^4 = 1$

$2^4 = 16 = 1+3+5+7$

$3^4 = 81 = 1+3+5+7+9+11+13+15+17$

3^4 can also be expressed as $3^3 \times 3$. Accordingly-

$3^4 = (7+9+11) \times 3$

Or 21+27+33

$4^4 = 256 = 1+3+5+7+9+11+13+15+17+19+21+23+25+27+29+31$

With further increase in the value of the base-numbers, this series will grow very long, and thus become non-practicable.

Value of the 5th power of the natural numbers

$1^5 = 1$

$2^5 = 32 = 5+7+9+11$

$3^5 = 243 = 19+21+23+25+27+29+31+33+35$

With further increase in the value of the base-numbers, this series will grow very long and cumbersome, thus become non-practicable.

Value of the 6th power of the natural numbers

$1^6 = 1$

$2^6 = 64 = 13+15+17+19,$

Or $2^6 = 64 = 1+3+5+7+9+11+13+15$

For the higher numbers or for the higher exponential powers this method will become non-practicable.

Addendum – 3

The author Mr. Satya Prakash Verma, our beloved father, wanted to perform this experiment himself, but unfortunately, fate had different plans. COVID took him away from us abruptly and untimely on June 23, 2021. He might have had some intuition and added the steps for performing this experiment towards the end of the book. He was in the process of finessing the book before publishing it when he left this world. Our mother wanted to release this book on April 28, 2022, when he would have been completing 80 years on this planet. He had a scientific mind and also tried to apply for some patents but didn't have sufficient funds to create a prototype. He had a lot of ideas shaping up in his mind. Apart from being a great mind, he was also a loving husband and a doting father. He was well respected and liked by his friends, family, and colleagues. We are so proud of him. Following is the experiment in his own words. He mentioned this idea in his closing remarks.

- Akanksha, Ashutosh & Aditi

Measuring the Unidirectional Speed of Light

The Necessity:

At the present, the speed of light is normally measured by making a light beam to reflect back and forth several times between a set of mirrors. Measuring the speed of light in a single direction is not felt necessary because, based on *Michelson-Morley's experiment*, it is believed that light-waves travel in any direction at a constant speed, even in the reference frame of the speed of any moving object. In case the light really travels at the same speed in any direction without being affected by the Earth's speed, then it shall take equal time to travel equal distances, whether measured in the direction of Earth's motion or the opposite of it. However, this idea has never been physically verified by measuring the speed of light in both of these directions separately and independently of each other. However, direct verification of the aforesaid belief that the speed of the light remains unaffected by the speed of the earth seems necessary to establish the truth.

The main aim of Michelson-Morley's experiment was to detect the hypothetical "Aether," through which the ripples of light are believed to propagate. Michelson-Morley predicted that two different light beams traveling perpendicular to each other between two different sets of mirrors placed 11 meters apart will travel through a *fixed distance* of 11 meters. The scientist-duo *assumed* that the speed of the beam traveling perpendicular to the Earth's motion would not at all be affected by the Aether; however, the speed of the second beam, which will be made to travel in the direction of the Earth's motion, would surely change in the reference frame of the moving Aether, accordingly, both the beams will travel between the *fixed distances but with different speeds.* The proposed experiment differs from the aforesaid experiment: "In this experiment, the speed of a light beam, which is made to run in the direction of the Earth's motion, is compared with the speed of the other beam running opposite of the Earth's direction. Although the distance between the two end-points between which the light beams are made to run would never vary, however, for the first beam, these end-points will also move with the Earth in the same direction, whereas for the beam running in the direction opposite of the Earth, these end-points will move in the direction opposite of the beam. Therefore, these two light-beams *would run at a fixed speed but between two different distances, not between the fixed distances.* A sufficiently long distance of three (3) kilometers between the two end-points has been chosen in the proposed experiment so that the effect of the Earth's speed, if any, could be noticed very clearly.

Although such a measurement is not considered feasible, however, such a measurement is definitely possible by the method disclosed below: –

The Required Equipment/Material

1. *Cesium-133 Atomic clocks* (alternatively, *chip-scale atomic clocks*) – 3 nos.

2. Light-sensitive screens made of photoelectric cells- 2 nos. (1 of these screens shall have a wide central opening)

3. A light-sensitive electronic device for very precise recording of the time instant when the laser pulse would hit the aforementioned screen – 2 nos.

4. 6" to 10" diameter pipeline with black interior- Length 3 kilometers + approximately 20 to 30 meters extra.

5. Electronic switching device with the facility to precisely record the time of switching-on the laser pulse: Preferably 2 nos. (Only a single device will also serve the purpose)

6. A composite source of Laser that can emit two parallel laser pulses simultaneously and precisely at the same time-instant; out of these two pulses, the main pulse is emitted from the center, whereas the auxiliary pulse from about 2 inches away from the main pulse.

Method:

Lay the pipe exactly in the direction of Earth's motion in its orbit, i.e., West to East, at any convenient elevation; care shall be taken to ensure that the pipe is laid in a perfectly straight line so that the laser beam may pass through it without hitting its interior surface. Next, install the source of the laser and both the screens in the manner as depicted in the drawing given below:

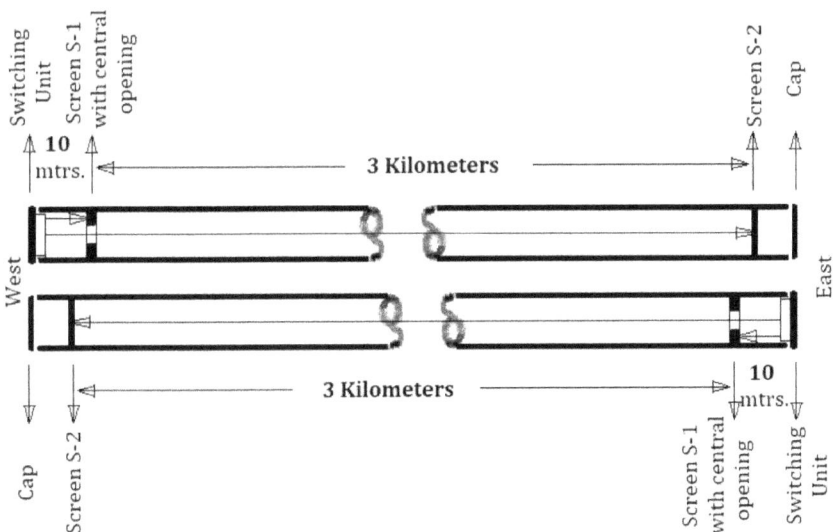

Next, connect the atomic clocks, one each to (a) switching unit, (b) screen S-1, and (c) screen S-2, taking care that 1) All the clocks shall show the same time, 2) Lengths of all the wires connecting each above device to the respective atomic clocks shall be exactly equal in length, so that time lost by the signal to reach the clock, shall be equal in all the three cases & 3) Assembly of switching unit and screens, etc. shall be done in the complete darkness.

In order to measure the speed of the light in the direction of the Earth's motion, i.e., from West to East, the switching unit shall be installed at the western end of the pipe. Immediately after putting on the switch, both the pulses of the laser would start their journeys exactly at the same time instant simultaneously. Let the time of switching on these pulses be t_0. At this instant of time, both the laser pulses would shoot off at the speed of 299,792.458 km per sec or say 300,000 km per sec, whereas the pipe assembly, along with both the screens, would also keep moving in the same direction with a speed of 30 km. per sec. Accordingly, when the auxiliary pulse would hit the screen S-1, this screen placed at a distance of 10 meters from its source would also move ahead by a distance of 1 millimeter, i. e., the light would have to travel a distance of 10.001 meters to reach this screen. Let the time recorded at the first screen S-1 be t1. Accordingly, the time-period taken by the auxiliary beam to reach screen S-1 would be $(t_1 - t_0)$.

Similarly, the main pulse would first travel past screen S-1 through its central opening; next, it would hit the screen S-2 at a time instant t2. Accordingly, the second laser beam would travel the distance between screen S-1 and the onward-moving screen S-2 in a time-period of (t2 – t1). Now, since the screen S-2 is also traveling along with the Earth, this beam would move a little extra distance over and above the fixed distance of 3 kilometers existing between both of these screens. The second beam would take a time of about 10,000 nanoseconds to travel the distance of 3 kilometers; however, by this time, screen S-2 would also move ahead by approximately 30 centimeters; in addition, it would also move ahead by 0.463 millimeters approx., with the spin of the earth, i.e., a total distance of 30.0463 centimeters over and above 3 kilometers. Accordingly, this pulse of the laser, in order to cover this extra distance, will take an extra time-period of little more than 1 nanosecond. Similarly, this laser-pulse, in order to travel from its source to the screen S-2, will take a time period of t2 – t0. The speed of light in the direction of Earth's motion can now be calculated on the basis of 3 different sets of readings obtained as above.

In order to verify the correctness of the idea that light travels at the same speed in any direction, the assembly of the equipment illustrated above shall now be reversed, and the said experiment may be repeated to measure the speed of light in the direction East to West.

In this case, laser-beams will move from East to West, whereas both the screens S-1 and S-2 will continue to move from West to East as before. Accordingly, by the time the main laser pulse hits the screen S-2, the said screen S-2 would also travel through a very small distance in the direction opposite to that of this laser pulse; therefore, this time, the laser-pulse would have to travel through a distance little lesser than 3 km., accordingly, it will take a time period of little lesser than 10,000 nanoseconds to reach the screen S-2. Now, since the said laser beam had to travel through a distance of a little lesser than 3 Km., the screen S-2 will move in the opposite direction, by a distance of a little lesser than 30 centimeters. Accordingly, this pulse, after passing through screen S-1, will take a time period of a little more than 9,999 (10,000 − 1) nanoseconds. Accordingly, in this direction, the laser would ostensibly appear to move at apparently a higher speed as compared to its normal speed.

Conclusion –

In case we actually obtain the readings as predicted above; then it would be inferred that-

1. Even the light in these two directions ostensibly propagates at different speeds. *However, in case the speed of the Earth and the direction of its motion are both considered, i.e., the actual distances traveled by different light-beams are taken into account, then the speed of light in both of these directions would be found to be exactly the same, as was envisaged by Einstein.* Even then, *this experiment will prove without any doubt* that though light travels with constant speed in any direction, *its speed, when viewed in the reference frame of the speed of other moving objects, would seem to have ostensibly changed in the reference frame of the speed of any moving object that moves at whatever speeds.*

2. Such a result would also prove without any doubt that - *"Neither the objects do contract at the speed of light, nor the time dilates or comes to a total halt at this speed."*

3. The result obtained in this experiment would further indicate that light does not necessarily propagate in waves; in case, as predicted in this book, the light propagates in the form of a shower of vibrating or spinning particles, even then a similar result would be obtained, in that case, no medium would necessarily be required for the propagation of the light.

×   ×   ×

Note: -

1. Since the Earth takes approx. 24 hours to rotate through an angle of 360°, the screen S-2, during a time period of 10,000 nanoseconds, i.e., the time period required by the light to travel through a distance of three (3) kilometers, would tilt through an angle of approx. $4 \times 10\text{-}8$ degrees. Tilting of screen S-2 by such a small angle would probably, not affect the result of this experiment in any way.

2. Probably, the unidirectional speed of the light beam propagating for a short distance can also be measured with the help of the Attosecond Technology (please see Chapter-3). However, it would be difficult to cognize the difference in the time taken by the two individual light beams running between short distances; one of them running in the East-West direction and the other one in the opposite of it, for such a short distance, unless a fraction of 1/10,000 nanoseconds could also be measured.

3. In November 2007, the speed of light was measured at the Advanced Laboratory, Physics, Wisconsin, USA, where, a laser pulse was first divided into two parts, one moving on a short distance and the other one on a comparatively larger path. The pulse moving on the short path was directly sent to a photomultiplier tube, whereas the other one was sent on a much longer path, at the end of this path this pulse was directed to the photo-multiplier tube. The speed of light was measured on the basis of the delay noted between both of these signals. This method can also be used to measure the unidirectional speed of the light, over a short distance.

4. The speed of light as well as that of the Earth both, can be measured by the experiment described in the end of Chapter- 4, by using atomic clocks and a light-sensitive screen.

Few words from Papa's dear friends, colleagues and family members:

In a very simple and easy language, the difficult and complex subject of science - "Mysteries of the Universe" has been presented by my dearest younger brother Late Shri Satya Prakash Verma 'Satyendra' in front of the common readers with complete truth, integrity, and fairness. I believe that this is one of the best work among the books written on this subject.

- Mr. Dharm Prakash Verma

"Why do you cry when he goes to his heavenly abode."
Hundreds of salutes to the holy memory of respected Shri Satya Prakash Verma ji, who was of saintly nature and had goodwill towards everyone. He has been a very honest, hard-working engineer full of technical knowledge. I remember when he remained without sleep for twenty days but did not let any subordinate or co-worker get hurt. He was rich in godly qualities like fearless and without rancor. May the Almighty God always give him a place at his feet.

- Vidya Sagar Malik.

He was my elder brother rather than my best friend. I have no words to describe his personality. He was always calm, quiet, and very polite. He had very deep theoretical & practical technical knowledge. I very well remember that he has made a classical musical guitar for Respected Roy dada totally on his own. He was very interested in old filmy & private songs as well deep knowledge of them. Also, I never found him excited. Again, I pray to God to rest his soul in peace.

- Prakash Manke

My life was blessed through friendship with Late Shri S P Verma ji, who knew my heart and mind as well. The values of our friendship are beyond measure. My close association started with him in 1982 for a good social cause of upliftment of education of children in the remote area where we lived at that time. He was

President of an Education Society. His contribution in this field can never be forgotten and will always be remembered. He was also an eminent Power Engineer. His excellent professional knowledge, honesty, commitment, dedication, and hard work won many awards for him in his professional career. There are many examples that are quoted for his honesty. He had an innovative mind and was normally busy in doing new experiments related to science. He had a keen interest in Astronomy. He was also fond of writing articles related to science. Shri Verma ji was an excellent human being, a very simple personality, and true by heart. He was very helpful to colleagues and loved by mostly all. His sudden and unexpected demise is a great personal loss to me. May almighty God bestow peace to the departed noble soul.

- Namdeo K Rupwani

Paying humble tributes to the most respected Late Shri S. P. Vermaji on the occasion of his first death anniversary, we are sharing here the unforgettable moments spent in his company - It is a matter of those days when in 1976-77, many engineers were posted for the construction and operation of 120 MW units in Amarkantak Thermal Power Station, Chachai, MP. At the same time, Late Vermaji also got posted in Chachai. I also got the opportunity to work under him. We were all impressed by his technical skills, sweet demeanor, and simplicity. He was truly an 'Ajatashatru' (One who is without enemies) and believed in taking everyone along. During his tenure, in the face of adversity, he always emerged victorious with his patience. It was the good fortune for both of us that he was also our neighbor in the old B-type quarter of Chachai. He was always ready to help. He was a person of calm, simple, and soulful nature. He had a special attachment to his family. He was never found to be angry. After retirement, in Noida and Jabalpur too, he always met with the same affinity. He was particularly interested in music and sports and had good knowledge about them. He himself was a good player of table tennis and had a good collection of old film songs. We will never forget the time spent with him.

- Sudhir and Swati Rege

I first met Vermaji in 1967 after my transfer from Jabalpur to Sarni. We were both reporting to the same manager and, co-incidentally, in the engineers' hostel, I was allotted a room adjacent to his room. This was the start of our friendship, which continued for almost 55 years.

Vermaji was of a very quiet and simple nature, he was a man of few words and liked minding his business.

There are many fond memories which I would not be able to share here. All I would like to say is I have lost my dear friend and brother. God, please take him into your care and provide him with peace.

- L.N Gupta

Papa - You were the most loving, caring, and forgiving person I have known in my entire life. You loved unconditionally. You have always inspired me to do the right thing no matter what. You taught us not to act the way the other person would have behaved with us but to always behave in a loving, sympathetic manner. You were the one I always confided in and looked up to you for guidance. I would tell you things and gauge by your reaction whether what I did was right or wrong. Your faith in me always gave me confidence and inspired me to dream big. You empowered me by believing in me and showing me how to do things instead of overprotecting and making me weak in the process the way many other parents do. I always knew that I could come to you with any and all of my problems without worrying about being judged. And when you and Mummy were by my side, I never had to worry about what the world thought of me or my actions. You were the most honest, helping, knowledgeable and selfless person around. I wish you did not have to suffer in your last few days and would have done anything in my power to save you. You left this earthly world on 23rd June 2021, but I know your soul will always be out there watching over us.

Love you so much Papa. You are the best!

- Mini

Words are not enough to express how much we loved our father and explain how brilliant he was; his concepts on almost all subjects were very clear, he possessed knowledge in almost all fields, and he was always prepared to discuss any topic; he was logical and was able to convince everyone with his reasonings. By education, he was a mechanical engineer; however, he was always interested in exploring other areas as well; by his own learning, he developed good knowledge in homeopathy, and he helped many with his knowledge. He was excellent in his job; however, if he would have chosen to become a doctor, he could have done wonders in the medical field as well.

His heart was filled with love for all human beings as well as he was very sensitive about the other living creatures on earth. He used to say that only your love for others and good doing is the correct way to reach God; though many people took benefit of his simplicity, I never saw him talking negatively about anyone. He

407

believed in doing good for everyone and never expected anything in return.

His love for his family and friends needs no explanation. He touched everyone's heart in some or the other way. It is rare to find a human being as selfless and as humble as my father was, most people don't know his difficult part of life as he never sought help from anyone no matter what. He always believed in giving and he never saved anything for himself, that is why his earnings cannot be calculated in terms of wealth but the love and respect he gained all over his life is his actual earnings.

We are proud of our father, and he will always be loved and respected in our hearts.

- Ashutosh Verma

www.ingramcontent.com/pod-product-compliance
Lightning Source LLC
Chambersburg PA
CBHW050502210326
41521CB00011B/2285